Why Guattari? A Liberation of Cartographies, Ecologies and Politics

This book examines Félix Guattari, the French psychoanalyst, philosopher, and radical activist, renowned for an energetic style of thought that cuts across conceptual, political, and institutional spheres.

Increasingly recognised as a key figure in his own right, Guattari's influence in contemporary social theory and the modern social sciences continues to grow. From the ecosophy of hurricanes to the micropolitics of cinema, the book draws together a series of Guattarian motifs which animate the complexity of one of the twentieth century's greatest and enigmatic thinkers. The book examines techniques and modes of thought that contribute to a liberation of thinking and subjectivity. Divided thematically into three parts – 'cartographies', 'ecologies', and 'micropolitics' – each chapter showcases the singular and pragmatic grounds by which Guattari's signature concepts can be found to be both disruptive to traditional modes of thinking, and generative toward novel forms of ethics, politics, and sociality.

This interdisciplinary compendium on Guattari's exciting, experimental, and enigmatic thought will appeal to academics and postgraduates within Social Theory, Human Geography, and Continental Philosophy.

Thomas Jellis is a British Academy Postdoctoral Fellow at the School of Geography and the Environment at the University of Oxford, and a Research Fellow at Keble College.

Joe Gerlach is Lecturer of Human Geography at the School of Geographical Sciences, University of Bristol.

JD Dewsbury is Professor of Human Geography at the University of New South Wales, Canberra, Australia.

Routledge Studies in Human Geography

This series provides a forum for innovative, vibrant, and critical debate within Human Geography. Titles will reflect the wealth of research which is taking place in this diverse and ever-expanding field. Contributions will be drawn from the main sub-disciplines and from innovative areas of work which have no particular sub-disciplinary allegiances.

Living with the Sea
Knowledge, Awareness and Action
Edited by Mike Brown and Kimberley Peters

Time Geography in the Global Context
An Anthology
Edited by Kajsa Ellegård

Consolationscapes in the Face of Loss
Grief and Consolation in Space and Time
Edited by Christoph Jedan, Avril Maddrell and Eric Venbrux

The Crisis of Global Youth Unemployment
Edited by Tamar Mayer, Sujata Moorti and Jamie K. McCallum

Thinking Time Geography
Concepts, Methods and Applications
Kajsa Ellegård

British Migration
Globalisation, Transnational Identities and Multiculturalism
Edited by Pauline Leonard and Katie Walsh

Why Guattari? A Liberation of Cartographies, Ecologies and Politics
Edited by Thomas Jellis, Joe Gerlach, and JD Dewsbury

For more information about this series, please visit: www.routledge.com/Routledge-Studies-in-Human-Geography/book-series/SE0514

Why Guattari? A Liberation of Cartographies, Ecologies and Politics

Edited by Thomas Jellis, Joe Gerlach, and JD Dewsbury

Routledge
Taylor & Francis Group
LONDON AND NEW YORK

First published 2019
by Routledge
2 Park Square, Milton Park, Abingdon, Oxon OX14 4RN

and by Routledge
605 Third Avenue, New York, NY 10017

First issued in paperback 2020

Routledge is an imprint of the Taylor & Francis Group, an informa business

British Library Cataloguing-in-Publication Data
A catalogue record for this book is available from the British Library

Library of Congress Cataloging-in-Publication Data
A catalog record for this book has been requested

ISBN 13: 978-0-367-72994-3 (pbk)
ISBN 13: 978-1-138-18349-0 (hbk)

Typeset in Times New Roman
by Apex CoVantage, LLC

Contents

Figures

Contributors

Manola Antonioli holds a PhD in Philosophy and the Social Sciences from EHESS (Paris) and previously oversaw the seminar series at the Collège International de Philosophie. She currently teaches the philosophy of architecture and urban studies at the Ecole Nationale Supérieure d'Architecture of Paris, La Villette. She has recently edited four books: *Théories et pratiques écologiques* (Presses Universitaires de Paris Ouest, 2013); *Paysage variations* (Éditions Loco, 2014); *Machines de guerre urbaines* (Éditions Loco, 2015); and *Biomimétisme* (Éditions Loco, 2017).

Christoph Brunner is Assistant Professor for Cultural Theory at Leuphana University Luneburg. In his research he deals with the relations between art, media, and activism investigating the aesthetic politics of social movements, in particular. He is director of the ArchipelagoLab for Transversal Practices, member of the SenseLab Montreal, and part of the European Institute for Progressive Cultural Policies (eipcp) and co-editor of *transversal texts*. He has published in *Third Text*, *Conjunctions*, *Inflexions*, *Open!*, and *Fibreculture*, among others.

Rebecca Catarelli is currently serving as Chief Operating Officer at Marlboro College, an experimental liberal arts institution in Vermont, USA, committed to helping undergraduate students pursue self-designed study that emphasises interdisciplinarity and individual expression. Having transitioned to the administrative side of academia several years ago, this will likely be her only publication of this type, which makes her feel both wistful and relieved. Rebecca continues to write experimentally and is currently working on a book that considers the instability of truth and experience through a narrative entwining motherhood, generational trauma, and the enigmatic nature of giant squid.

David B. Clarke is Professor of Human Geography and at Swansea University, where he previously served as Head of Department (2015–2018). He is a Co-director of the Centre for Urban Theory. His substantive research interests include work on cities, cinema, consumer culture, and value. He is the author of *The Consumer Society and the Postmodern City* (Routledge); editor of *The Cinematic City* (Routledge); and co-editor of *Jean Baudrillard: The Disappearance of Culture* (Edinburgh); *Jean Baudrillard: from Hyperreality* *to Disappearance* (Edinburgh); *Jean Baudrillard: Fatal Theories* (Routledge); *The Consumption Reader* (Routledge); and *Moving Pictures/Stopping Places: Hotels and Motels on Film* (Lexington).

JD Dewsbury is Professor in Human Geography at the University of New South Wales, Canberra, Australia. His research focuses on the performative and non-representational nature of social life, in particular, on how we view space, environment, affects, human agency, and objectivity. Using post-continental philosophy his research looks at understandings of habit, materiality and politics, the relationship between ontology and events, and the impact of assemblage theory and affect in research practice. Although situated in human geography, he has produced successful collaborations with performance studies, political theory, sociology, and environmental studies. His research has previously appeared in the journals *Environment & Planning D, Cultural Geographies, Transaction of the Institute of British Geographers* and *Performance Research*, and he is currently completing a monograph *Performative Spaces: Events, Materiality, and Subjectivity* (SAGE).

Marcus A. Doel is Professor of Human Geography at Swansea University, where he is also Deputy Pro-Vice-Chancellor for Research and Innovation, and Co-Director of the University's Centre for Urban Theory. Marcus is the author of *Geographies of Violence: Killing Space, Killing Time* (Sage) and *Poststructuralist Geographies: The Diabolical Art of Spatial Science* (Edinburgh University Press), and the co-editor of a number of books, including *Jean Baudrillard: Fatal Theories* (Routledge), *The Consumption Reader* (Routledge), and *Moving Pictures/Stopping Places: Hotels and Motels on Film* (Lexington). He has written widely on poststructuralist spatial theory, having published over 100 journal articles and book chapters.

Sasha Engelmann is Lecturer in GeoHumanities in the department of Geography at Royal Holloway University of London, where she co-directs the GeoHumanities Creative Commissions programme. She collaborates with creative practitioners to explore questions of environmental sensing and politics. Since completing her doctoral research as a creative ethnographer in collaboration with Studio Tomás Saraceno in Berlin, she has become an active member of the Aerocene community. She is currently preparing a book manuscript provisionally titled *Elemental Lures*.

Gary Genosko is Professor at the Faculty of Social Science and Humanities at the University of Ontario. He works on communication and cultural theory, subcultures in the digital underground, and whistleblowers. His philosophical interests includes the work of Félix Guattari and Jean Baudrillard. Post-media, communication modelling, critical semiotics, and media ecology play important roles in his writing. He is the editor of *The Guattari Reader* (Blackwell), and has also written *Félix Guattari: An Aberrant Introduction* (Continuum), *Félix Guattari: A Critical Introduction* (Pluto Press), and *The Reinvention of Social Practices* (Rowman and Littlefield).

Joe Gerlach is Lecturer in Human Geography at the School of Geographical Sciences, University of Bristol. Cutting across cultural and political geography, his research interests are centred conceptually on the notion of micropolitics and the geophilosophies of Spinoza and Guattari. Empirically, he has written on geographical matters in both critical-cartography and in nature-society politics in Latin America, with research published in *Progress in Human Geography, Transactions of the Institute of British Geographers, Environment and Planning A, Cultural Geographies,* and *GeoHumanities*.

Maria Hynes is a Senior Lecturer in the School of Sociology, Australian National University. She has an enduring interest in forms of experimental thought and practice, particularly those at the intersection between art and science. Her research pursues an ethico-aesthetic conception of the social sciences, in areas relating to affective and biopolitical forms of power, practices of creative resistance and anti-racism. She has published in a variety of journals, including *Angelaki, Environment and Planning A & D, British Journal of Sociology, Sociological Review, Performance Review, Parallax, Culture Machine, Borderlands, Fibreculture,* and *Continuum*.

Thomas Jellis is a British Academy Postdoctoral Fellow at the School of Geography and the Environment at the University of Oxford, and a Research Fellow at Keble College. He has written on geography's relations to experimentation, art, and minor theory. His current research seeks to trace a geo-history of exhaustion.

Anja Kanngieser is a political geographer and sound artist. They hold a Vice Chancellors Research Fellowship in Geography, University of Wollongong, Australia. They are the author of *Experimental Politics and the Making of Worlds* (2013), and have published in interdisciplinary journals including *South Atlantic Quarterly*, *Progress in Human Geography* and *Environment and Planning D*. http://anjakanngieser.com/

Michele Lancione is an urban ethnographer and activist interested in issues of marginality and diversity, homelessness, and radical politics. His works have been published in journals like EPD, EPA, Transactions of the IBG, Urban Studies, and IJURR, among others. His first edited volume is *Rethinking Life at the Margins* (Routledge, 2016) and his works include a documentary film around forced evictions in Bucharest, Romania (www.ainceputploaia.com). Michele is also one of the founders of the forthcoming publication, the *Radical Housing Journal*. He is based at the Urban Institute and the Department of Urban Studies and Planning, University of Sheffield. You can get in touch @michelelancione.

Andrew Lapworth is a Lecturer in Historical and Cultural Geography at the University of Bristol, UK. His research interests lie in continental philosophy (especially the thought of Deleuze, Simondon, Guattari, and Whitehead) and its implications for geographical thinking around ethics, politics, aesthetics, and subjectivity; contemporary encounters of art, technology, and science; and cinematic geographies and the politics of thought. His research has previously appeared in the journals *Theory, Culture & Society*, *Cultural Geographies*, *Transformations: Journal of Media and Culture*, and *The Journal of Urban Cultural Studies*.

Mahoro Murasawa, born 1968, is a Professor of Sociology of Ryukoku University. He works especially on Gabriel Tarde and Félix Guattari, and is interested in the problematic of ecology and subjectivities. His latest publication is *Life, Mind and World: Dialogue with Hisao Nakai* (Kawade-shobo) in 2018.

Stéphane Nadaud, born 1969, is a French Pedospychiatrist and Philosopher. He works on Guattari and the political and philosophical problematic of subjectivities. He edited Guattari's *The Anti-Œdipus Papers* (Semiotexte) and *Qu'est-ce que l'écosophie* (Lignes). Stéphane and Mahoro are friends 'beyond frontiers' and regularly write together in both French and Japanese. They hope, one day, to be able to walk the Shikoku pilgrimage together.

Anne Querrien, born 1945, studied sociology and political science, was an activist in the French student union and in the organisation created by Félix Guattari in 1965: *Fédération des groupes d'études et de recherches institutionnelles*. A student in Nanterre University in 1968, she participated in the 22nd March Movement. She participated in the *Centre d'études et de recherches institutionnelles* led by Félix Guattari, and followed Deleuze's lessons in Vincennes. She worked in the Ministry for Public Works and Ecology, as the editor of the journal *Annales de la recherche urbaine*. Today she is one of the editors of *Multitudes* and *Chimeres,* created by Deleuze and Guattari.

Tom Roberts is a geographer and Research Associate at the School of Physical, Environmental and Mathematical Sciences at the University of New South Wales Canberra. Tom's intellectual background is in social and cultural geography, with expertise in theories of affect, new materialism, and post-humanist thought. His research addresses the ontological status of matter within geography and cognate disciplines, in order to better understand the kinds of material processes implicated in contemporary spatial experience. Tom is especially interested in 'processual' materialism and has published work on a variety of process philosophers, including Félix Guattari, Alfred North Whitehead, Gilbert Simondon and Gilles Deleuze.

Arun Saldanha is Professor in the Department of Geography, Environment, and Society at the University of Minnesota. He is author of *Psychedelic White: Goa Trance and the Viscosity of Race* (Minnesota, 2007) and *Space After Deleuze* (Bloomsbury, 2017), and co-editor with Hoon Song of *Sexual Difference Between Psychoanalysis and Vitalism* (Routledge, 2013), with Rachel Slocum of *Geographies of Race and Food: Fields Bodies Markets* (Ashgate, 2013), and with Jason Michael Adams of *Deleuze and Race* (Edinburgh, 2013). Arun is working on a theoretical book on the materialities of race and the edited collection *Prince from Minneapolis*.

Nina Williams is a cultural geographer influenced by non-representational theory, post-humanist thought, and process ontologies. She lectures in social and cultural geography at the University of Bristol and the University of New South Wales, Canberra, Australia. Her work engages with theorisations of ethico-aesthetics, minor creativity, and the processes of subjectification, particularly as they are understood in the philosophies of Félix Guattari, Gilles Deleuze, and Henri Bergson. She draws upon these conceptual starting points to develop experimental and collaborative techniques of research in the contexts of art and curation; walking and mapmaking; audio technologies; and fashion and style.

Acknowledgements

This book has been in the making for some time now. We would like to start, therefore, by thanking everybody involved for their patience and enthusiasm, not least that of Ruth Anderson, Priscilla Corbett, and Faye Leerink at Routledge, and Anandan Bommen for managing the project to completion. Thomas Jellis and Joe Gerlach are grateful to the Economic and Social Research Council for initial support, and subsequently to the British Academy for the award of postdoctoral fellowships which sustained this project; grant numbers pf160023 (Jellis) and pf140034 (Gerlach). Some of the ideas herein were first essayed at the Annual Meeting of American Association of Geographers in New York in 2012 during sessions entitled 'For Félix: Transversal Geographies'. We are grateful to all the participants, many of whom feature in this collection. In the intervening period, we are grateful to audiences at the School of Geographical Sciences, University of Bristol, and Nottingham Contemporary. To that end, this book is the (sometimes frantic) culmination of conceptual and transversal energies of many bodies. We would like to thank those whose ideas and generosity have made themselves felt in the composition of the text. In particular we extend our gratitude to colleagues at the University of Oxford, namely Andrew Barry (now University College London), Maan Barua (now University of Cambridge), Marion Ernwein, Ian Klinke, Richard Powell (now University of Cambridge), and Sarah Whatmore. We are grateful, also, to Hertford College, Jesus College, Keble College, and Mansfield College for providing institutional support, to which end we extend particular thanks to Ali Rogers, Colin Clarke, and Patricia Daley. Likewise, thanks to colleagues at the University of Bristol, especially, Keith Bassett, Maria Fannin, Andrew Lapworth, Merle Patchett, James Palmer, and Nina Williams. Sincere thanks too to colleagues Down Under, Maria Hynes (Australian National University), Thomas Keating, Tom Roberts, and Scott Sharpe (University of New South Wales). And thank you to those who have at various points have delivered well-timed pep talks or just reminded us to stay the course: Ben Anderson, James Ash, Jane Bennett, Christoph Brunner, William Connolly, Paul Harrison, Hayden Lorimer, Erin Manning, Brian Massumi, Anne Sauvagnargues, Nigel Thrift, and John Wylie. Andrew Goffey has been very supportive, not only with invites to workshops ('Bad Atmospheres and Toxic Positivity' in 2015; 'Winter Detox' in 2016) but for talking Guattari no matter where or when (not to mention all the translations). We also extend our sincere thanks to Derek McCormack for getting us going with Guattari, and whose generosity in both theory and mentorship has ignited and sustained our own, often inadequate, thought experiments, and a friendship beyond that.

Last, but never least, thanks to Suwita, Soohyun, and Noah.

Foreword

Gary Genosko

Creating and sustaining Guattari studies

In 1994 when I began working on *The Guattari Reader* (1996), eventually published in 1996, Blackwell Readers were an interdisciplinary touchstone for the consolidation of theoretical contributions of both established and soon-to-be-established figures. My choice of Guattari was risky to the extent that I did not know if, only a few years after his death in 1992, that permission to translate could be negotiated from an estate or at least a centralised repository. Rumours were circulating about the executors, and I consulted both Brian Massumi, Michael Hardt, and Timothy Murphy for advice about this situation: who was the literary executor of the material deposited into the French government's Institute for Contemporary Publishing Archives?

It turned out that the permissions process was a tangled mass of string that took some time to unravel. The diversity of copyright holders is nothing new to editors, but the Guattari children, in particular his son Bruno, Félix's contacts in the magazine and publishing worlds, including colleagues at Éditions Recherches and *Chimères*, not to mention *Le Monde* in the pages of which Guattari published his final public statement a few months after his death, were all helpful and flexible. Anne Querrien, Florence Pétry, and Maurice Nadeau, editor of *Quinzaine Littéraire,* were particularly helpful. There was nothing quite as pleasing as Nadeau's scrawled note in the margins of my request for permission to translate: *D'accord, gratuit*. The journal founded by Deleuze and Guattari *Chimères* sent me a half dozen copies of the journal to send around to interested Canadian parties as my only cost.

At the time I wanted Balthus' painting *The Street* to be on the cover, but the Balthus estate would not cooperate. This precipitated some tense faxes back and forth with Blackwell until we settled on the Marcel Duchamp image. But whenever I look at the cover I still think it was a compromise on two grounds. First of all, Guattari discussed Duchamp, of course, but focused on the later work *Etant donnés*, and not so much the early futurist nudes. Guattari has written a wonderful essay, "Cracks in the Street", that put the Balthus painting on my radar. Guattari's heterogeneous writings on art remain uncollected. And the Balthus estate remains intractable. As editor, I was perhaps too conservative, or at least in the grip of modern art. Why not a George Condo or, even better, a David Wojnarowicz? A still from one of the Sarenco films for which Guattari received a credit?

The organisation of *The Guattari Reader* was based on the principle that the anti-oedipal approach could be introduced through the position of Guattari on the anti-psychiatry social movement, especially its European versions and variations, and that the emergence of institutional analysis and schizoanalysis found in this manner a historical and sociological footing. The first two sections situated Guattari in relation to Italian alternatives to psychiatry, as well as British experiments and the critique of psychoanalysis, giving way in the second part to positive statement on schizoanalysis and contemporary theoretical stalemates such as postmodernism and post-media practices. The third section was a sprawling interview in which Guattari refined his approach to institutions and organisational politics. I devoted a section to the role of semiotics in Guattari's thought, linking it to what he learned from Foucault's theory of power, crossing a microphysics of power with molecular politics. The fifth section included a number of papers and interviews Guattari had given on queer subjectivities, and the final and sixth section dealt with ecologies, typology of capitalism, prospects for renewing communism, and the remaking of social practices on molecular scales. It was this final essay, first appearing in *Le Monde* after his death that still inspires me. Its tone has never left me, and the title of my new collection of career-spanning essays, *The Reinvention of Social Practices* (2018), comes from it. Much in evidence is the unmistakable Guattarian tone of muted hope for the post-media era that would surpass media induced passivity, loneliness, individualism, the ravages of capitalism, and eludes the persistent molecules of fascism.

While I was working on *The Guattari Reader* I received a tragic message from Jacques Pain, from Université de Paris X. He simply noted at the bottom of a letter that François Tosquelles, the great Catalan red psychiatrist, had passed away on 25 September 1994. Tosquelles inspired Guattari to rethink the doctor–patient–institution relationship and to establish institutional analysis as a politically relevant alternative to psychiatry and psychoanalysis.

I edited *The Guattari Reader* as an independent scholar. I relied on favours because of my precarity; I cajoled, begged, borrowed, traded. Almost everything was still photocopied and exchanged by snail mail as it was not until 1995 that email became unavoidable, at least for me. I enlisted a team of translators; some more polished than others, and set about editing in earnest. The files I carried around were like paving stones, and when I occasionally set them down, beneath them was a dune-like stretch of sand, hardly a beach.

Global Guattari

I visited the Guattari archives at the Institute for Contemporary Publishing Archives in the fall of 2000 before the collection was completely inventoried. What I could consult, and what was held back by special request in accordance with the massive intellectual biographical project of François Dosse (2010), at that time already underway on Deleuze and Guattari, revealed a few surprises.

These provided the foundation for my efforts to globalise Guattari studies first by situating his numerous and lengthy visits to both Japan and Brazil. The impressive array of short Japanese texts on a variety of topics composed by Guattari, including a number of interviews, during his visits there in the 1980s, provided the basis for a chapter in my next book, *Félix Guattari: An Aberrant Introduction* (2002), as well as a few small translations. Later I would revisit the relationship between Guattari and Japan and his attempt to grasp the Japanese unconscious, not in the secret universe of each person, but as a social or extra-psychical entity perfusing contemporary modern culture. My later explorations were made possible by the translator and editor Drew Burk at his imprint Univocal (now University of Minnesota Press), and my colleague Jay Hetrick, under the title *Machinic Eros* (2015).

Guattari developed a sense of the singularity of Japan's "mutant creationism" during his visits there over the course of the Bubble Economy of the 1980s up to its collapse in the early 1990s. Guattari visited Japan at least eight times during the period between 1980 and 1992. These visits can be clarified in the following manner: early visits involved dialogues with Tetsuo Kogawa (fall 1980 and spring 1981) and Kuniichi Uno (University of Kyoto in 1983 and Uno's La Borde visit in 1984-1985); mid-decade dialogues with Min Tanaka (summer 1985) as well as appearances at pirate micro-FM radio station Radio Homerun and a pilgrimage to the Sanya neighbourhood in Tokyo in honour of the assassinated documentary filmmaker Mitsuo Sato, followed by a February 1985 visit to Tokyo where Guattari writes "Tokyo, the Proud", and a June 1986 visit to the Yayoi Kusama show to which he contributes a catalogue essay in anticipation of the major Centre Pompidou exhibition the following year of *Japon des avant-gardes* (staging a dialogue with Akira Asada). The later visits become strange as they are sponsored by the Seibu department store (1987), and the large-scale Japan Institute of Architects event (encompassing Guattari's dialogue with Shin Takamatsu and the Nagoya urban planning presentation), followed in the same year by the Keiichi Tahara exhibition and catalogue essay in Paris. There was one more visit in the summer of 1992 as Guattari's old friend, painter Imaï Toshimitsu, was hospitalised for leukaemia treatment (he passed away in 2002), and he planned to realise a film with photographer Keiichi Tahara.

During the 1980s, Japanese translations of Guattari's single-authored books began to appear, as well as documents of his activities in Japan. Kuniichi Uno and Masaaki Sugimura are key figures in these efforts. Guattari's repeated visits to Japan were submersive experiments into a machinic interconnectedness striated by animist tendencies that he deployed in a panoply of interfaces – 'collaborations' and different kinds of writing (short pieces for fashion magazines, film scripts, art criticism) for likely and unlikely audiences. During his visits to Japan, Guattari participated in the transits of machinic subjectivities and objectivities and joined some of their assemblages through his engagements, trying to understand the machinic eros of Japanese culture, that is, the desire to be in the thick of things, immersed in the pop cuteness, and stuck in refrains of gaming.

Sugimura's Guattari Studies Group at Ryukoku University in Kyoto recently in 2018 translated my Pluto volume, *Félix Guattari: A Critical Introduction* from the Modern Masters Series (2009) into Japanese. The spirit of this group, with its study, translation, and graphical (screen prints) investigations of Félix is an antidote to the 'big Deleuze' brand and the Guattari-phobia so prevalent therein.

Further, Guattari was instrumental in encouraging cultural exchanges between France and Japan in the mid-80s as well, through his involvement in the Centre Georges Pompidou in Paris. Some of his visits to Japan were sponsorships by organisations that could only have occurred during the bubble economy, and some of his conversations with public intellectuals around the infantile character of Japanese capitalism and artists like Buto dancer Min Tanaka, concerning animal and vegetal becomings, gave rise to fraught dialogues, but did not prevent collaborations like Tanaka's performance at Clinique de la Borde, captured in the film by partner Joséphine Guattari and friend François Pain, *Min Tanaka à La Borde* (1986).

A further surprise was Guattari's self-definition as a writer of philosophically-inspired fictions. Some of his unpublished theatrical dialogues were based on ancient Greek philosophical figures, such as Socrates and Parmenides. Others were more overtly Sartrean in formation. Guattari's notes for a made for television film about, or rather, *by* a molecular Kafka, would come to garner my attention given his curatorial interests in the Kafka centenary in 1984. Guattari never stopped working on Kafka. He loved Kafka's understanding of bureaucratic perversion in all of its micro-fascistic power rendered in static forms, procedures, protocols, and hierarchies, alienating those encountering it. The problem is the curious pleasure bureaucracy affords to the growing ranks of those who take delight in administration. Kafka's vision was not sombre and sad, but full of humour. The "great paranoid bureaucratic machine" and the "little schizo machine" are in the same assemblage. The suits proliferate, and office machines mutate. This happens together, not apart. Guattari could not quite plug himself into arts administration, and gave to curator Yasha David full control of the Kafka exhibit. Unfortunately, it would not travel to Japan as Guattari had hoped.

When I first stepped into the foyer of the Institute for Contemporary Publishing Archives, then still in Paris, a copy of the original Portuguese volume *Guattari Entrevista Lula* was on display. Guattari's visit to Brazil in 1982 and his interview with Lula da Silva, then head of the Workers' Party, was an important political and personal marker for Guattari as he had threatened to relocate to Brazil during a lengthy period of political greyness in France. In 2003 I published a translation of this interview, and introduced it by recounting the events leading up to Lula's election as President of Brazil in 2002, 20 years after the interview took place. *The Party Without Bosses* was an amalgam: translated interview, up-to-date report on Lula's election as president, and reflection on the Guattari-Lula relationship. Originally, the book included a set of scholarly essays on Guattari, but publishing in Canada at the time was undergoing a seismic shift with the incursion of big box stores that were over-ordering and holding ransom the unsold books from small presses like my publisher, Winnipeg's Arbeiter Ring. Semiotext(e) later picked up this trail with the translation of Suely Rolnik's compendium of Guattari's talks and writings during his visits to Brazil, *Molecular Revolution in Brazil* (2008).

Scholarly scaffolding

English translations of Guattari's writings have lagged behind the early efforts by Japanese scholars, and the many editions of collections about his activities in Brazil. There are still important missing pieces: a consolidation of the two different books originally published in 1977 under the title of *The Molecular Revolution*, and a very early and in some ways isolated translation of selections from these by Penguin house translator Rosemary Sheed in 1984. Translators continue to tackle new volumes of occasional writings such as *Qu'est-que l'écosophie* (2013). There are many isolated articles and interviews that remain untranslated.

I recently worked on Guattari's homage to French psychoanalyst Françoise Dolto (see Guattari, 2018). Guattari published "A Game of Scrabble with Lacan" in *Le Monde* (August 28, 1988) four years before his death. Dolto passed away on 25 August 1988, so Guattari's response was immediate. Among the many articles he placed in the newspaper's pages, this homage to Dolto stands apart as it does not use Guattari's theoretical vocabulary. Instead, it is a personal reflection on Dolto's important innovations and contributions to child analysis. Part obituary, part muted intellectual biography, and certainly part reconciliation with the Lacanian tradition in which he was trained, Guattari recounts in outline Dolto's personal and professional trajectory.

Above all, Guattari deploys the figure of the game of Scrabble as Dolto's characterisation of how beyond life she would rejoin Lacan and engage him in a friendly game, letting the board serve as an integrative device for expressing mutual respect, without giving up the challenge of balancing lexical meaning and absurdity. With no room for caustic remarks about linguistic imperialism, Guattari simply lets the letterati sort it out among themselves across the board.

"The Guattari Effect" conference at Middlesex University took place in 2008 and gave rise to an important volume in 2011 of selected proceedings and translations. This event, at the impetus of the indefatigable guattarian, Eric Alliez, remains the major English-language site of reflection on Guattari's legacy, even though a few key figures were absent: Janell Watson (2009) and Maurizio Lazzarato (2014), both of whom have published important studies on Guattari. It should also be noted that the title changed in 2018 from *Deleuze Studies* to *Deleuze and Guattari Studies* of the flagship philosophical journal, after 10 years of publication marks a decisive gesture of inclusion, but it has not been yet worked out on the ground in the multiple annual conferences.

There is so much exciting work being done in the field at the moment. Research into the legacies of institutional analysis will help to contextualise Guattari's formulation of schizoanalysis by giving it a clinical foundation; Guattari's film studies, hitherto based on his theory of minor cinema, are front and centre after the translation of his script for a science fiction film, *A Love of UIQ* (2016). Although he worked on a number of different film proposals, this script remains unfilmed, yet the presentations by translators Silvia Maglioni and Graeme Thomson in their filmic response *In Search of UIQ* point in the direction of an eventual full realisation. The enduring influence of Guattari's *The Three Ecologies* (2000) continues to be felt in numerous books renewing and expanding his unique theorisation of ecosophy. There is even talk of a new translation.

What remains to be accomplished is a comprehensive review of Guattari's aesthetics. Schizoanalysis emerges from a series of criticisms aimed at structuralist versions of linguistics, psychoanalysis, and Marxism, the closed interpretive grids and scientific pretensions of which are overcome by aligning practical institutional analytics with aesthetic practices. By gathering and analysing Guattari's writings on art a greater understanding of the aesthetics of the schizoanalytic project may be gained. Examples include writings on architecture, cinema, painting, photography, theatre, and poetry. As Guattari's writings on art remain uncollected, the task would require bringing together for the first time his contributions. A number of contemporary critics and philosophers have partially situated Guattari's art writing within aesthetic movements such as modernism, relational aesthetics, and political art, but the big picture remains to be developed and elaborated. Many of Guattari's own fictional texts remain untranslated and languish in obscure locations.

There is a pressing need to gather together and translate his seminars, many of which are available in French from journals such as *Chimères*. Indeed, his case notes and inventions at La Borde remain largely untapped, including the formation of the Club de la Borde, glossary of constitutive definitions, and a more fine-toothed understanding of the developments of 'the grid', his conception of the rotating schedule of tasks and how it was adapted along its trajectory. With so few of Guattari's case studies available, his clinical practice remains elusive. Of course, we know that his first work as a psychotherapist, working alongside Jean Oury, treating a schizophrenic patient known by the initials R.A., involved the creative use of a portable tape recorder, which seemed most effective when it was turned off.

The pool of Guattari translators in the English world is growing, with Andrew Goffey and Taylor Adkins doing some heavy lifting. Philosophy lists remain open to Guattari's unclassifiable writings. Edited collections of secondary literature like this one are beginning to emerge in the spirit of the effort I initiated in a volume devoted to the study of Guattari's thought in my three volume collection, *Deleuze and Guattari: Critical Assessments of Leading Philosophers* (2001), and a special issue of *Deleuze Studies* (Félix Guattari in the Age of Semiocapitalism) (2012).

I am often asked about the status of schizoanalysis today. In answering I usually fall back on Jean-Claude Polack's assessment: there are no practising schizoanalysts, at least in France. This leaves only one solid option: what about Brazil? The exemplary theory and practice of Peter Pál Palbert with outpatients in his travelling theatre troupe, Ueinzz Theatre Company is a peripatetic artistic cartography and mobile micro-institution. If there is an answer to the question, this is part of one.

We are only beginning to appreciate the presence of Guattari's thought and how it may contribute to a critical understanding of our present situation. The transdisciplinary uptake of Guattari's thought has been slow, perhaps fittingly so as this pace has provided time for extended enjoyment, unlike the hurried industry that has grown up around Deleuze. Franco Beradi (2008) more than anyone else has helped readers to grasp the potential of Guattari's visionary cartography – his techno-nomadism, production of subjectivity as the real, creation of new forms collective assemblage, and release of the molecules swimming beneath heavy molar formations. Berardi's books are like a semiochemistry kit that readers can set up and try their hands at mixing, heating, cooling, pouring, accelerating, and capping. What's best: there isn't an app for this.

I have spent more than 20 years trying to answer the question: why Guattari? I may be getting close to something akin to an answer.

References

Alliez, E. and Goffey, A. (eds.) (2011) *The Guattari Effect*. London: Continuum.

Berardi, F. (2008) *Félix Guattari: Thought, Friendship, and Visionary Cartography* (trans. G. Mecchia and C. J. Stivale). London: Palgrave.

Dosse, F. (2010) *Gilles Deleuze and Félix Guattari: Intersecting Lives* (trans. D. Glassman). New York, NY: Columbia University Press.

Genosko, G. (ed.) (1996) *The Guattari Reader*. Oxford: Blackwell.

Genosko, G. (ed.) (2001) *Deleuze and Guattari: Critical Assessments of Leading Philosophers*. London: Taylor & Francis.

Genosko, G. (2002) *Félix Guattari: An Aberrant Introduction*. London: Continuum.

Genosko, G. (2003) *The Party without Bosses*. Winnipeg, MB: Arbeiter Ring.

Genosko, G. (2009) *Félix Guattari: A Critical Introduction*. Pluto: London. [trans. into Japanese by M. Sugimura and M. Matsuda. Tokyo: Hosei University Press, 2018.]

Genosko, G. (ed.) (2012) Special Issue of *Deleuze Studies*, "Félix Guattari in the Age of Semiocapitalism". Edinburgh: Edinburgh University Press.

Genosko, G. and Hetrick, J. (eds.) (2015) *Machinic Eros: Félix Guattari's Writings on Japan*. Minneapolis, MN: Univocal.

Genosko, G. (2018) *The Reinvention of Social Practices: Essays on Félix Guattari*. London: Rowman & Littlefield International.

Guattari. F. (1984) *Molecular Revolution* (trans. R. Sheed). London: Penguin.

Guattari, F. (2000) *The Three Ecologies* (trans. I. Pindar & P. Sutton). London: Athlone Press.

Guattari, F. and Rolnik, S. (2008) *Molecular Revolution in Brazil* (trans. K. Clapshaw and B. Holmes). Los Angeles, CA: Semiotetext(e).

Guattari, F. (2013) *Qu'est-ce que l'ecosophie*. Paris: Lignes.

Guattari, F. (2016) *A Love of UIQ* (trans. G. Thomson and S. Maglioni). Minneapolis, MN: Univocal.

Guattari. F. (2018) A Game of Scrabble with Lacan (trans. G. Genosko and I Ariss). *The Lacanian Review*, 4: 118-121.

Lazzarato, M. (2014) *Signs and Machines: Capitalism and the Production of Subjectivity* (trans. J.D. Jordan). Los Angeles, CA: Semiotext(e).

Pain, F. and Guattari, J. (dirs.) (1987) *Min Tanaka à la Borde*. Available online at: https://youtu.be/VgErye7jXbI

Watson, J. (2009) *Guattari's Diagrammatic Thought: Writing Between Lacan and Deleuze*. New York, NY: Continuum.

Introduction

For better and for worse

Thomas Jellis, Joe Gerlach, and JD Dewsbury

Pierre-Félix Guattari (1930–1992) remains something of an enigma. Variously described as psychoanalyst, political activist, and 'Mister Anti', and more besides, Guattari was involved in a range of ways in the intellectual tumult of post-war France. And yet, unlike many of his contemporaries, often know as *les soixante-huitards*, his work has not generated a significant body of secondary literature. As Goffey (2017) notes, this has resulted in a tendency for commentary, where it exists, to have a somewhat 'introductory flavour' to it.

Perhaps befitting of his character, there is no orthodox biography of Guattari. There is, of course, François Dosse's (2011) *Intersecting Lives*, which despite the "evident theoretical confusions and some unfortunate mistakes" (Alliez and Goffey, 2011: 7) draws out some important connections between Guattari's life and his thought. Much of the most interesting material about Guattari in the book emerges from 49 interviews conducted by Virgine Linhart for her own biography of Guattari, later abandoned (see Osborne, 2011). Inevitably, by developing the book around the intersecting lives of Guattari and his erstwhile collaborator, Gilles Deleuze, there is a danger that Guattari can only be understood in conjunction with his arguably more acclaimed friend. Franco 'Bifo' Berardi's (2008) *Félix Guattari: Thought, Friendship and Visionary Cartography* is a very different kind of biography, akin to a set of reflections on their relationship and describes how Bifo encountered Guattari's thought, written over the course of 10 years or so.[1] As Bifo (2008: 5) puts it, the book seeks to "reconstruct the rhythmic map of Félix-thought". There is, too, Guattari's own fragmentary autobiography, which was published as *Ritournelles*.[2] This 'internal monologue' was completed in 1992, just months before he died, with the help of his friend the artist Gerard Fromanger. The original text was edited down from 300 pages to 80, and described by Guattari as memory fragments (see Dosse, 2011: 429); it returned to themes raised in a novel tentatively called *33.333*, which referred to his life and birthday (30 March 1930), that never saw the light of day.

Rather than trace the course of Guattari's life, then, we call attention to these texts which are better placed to do so, not least because there are difficulties, not to mention tensions, in trying to recount a life of somebody who did not believe that the subject coincided with the individual. As Frida Beckman notes, "to focus on and give a chronological account of a life . . . inevitably delimits all the things that it expressed" (2017: 9). She suggests approaching biography through foregrounding "the intensities and multiplicities that run through an individual". For our purposes here, rather than focus on Guattari's life, if such a story could be told in all its detail, we want to chart an intellectual trajectory of sorts. In short, we want to highlight the intensities and multiplicities that animated Guattari's thought, as much as the after-lives of such thinking.

In a diary entry penned when he was 23, Guattari exclaimed: "I WANT TO WRITE A BOOK".[3] Some 19 years later, in 1972, Guattari published not one but two books. *Anti-Oedipus*, co-authored with Deleuze (Deleuze and Guattari, 2004a), and *Psychoanalysis and Transversality* (Guattari, 2015a), a collection of essays, were to be the first of many. Over the next 20 years, Guattari co-authored three further books with Deleuze: *Kafka: Towards a Minor Literature* in 1975 (Deleuze and Guattari, 1986a); the second volume of *Capitalism and Schizophrenia*, the follow-up to their first collaboration, *A Thousand Plateaus* in 1980 (Deleuze and Guattari, 2004b); and *What Is Philosophy?* in 1991 (Deleuze and Guattari, 1994).[4] He also wrote with others, producing *Communists Like Us: New Spaces of Liberty, New Lines of Alliance* (Guattari and Negri, 1990) and *Molecular Revolution in Brazil* (Guattari and Rolnik, 2008). But, and crucially, he wrote a number of books on his own, including *Molecular Revolution* in 1977 (Guattari, 1984), *The Machinic Unconscious* in 1979 (Guattari, 2011); *Schizoanalytic Cartographies* (Guattari, 2012a) and *The Three Ecologies* in 1989 (Guattari, 2008); and *Chaosmosis* in 1992 (Guattari, 1995).[5] In addition to these texts, other less well-known pieces have also come to light. For instance, *Lines of Flight* (Guattari, 2016a) was not discovered until after Guattari's death, and a screenplay, *A Love of UIQ* (Guattari, 2016b), has only recently been published. Moreover, there have been a series of books collecting his shorter pieces, such that in addition to *Psychoanalysis and Transversality*, which covered his work between 1955 and 1971, there has been *Chaosophy* covering texts and interviews from 1972 to 1977 (Guattari, 2009a); *Soft Subversions* for those between 1977 and 1985 (Guattari, 2009b); and, hopefully, *The Winter Years* for the final stretch, 1986–1992.[6] Other edited collections include *The Guattari Reader* (Guattari, 1996), *The Anti-Œdipus Papers* (Guattari, 2006), and *Machinic Eros: Writings on Japan* (Guattari, 2015b). A number of books remain untranslated.[7] Guattari's final text, written just a few weeks before his death, was published posthumously shortly afterwards and later translated as 'Remaking Social Practices' (see Guattari, 1996: 262–272). His next 'project' may have centred on war, after a series of dialogues with Paul Virilio, material which only exists in archival boxes (and may well remain so, now that both thinkers have passed away) (see Dosse, 2009; 2011).

Two remarks might be made at this point. Firstly, we might note that Guattari published very little in the 1980s. As Dosse (2011: 423) writes, "in the mid-1980s, the indefatigable Guattari, ever in search of new ideas, lost his footing". Guattari referred to this period of deep depression as the winter years, in which there were disputes at La Borde, the death of his mother, several cardiac episodes, and stretches where he was described as catatonic. Secondly, we might note the disjuncture between the points at which Guattari's work was originally published and when it became translated into English. Nearly all the translations have been undertaken since Guattari's death, and much of this in the last decade or so. Furthermore, translation has proceeded such that Guattari's earlier work is appearing a good deal after his later work. This has complicated the reception of his ideas and also pointed to the porosity of texts, as concepts travel between solo and collaborative endeavours. A case in point is *The Machinic Unconscious*. The temptation is to read this book through Guattari's contemporaneous work with Deleuze, to such an extent that the former "may be read as a workbook for *A Thousand Plateaus*" (Genosko, 2012: 167). Yet for Guattari, it was important not to see in this "a business of paternity relating to the ideas advanced" (Guattari, 2011: 333n.4).

The question of how to distribute "the weight of authorial input and authority correctly" continues to bewitch readers of Deleuze and Guattari (Genosko, 2012: 158). The aim, if we follow the pioneering work of Gary Genosko is not to 'separate' Guattari from his co-authors. Indeed, for all that Guattari has been occluded, we seek to avoid "an exegetic practice that would simply 'rebalance' the prevailing doxa of Deleuze studies" (Alliez and Goffey, 2011: 9). But it is worth noting just how imbalanced the relationship has become. This has manifested itself in at least three ways. Firstly, there have been attempts to either pull the two apart completely, as Žižek (2004: 20) has done – such that he is apparently able to identify the 'Deleuze proper' from the 'guattarized books' – or, in Badiou's (2000) case, to simply to ignore their collaborations. Deleuze was clearly uncomfortable with the way in which commentaries tended to focus on his work at the expense of Guattari and he drew attention to the futility of trying to "disentangle inseparable elements and identify who did what" (Deleuze, 1995: 7). Secondly, and more commonplace, is the use of parentheses, so that the authorship is listed as 'Deleuze (and Guattari)', qualifying Guattari's role or authority. In such accounts, Guattari is an after-thought. Thirdly, and most routinely, is the simple ordering of the two names – philosopher first, idiot savant second. This only becomes evident once you realise that it is still jarring to read of Guattari and Deleuze (Genosko, 2012: 167). One of the aims of this collection, alongside the few secondary texts on Guattari (Alliez and Goffey, 2011; Elliot, 2012; Genosko, 2002; 2009; 2018), is to showcase just how important Guattari is in his own right.[8]

In his own words, he had "everything in my head, nothing in the pocket" (Guattari, 2006: 400). This is corroborated by the sheer range of topics he engaged with, from "[o]ceanic anthropology and contemporary art to institutional therapy and autonomist political praxis" (Alliez and Goffey, 2011: 1), but it was also much too modest. Guattari wrote a great deal. He was always on the move. As Deleuze (2007: 237) said of him: "He jumps from one activity to another, he sleeps little, he travels, and he never stops. He never pauses". In many ways, Guattari's thought followed a similar rhythm. He jumped from one idea to the next, in ways that can be difficult to follow. And part of this movement of thought meant for a very different language. In effect, Guattari (2009b: 21) "had had to forge [his] own language in order to confront certain questions, and to forge a language means to invent words, keyterms, carrying-case terms". On this he was unapologetic:

"I understand that this annoys some people but, in the end, that is not problem" (2009b: 21–22). Indeed, his speed of thought and his jargon machine (Gerlach and Jellis, 2015a; Goffey, 2012), or what Stivale (2009) has decribed as Guattari's "renowned supposed 'difficulty'", may be one of the reasons for why Guattari is so intriguing. Indeed it is one of the questions which animates this book.

Why Guattari?

To pose the question 'why Guattari?' is not without risk. It invites accusations of a rank individuation in the form of homage, and indeed in the guise of hagiography. Veneration is, of course, vehemently antithetical to a Guattarian ethos, and moreover, not possible if one takes seriously Gilles Deleuze's (in Deleuze and Parnet, 2006: 12, emphasis added) account of his friend, "one whose proper name denoted something which was happening and *was not a subject*". Indeed, geographer Maria Fannin (2015: 173) addresses the pitfalls head-on, noting, "it would be unfortunate to see the philosophers who were most critical of orthodoxy in their own work themselves become the names associated with the new orthodoxy". The question 'why' also tempts a petulance in the hair-trigger rebuff, *why not* Guattari? And, predictably, why not *with* Gilles Deleuze? More egregious still, it risks the insistence that Guattari and his work can only matter via an umbilical dependence to 'the conjuncture'; a retrofitted context or pre-prepared problem enjoining suspiciously well-formed historical and geographical constraints. To stake Guattari's significance on a contrived tethering to ostensible urgencies of the present is to fall, hopelessly so, into the conceit of a social science itself impelled by both deadened notions of 'the empirical' and by financier reveries of impact.

Contrary to his mischievous caricaturing as Mister Anti, Guattari was no grumpy antagonist of disciplinary lines in geography and in the wider social sciences. In respect of the former, his proclivity toward mapping is of particular repute. In respect of the latter field of enquiry he harnessed, sympathetically, its diagrammatic and modelling tendencies, but without recourse to associated forms of behaviouralist or positivist intent. Indeed, reflecting on his analytical turn toward models, Guattari (2012a: 3) remarked:

> all modelling systems are valid, all are acceptable, in my opinion. This is solely to the extent that their principles of intelligibility give up any universality pretention and admit that they have no other mission than to contribute to the cartography of existential Territories, implying sensible, cognitive, affective, aesthetic, etc. Universes, for clearly delimited areas and periods of time.

This caution in hand, Guattari developed concepts that would, *inter-alia*, "protect schizoanalysis from every temptation to give in to the ideal of scientificity that ordinarily prevails in the 'psy' domains, like a collective Superego" (2012a: 32). It is therefore not science and the universe that are targeted by Guattari, but modes of scientificity and universality instead. Put differently, any tactic, technique, semiosis, or science that territorialises the subject or the unconscious – and conversely any subject or unconscious that reterritorialises those self-same techniques and semiotics – are matters of reprove for Guattari. Throughout his career, it was precisely this gridding of existence that Guattari both found intolerable and which he engineered as the transversal line of critique. It is in this transversality through which Guattari asserts the primacy of the line over the point: an arrangement that is arguably contorted – in the contemporary social sciences – in vice versa fashion. In this sense, the question why Guattari precipitates a rather obvious and perhaps unseemly answer, namely, that modes of social enquiry remain enraptured to the 'point' when instead they should be attending to the line. Whilst undoubtedly there are manifold exceptions to this critique, the place of the problem and the role of empirics have too often been reduced, paradoxically, to a range of a-priori points and markers. Empiricism, such as it is, trades in speculation for pre-determination. This claim is a reverberation of Guattari's own concern for the paralysis of subjectivity, to the extent that, "it loses the taste for difference, the unpredictable, and for the singular event" (Guattari 2015b: 98). We might already argue that the appetite for difference was lost long ago, preceding even Guattari's own demise. To that end, and to repurpose a Guattarian (and indeed Deleuzian refrain), not only do we have the unconscious and the problems we deserve, but moreover the social science we deserve too. "Does the unconscious still have something to say to us?" asks Guattari (2011: 9). Yes, undoubtedly so. How can it ever not have something to say, lest Freud lay down further road blocks? In which case, the more pressing question, it might be argued, is 'does the social still have something to say to us'? Only, perhaps, if we get the social right. Guattari (2009b: 115–116) himself exhorts,

> [f]irst of all we must stop claiming that there is no more 'social', that it no longer exists and that nobody gives a damn. We should at least try to recognize the nature of the phenomena we're dealing with, try to recentre the focus where politics has migrated, where the situation has become critical, difficult to get a grasp on, to attach a meaning to.

The response to this provocation by the social will be conditioned by collective, micropolitical capacities to generate and resingularise cartography, ecology and politics in the push toward a liberation of thought and theorising. The response – any response – is dependent, too, on a resingularisation of institutional and industrial planes of consistency; a retooling of 'royal science', of publishing regimes also, and a facing-down of attempts to foreclose theory and theorising in geography and beyond.

Guattari is an obvious ally in addressing this challenge, but perhaps for reasons not entirely self-evident. On the one hand, and indeed, Guattari clearly affords considerable conceptual verve with which to counter the microfascist tendencies that progressively breach surficial intellectual labour, not least in the semiotic subjugation of theory in favour of a hackneyed empiricism in the social sciences. As an aside, and to echo, strangely, a Spinozist puzzlement for collective desiring toward servitude (in spite of ourselves), why, now, does academia desire its own atrophic decline in thought? Returning then, and abruptly, to Guattari's own theorising, the vitalist energy of his work, of course, invokes a turn to molecular registers of existence and likewise a simultaneous attentiveness to affect. On the other hand, and this is perhaps the 'tactic' in Guattari's work somewhat underplayed analytically, that whilst his thinking makes clear that the molecular is the 'make or break' of existence, he nonetheless refuses to decry the significance of molar spaces, forms, and assemblages. In this respect, Guattari is arguably less dismissive, nay less supercilious than Deleuze in his countenancing and subsequent diagramming of the molar. To wit, in relation to the couplets molar/molecular, macro/minor, absolute/relative and likewise tensed to their respective, associated nouns, 'revolution', 'politics', and 'unconscious', Guattari stressed both the immanence and significance of what can only be described crudely and quantitatively as the 'larger' of the values; molar, macro, absolute. Desire, for example, is always tied into and generative of social and asocial fields, not that such fields necessarily entail a greater scalar concern than that of molecular fields.

Guattari's work, as such, evokes a certain pragmatism that can be held in creative tension with the generative impracticality of his philosophy (Gerlach and Jellis, 2015a). That pragmatism is felt in the wisp of Guattari's gentleness in theory. Speculating on the viability of a molecular revolution, Guattari (2009a: 276) is conciliatory, remarking it "can only develop in parallel with [a] general, political crisis". This position might appear remarkably conservative for someone with Guattari's micropolitical credentials. In part, of course, it is a reflection of Guattari's own militant agitating and activist commitments. At the same time, this is not to disavow an unremitting insistence, on Guattari's part, that it is affect, and its attendant capacity, "to invade, disorient, and breakdown subjectivation" (Genosko, 2018: 150) that makes the difference. To that end, if one refuses to pose the question, 'why Guattari?' the social sciences will continue to be hobbled by the erroneous tethering of affect to emotional anchor points, themselves glued to all-too-stable cultural matrices. Furthermore, it places in abeyance the critique of grounding, and the insistence on rooting social theory and social enquiry in contrived a-priori grounds. It is in short, a myopic take on what has happened, and what is happening, namely that, "[c]ontemporary human beings have been fundamentally deterritorialised. Their original existential territories – bodies, domestic spaces, clans, cults – are no longer secured by a fixed ground; but henceforth they are indexed to a world of precarious representations and in perpetual motion" (Guattari 2015b: 97). How then to respond? Compromise, as JD Dewsbury (2015: 156) puts baldly, is off the table, noting that Guattari's work itself is uncompromising and that his ideas, "push us to let go of our difference of approach, to embrace instead the ephemeral and plural differences that emerge in the singularity of the research encounter itself, whether that be with a book, person, art work or thought". Through, and sometimes with, Guattari, then, are the fleeting spaces and moments in which to cultivate aberrant movements and machines that unpick stratification, and in which to untether from a universal time.

Impractical liberation

> This desire for a fundamental liberation, if it is to be a truly revolutionary action, requires that we . . . overturn the notion of the 'individual', . . . our sedentary selves, our 'normal social identities', in order to travel the boundaryless territory of the body, in order to live in the flux of desires that lies beyond sexuality, beyond the territory and the repertories of normality.
>
> (Guattari, 2009a: 209–210)

Elsewhere we have written of Guattari's 'impracticality' in philosophy (Gerlach and Jellis, 2015a; 2015b), the thematic injunction of which (and perhaps we should have known better) has been misunderstood in some quarters, sometimes willfully so. Such a claim does not – and would never – claim that Guattari was not engaged in all kinds of practical activities. Indeed, we might reflect on just how important his various practices and collaborations were for his thinking. Consider the various *institutional* contexts in which Guattari operated (see Goffey, 2015; 2016), and how "his thinking [was] always concerned with the specificity of concrete situations, even if some of his texts . . . seem to veer into the realms of extreme abstraction" (Goffey, 2016: 41). Crucially, one such institutional context that Guattari never found a home in was academia. He did not obtain a doctorate and never held an academic post. Indeed, much of his work took place outside the 'accredited structures' of academic endeavour (Goffey, 2017), in large part through his connection to collaborative group practices. Guattari was a member of many groups during the course of his life, too many to list here, but perhaps best known – as Anne Querrien (this volume) notes – were FGERI and their journal *Recherches*, and CERFI.[9] In addition, and crucially, Guattari spent much of his life in the clinical setting of La Borde, near Blois, which was much more than just a place of work for him (see Polack and Sabourin, 1976).

Although Guattari (1996) advocated a toolbox approach to the development of concepts, it was not a case of anything goes; concepts had to 'do' work. To talk of the impractical, then, is to underscore the analysis that Guattari demands from us, such that the thinkers, theories, and concepts we engage with are never rendered stable (what he sometimes refers to as 'metamodelisation'). To undertake analysis in a Guattarian fashion runs counter to the commonplace practice in the social sciences of parachuting in theory "for credibility, where there is no disruption", where theory is figured as "explanation, as analytical panacea" (Gerlach and Jellis, 2015b: 180). Put differently, "invoking the impractical is quite simply a means of disrupting or extending what constitutes application" (2015b: 180; see also Doel, 2015; Dewsbury, 2015).

It is from such questions around the disruption of application that we turn to a Guattarian liberation of cartography, ecology, and politics – and indeed of thought. 'Liberation', as a term, as a vocation, has its detractors. Guattari was, himself, all too wary of the false promise of a liberation towing an entourage of heroic and all-too-eager macropolitical tropes in its wake. If one were to accentuate, for a moment, the Spinozist inflection in Guattari's thought, in order to arrive at liberation, one needs first to resingularise the notion of liberation itself. That is to say one might acknowledge that liberation has nothing to do with freedom, and even less to do with freedom of will. It is in that moment of acknowledgement – tacit, stoic, somatic, transversal – that one is untethered from an inadequate understanding of one's affective capacity (individuated and collective), and indeed sensitised toward the micropolitical potential of liberation. It liberates liberation from a grid that pretends deliverance, but that otherwise belies a stubborn attachment to a fixed signifying referent; a conceit identified by Guattari and Rolnik (2008) as an abusive form of utopia. Liberation is not the recuperation of a primordial innocence in thought, but an experimental disposition toward a radical indeterminacy which itself refuses grounding in all senses. Once tied to repressive groupings, liberation, as per affect, is no more. In this respect, and precisely because he insists on the 'primacy' of assemblage (*agencement*), Guattari ventures far beyond Spinoza's disquisition on the emancipation of the self, and of the divestment of an individuated mind's inadequacy. Liberation, moreover, is diagrammatic; it multiplies functioning, tentatively sketching *toward* a futurity – a virtuality – without arriving there completely. It is a resingularisation affirming the right to singularity of all mutant productions of enunciation, always operating on the outside of the dominant fields of reference. Liberation, in sum, finally, all told, is disorientation – a 'schizogeography', a collective – and crucially – minoritarian enunciation of multiple and excessive earths.

Disorientation, as a motif, draws attention to the danger of the converse: of too much orientation. Just how prescriptive must be the directions before they are found out for their microfascist intent? To that end, what this book *is not*, clearly, is a user's manual for liberation. Similarly, this book, then, is not simply a celebration of Guattari's thought, nor an exegesis. We invite readers to find their own relays, mutations, and questions in his work – in short, to think-with Guattari. As such, and like Daniel Smith (1997: xiii), we hope that this book "is not a mere collection of articles" but "develops a series of concepts like so many motifs that appear and reappear in different essays, which enter into increasingly complex relationships with each other". One gets to know these concepts, "not through representation but through affective contamination. They start to exist in you, in spite of you" (Guattari, 1995: 92). By way of generative constraint, the book is organised around three conceptual nodes, themselves at the heart of a Guattarian vocabulary, namely, 'cartographies', 'ecologies', and 'micropolitics'.[10] Indeed, each of these nodes functions as something of a refrain across Guattari's writings yet disavow any 'mapping onto' some kind of chronological progression of his thought. In introducing each of these nodes by way of an intervention, we seek to open up the problem of their choice (why cartographies? why ecologies? why politics?) rather than offering – and thus to be encumbered by – a definite, and reassuring, answer as to their appropriateness.

Notwithstanding the transversal energies that undermine any sense of mutual exclusivity between these nodes, this geo-Guattarian triptych actuates a micropolitical 'demand' to liberate these terms from established mooring points. In responding to this demand (one intended to be more gentle in tenor than perhaps it appears and feels in writing), the resingularising of these conceptual and thematic nodes point, simply, to Guattari's overriding ambition: to map the production and transformation of subjectivity, in order that experimental cartographies, ecosophies, and micropolitics might be generated from that schizoid mapping; that very liberation of a generative unconscious.

Notes

1 As Bifo notes at various points, he thought about writing this book a few days after Guattari's death and kept writing for almost 10 years (the original, in Italian, came out in 2001).
2 *Ritournelles* was published initially in two parts in La Nouvelle Revue Francaise in January 1999 and then in the next issue, in April (Guattari, 1999a, 1999b). It was later re-published by Lume (Guattari, 2007a).
3 See Dosse (2011: 530n.67).
4 Two further books were published by Semiotext(e), *On The Line* (Deleuze and Guattari, 1983) and *Nomadology: The War Machine* (Deleuze and Guattari, 1986b), both of which went on to be developed in *A Thousand Plateaus*.
5 According to Paul Virilio, who was editor of Editions Galilée, Guattari had to be convinced into publishing *The Three Ecologies* as a book in its own right, rather than as an afterword to *Schizoanalytic Cartographies* (see Dosse, 2011: 391). Virilio was vindicated, as the latter was a huge success, while the former failed to attract a very large readership.
6 *Chaosophy* and *Soft Subversions* were first published in 1995 and 1996, respectively. Revised versions were then reissued in 2009 (see Dosse, 2009; Stivale, 2009).
7 The growing interest in Guattari's work is indicated not only by recent translations but also by new publications. A little-known French press, Editions Lignes, has issued a number of out-of-print and previously unpublished texts. Those not (yet) translated include *Soixante-cinq Rêves de Franz Kafka* (Guattari, 2007b), *De Leros à la Borde* (Guattari, 2012b), and *Qu'est ce que l'Écosophie* (Guattari, 2014). There is, at present, no complete translation of *Molecular Revolution* or *The Winter Years*.
8 Notwithstanding some hyperbolic claims that "Guattari is now a canonical author in his own right" (Derbyshire, 2018: 158).
9 FGERI stands for *Fédérations des groups d'études et de recherches institutionelles*, or the Federation of Study Groups and Institutional Research; CERFI stands for *Centre d'études, de recherche et de formation institutionnelles*, or Centre for Institutional Study, Research, and Training.
10 As we've noted elsewhere, this "is not to play down the importance of other terms that make up Guattari's thoroughly geographical vocabulary, which might include deterritorialization, re-territorialization, universes, diagrams, mappings, tracings, lines, points, figures, folds, chaosmos, earth, geo-philosophy and so on" (Gerlach and Jellis, 2015a: 143n.4). Indeed, all these terms feature in various ways throughout this volume.

References

Alliez, E. and Goffey, A. (eds.) (2011) *The Guattari Effect*. London: Continuum.
Badiou, A. (2000) *Deleuze: The Clamor of Being* (trans. L. Burchill). Minneapolis, MN: University of Minnesota Press.
Beckman, F. (2017) *Gilles Deleuze*. London: Reaktion Books.

Berardi (Bifo), F. (2008) *Félix Guattari: Thought, Friendship and Visionary Cartography* (trans. G. Mecchia and C.J. Stivale). Basingstoke: Palgrave Macmillan.
Deleuze, G. (1995) *Negotiations: 1972–1990* (trans. M. Joughin). New York, NY: Columbia University Press.
Deleuze, G. (2007) *Two Regimes of Madness: Texts and Interviews 1975–1995* (trans. A. Hodges and M. Taormina; ed. D. Lapoujade). Los Angeles, CA: Semiotext(e).
Deleuze, G. and Guattari, F. (1983) *On the Line* (trans. J. Johnston). New York, NY: Semiotext(e).
Deleuze, G. and Guattari, F. (1986a) *Kafka: Towards a Minor Literature* (trans. D. Polan). Minneapolis, MN: University of Minnesota Press.
Deleuze, G. and Guattari, F. (1986b) *Nomadology: The War Machine* (trans. B. Massumi). New York, NY: Semiotext(e).
Deleuze, G. and Guattari, F. (1994) *What Is Philosophy?* (trans. H. Tomlinson and G. Burchill). London: Verso.
Deleuze, G. and Guattari, F. (2004a) *Anti-Oedipus: Capitalism and Schizophrenia* (trans. R. Hurley, M. Seem and H.R. Lane). London: Continuum.
Deleuze, G. and Guattari, F. (2004b) *A Thousand Plateaus: Capitalism and Schizophrenia* (trans. B. Massumi). London: Continuum.
Deleuze, G. and Parnet, C. (2006) *Dialogues II* (trans. H. Tomlinson and B. Habberjam). London: Continuum.
Derbyshire, P. (2018) Vicissitudes of psychoanalysis. *New Left Review*, 110: 151–160.
Dewsbury, J.D. (2015) Guattari's resingularization of existence: Pooling uncertainties. *Dialogues in Human Geography*, 5(2): 155–161.
Doel, M. (2015) Applied Guattari: From toxic theory to loopy thinking. *Dialogues in Human Geography*, 5(2): 167–171.
Dosse, F. (2009) Introduction to Chaosophy. In: Guattari, F. (2009a) *Chaosophy: Texts and Interviews 1972–1977* (trans. D.L. Sweet, J. Becker and T. Adkins; ed. S. Lotringer). Los Angeles, CA: Semiotext(e).
Dosse, F. (2011) *Gilles Deleuze & Félix Guattari: Intersecting Lives* (trans. D. Glassman). New York, NY: Columbia University Press.
Elliot, P. (2012) *Guattari Reframed: Interpreting Key Thinkers for the Arts*. London: IB Tauris.
Fannin, M. (2015) The spirit of Guattari? *Dialogues in Human Geography*, 5(2): 172–176.
Genosko, G. (2002) *Félix Guattari: An Aberrant Introduction*. London: Continuum.
Genosko, G. (2009) *Félix Guattari: A Critical Introduction*. New York, NY: Pluto Press.
Genosko, G. (2012) Deleuze and Guattari: Guattareuze & Co. In: Smith, D.W. and Somers-Hall, H. (eds.) *The Cambridge Companion to Deleuze*. Cambridge: Cambridge University Press. pp. 151–159.
Genosko, G. (2018) *The Reinvention of Social Practices: Essays on Félix Guattari*. London: Rowman & Littlefield.
Gerlach, J. and Jellis, T. (2015a) Guattari: Impractical philosophy. *Dialogues in Human Geography*, 5(2): 131–148.
Gerlach, J. and Jellis, T. (2015b) A lovesick note to the impractical. *Dialogues in Human Geography*, 5(2): 177–182.
Goffey, A. (2012) Translator's introduction: The artifice of jargon. On Guattari's style. In: Guattari, F. (ed.) *Schizoanalytic Cartographies* (trans. A. Goffey). London: Bloomsbury.
Goffey, A. (2015) Introduction to Guattari on transdisciplinarity. *Theory, Culture & Society*, 32(5–6): 125–130.

Goffey, A. (2016) Guattari and transversality: Institutions, analysis and experimentation. *Radical Philosophy*, 195: 38–47.
Goffey, A. (2017) Félix Guattari: Literary and critical theory. *Oxford Bibliographies*.
Guattari, F. (1984) *Molecular Revolution: Psychiatry and Politics* (trans. R. Sheed). London: Penguin Books.
Guattari, F. (1995) *Chaomosis: An Ethico-Aesthetic Paradigm* (trans. P. Bains and J. Pefanis). Sydney: Power Publications.
Guattari, F. (1996) *The Guattari Reader* (ed. G. Genosko). Oxford: Blackwell Publishers.
Guattari, F. (1999a) Ritournelles. *La Nouvelle Revue Française*, 548: 340–374.
Guattari, F. (1999b) Ritournelles (Fin). *La Nouvelle Revue Française*, 549: 314–352.
Guattari, F. (2006) *Anti-Oedipus Papers* (trans. K. Gotman; ed. S. Nadaud). Los Angeles, CA: Semiotext(e).
Guattari, F. (2007a) *Ritournelles*. Tours: Lume.
Guattari, F. (2007b) *Soixante-cinq Rêves de Franz Kafka*. Fécamp: Éditions Lignes.
Guattari, F. (2008) *The Three Ecologies* (trans. I. Pindar and P. Sutton). London: Continuum.
Guattari, F. (2009a) *Chaosophy: Texts and Interviews 1972–1977* (trans. D.L. Sweet, J. Becker and T. Adkins; ed. S. Lotringer). Los Angeles, CA: Semiotext(e).
Guattari, F. (2009b) *Soft Subversions: Texts and Interviews 1977–1985* (trans. C. Wiener and E. Wittman; ed. S. Lotringer). Los Angeles, CA: Semiotext(e).
Guattari, F. (2011) *The Machinic Unconscious: Essays in Schizoanalysis* (trans. T. Adkins). Los Angeles, CA: Semiotext(e).
Guattari, F. (2012a) *Schizoanalytic Cartographies* (trans. A. Goffey). London: Bloomsbury.
Guattari, F. (2012b) *De Leros à la Borde*. Fécamp: Éditions Lignes.
Guattari, F. (2014) *Qu'est ce que l'Écosophie*. Fécamp: Éditions Lignes.
Guattari, F. (2015a) *Psychoanalysis and Transversality: Texts and Interviews 1955–1972* (trans. A. Hodges). Los Angeles, CA: Semiotext(e).
Guattari, F. (2015b) *Machinic Eros: Writings on Japan* (eds. G. Genosko and J. Hetrick). Minneapolis, MN: Univocal Press.
Guattari, F. (2016a) *Lines of Flight: For Another World of Possibilities* (trans. A. Goffey). London: Bloomsbury.
Guattari, F. (2016b) *A Love of UIQ* (trans. S. Maglioni and G. Thomson). Minneapolis, MN: Univocal Press.
Guattari, F. and Negri, T. (1990) *Communists Like Us: New Spaces of Liberty, New Lines of Alliance* (trans. M. Ryan). New York, NY: Semiotext(e).
Guattari, F. and Rolnik, S. (2008) *Molecular Revolution in Brazil* (trans. K. Clapshow and B. Holmes). Los Angeles, CA: Semiotext(e).
Osborne, P. (2011) Guattareuze? *New Left Review*, 69: 139–151.
Polack, J.C. and Sabourin, D. (1976) *La Borde ou le droit a la folie*. Paris: Calmann-Lévy.
Smith, D.W. (1997) 'A Life of Pure Immanence': Deleuze's 'Critique et Clinique' project. In: Deleuze, G. (ed.) *Essays Critical and Clinical* (trans. D.W. Smith and M.A. Greco). Minneapolis, MN: University of Minnesota Press.
Stivale, C. (2009) Rethinking (with) Félix Guattari. In: Guattari, F. (ed.) *Soft Subversions: Texts and Interviews 1977–1985* (trans. C. Wiener and E. Wittman; ed. S. Lotringer). Los Angeles, CA: Semiotext(e). pp. 9–17.
Žižek, S. (2004) *Organs Without Bodies: Deleuze and Consequences*. London: Routledge.

Part I

Cartographies

Cartography, that most quintessential of spatial motifs, is not one readily associated with acts of liberation. On the contrary, it is 'oppression' to which cartography is welded; analytically, politically, and geographically. Our lives – collective, individualised, resingularised – are simultaneously composed and pulled apart by maps and mappings. Deterritorialisation and reterritorialisation, as it were, on repeat. Through lines, legends, and coordinates – perhaps to the point of tedium – geographers, philosophers, historians, and cartographers have called into question the power and potential of cartography. From mundane vectors to violent lines of flight, cartography's capacity for belligerence and recalcitrance is of planetary renown. To this end, Deleuze and Guattari's (2004) familiar description of geometry's scalpel-like intensity is both a condemnation of molar intransigence and an ambivalent meditation upon the micropolitical capacities of aberrant lines. Guattari, in particular, was one for drawing all manner of maps and schizoid cartographies. This mapping was not a hobbyist pastime. Guattari was no mere amateur cartographer. La Borde was as much a cartographic institution in a molecular assemblage as it was, and is, a psychiatric clinic set deep in the Loire valley. Guattari, cartographer-in-chief, turned to mapping as a technique for decoupling the unconscious from a Freudian referent-framework of caricatured meaning and desire. Cartography was also a tactic, for Guattari, of simultaneous problem prosing and self-preservation. On this, Genosko (2009: 11) suggests,

> Guattari's fundamental problem was always how to traverse multiple fields in his own life and work. His own auto-modelization across fields of social struggles, institutional commitments, analytic treatments of psychotics in group settings and other patients in private practice, without forgetting writing philosophy and fiction, gave rise to a unique analytic method. One of the hallmarks of this method is the proliferation of diagrams.

What seems remarkable is the ease with which Guattari appropriated a Euclidean geometry and other cartographic forms in the service of mental emancipation – that self-same geometry which sustained precisely the architectures of Freudian, Jungian, and Lacanian psychoanalysis. On this point, cartography for Guattari (2012: 149) is a method of, "re-mooring . . . social and analytical practices on the side of ethico-aesthetic paradigms". In harnessing the practice of re-mooring, or refiguring coordinates, the following chapters are an attempt to draw lines alongside Guattari and his schizoanalytic cartography, that is to say, a mode of cartography that inverts the ethics of tracing. As Guattari (2011: 184) explains; "rather than indefinitely tracing off the same complexes or same universal mathemes, [we] will explore and experiment with an unconscious in action". This would set in train a cartography of and for mental ecologies which is both autopoietic and rhizomatic. These rhizomatic energies enfold not into a representational chart of existence, neither of territory nor the unconscious. Such gridding would be tantamount to the mimetic, and therefore reductive, act of tracing. Instead, such energies capacitate the charging of non-representational cartographies: maps of affect, cartographic gestures of the virtual, and lines of flight bound not to points but to mutant coordinates. In sum, such rhizomatic energies suffuse an ontogenetic schizoanalysis, thus liberating cartography from insincere, almost laughable, ideological memes, ones propagated, in turn, by those who have so little to lose from following maps that chart institutional pathways and tenure tracks to a comforting and comfortable telos. Think differently. Map differently. The liberation of cartography, then, lies not in the desiring toward accuracy, but instead hinges on the uncertain diagramming of disorientation: aberrant deterritorialisations of the unconscious, of geography, of space – of existence.

Marcus A. Doel and **David B. Clarke** – speaking, as they insist, always as geographers – open the lid on Guattari's schizoanalytic cartographies. Here they diagram the schizo not as an avatar of madness nor convenient conceptual foil, but as a continual splitting and splintering; a detethering from rationalities that themselves fold into a tyranny of signification. Pacing, breathlessly, through Guattari's semiologies whilst holding-hands (furtively) with Hjelmslev, Doel and Clarke grasp at an a-signifying semiotics, a deliberately tentative ungrounding for a geographical schizoanalysis. A geo-philosophy, of sorts. A constellation of unfoldings, enjoinings, mutant coordinates, lines here, there – and occasionally betwixt them all. In short, rather than decrying the 'schizo' as an aberrant figure of humanist imaginations, as "a *figure of madness* or of *reason unhinged*" (Doel and Clarke this volume: 30), it might be diagrammed instead as, "a *creative energy*: splitting, differentiating, deviating, mutating, evolving, explicating, implicating, etc." (ibid). Indeed, the 'etc.' might be the defining characteristic of the schizo, its Spinozist-inflected conative drive to more, more, more – or rather, to put it differently, to and, and, and . . . and . . . ?

In mapping the unconscious with Guattari, **Manola Antonioli** pauses to *think against the grain* – against the grain and granularity of time, territory, and geography itself. It is to move from a colonising cartography to a schizoanalytic cartography. It is a provocation to all – for, after Foucault, we are all cartographers now – to do more with our maps, to conceive of them as works of art, nay political action. Antonioli's invective castigates, gently, the act of cartographic superimposition in all its forms, psychoanalytic and geo-political. It is, in short, a clear statement of Guattarian intent; that the map exceeds fixation by semiotic tracing. As such, the injunction here is one of a beckoning, a beckon to map and to eschew interpretation, to give rise to a schizoanalytic subjectivity whereby flows of signs intersect with machinic flows. Such a demand seems improbable given cartography's seductive invite to survey, and thus patronise – territorialise – what is laid out before us in either a map, or along the dreamy royal road to the unconscious. Sigmund Freud, as it were, navigating semi-conscious at the wheel. Yet as Antonioli suggests, in both attuning to, and actively proliferating machinic mutations actualised in cartography, one becomes witness to the deterritorialisation of existential fields – and to the proliferation of existential territories.

Walk, momentarily, with **Tom Roberts**, on a summer's evening as thresholds of intensity are crossed in the greening of trees. Do not stop, however, to pause in phenomenological prayer, for as Roberts (this volume: 54–55) argues, "the corporeality of the lived falls drastically short to the extent that it cannot think the incorporeal individuations that traverse and agitate the Territory from within". Glancing across to the process philosophy of Alfred North Whitehead, Roberts animates the importance of the incorporeal in Guattari's materialist diagramming of subjectivity, or rather in his 'cartographic materialism'. As Roberts admits from the outset, this is no easy task, not least in the task of prising an incorporeal individuation away from the conceit of finite, individuated forms. To that end, Roberts turns to Guattari's (and Deleuze's, let us not forget) fabulation of haecceity: a perpetual grasping *toward* becomings. Haecceity, tacitly or otherwise, draws us in to the speeds and slownesses, the consistencies and inconsistencies of things, themselves never capitulating to straightforward renderings of subject or object. Despite the gentle warmth of that summer's evening, there are nettles to be grasped (perhaps more firmly than haecceity would ordinarily permit), and Roberts does so in harnessing the manifold difficulty of Guattari's schizoanalytic cartographies, its four co-constitutive, unfolding domains, and its modulating (modelising) influences on the production of subjectivity. For all of Guattari's talk of subjectivity, Roberts admits both to a certain frustration toward the lack, otherwise, of 'subject' in his work, but also to an admiration of this absence. The stakes, after all, are considerable. Alongside a Guattarian incorporeality there can no longer be an "ethics of self-fashioning" (this volume: 51), a desiring toward the self, self-help, self-medication, no – as it were – cartography of the self.

How a map does its work of deterritorialisation is never given in advance, yielding not to representation, but instead to speculation (however frustrating or portentous this might appear). This speculative mapping is taken up in **Christoph Brunner's** examination of the valence of diagrams and diagrammatic practice both in Guattari's work, and that of cross-border activism in California. Brunner, in recognising Guattari's mischievous oscillation between the terms metamodelisation and cartography, nonetheless pitches both practices as core motifs of Guattari's interest in, and activism through, diagrams. Hinging on the artistic-activist work of Teddy Cruz and *Estudio Teddy Cruz*, the chapter draws attention to the virtual forces generated and perpetuated by diagrams. Here, modelisation is figured not as the tracing of fixed coordinates or assaying of general qualities, but a modulation – or transformation – of affects, complex relations, and unfolding territories. Apposite or otherwise, Teddy Cruz's timely political interventions into USA-Mexico border politics exceed the asinine rhetoric of either his 2016 US Presidential Candidate namesake (but for two consonants) or the incumbent

US President of whom Guattari had 'figured-out' his algae-like composition long before he entered the White House (see Guattari, 2008). Here, Cruz invokes not an overcoded tracing of a caravan-like migration, but instead insists on a cartography of transversal border relations, a context-specific diagrammatic urbanism.

Anne Querrien, a close friend of Guattari, having met him in 1965 and bearing witness to his "precocious love of philosophy" (Querrien and Boundas, 2016: 399), offers further testament to his manifold geographical thought. For Guattari, geography was not figured as an obeissance to cardinal directions on the compass, but underscored in large part by networks: of friends, activists, patients, and of course, of molecules. To that end, a Guattarian geography is, as Querrien argues, always at the same time a plural geography. This desiring for plurality in Guattari's own work – conceptual and clinical – much like all desiring-machines, is part autopoetic. However, as Querrien makes clear in composing a brief history of Guattari's clinical imbrications, that desire also emanated from and against a post-war consensus in French psychiatry steeped in monolithic techniques of therapy, and indeed bound similarly by monolithic understandings of institutional purpose. Querrien draws attention, therefore, to the scale of the task that befell Guattari, and indeed Jean Oury; one that arguably re-asserted itself long after Guattari's death during the Sarkozy presidency (2007–2012) and the threat of a return to an industrial complex in psychiatric care. Notwithstanding these considerable macropolitical challenges, Querrien animates Guattari's work as an unrelenting disruption of the spatiality and cartography of psychoanalysis, such that one can conclude that the geography of institutions is never settled. It would be misguided, however, to assume that this disruption takes the guise *only* of militant agitation. Here, Querrien insists on a 'smoothness' in both thought and technique, a gentleness, even, in working along lines of flight, harnessing all the while a pragmatic micropolitics and a schizoanalytic cartography to 'get things done'.

"If you cannot work with such abstraction, don't quote Guattari to bolster your argument" (Dewsbury this volume: 95). Here, **JD Dewsbury's** icy invective spotlights a social science increasingly mired in a reterritorialisation of theory and held, simultaneously, in rapt enamourment with readily packaged forms of empiricism. In refrains of lost time, Dewsbury laments a social science impoverished by a lack of connective tissue to the machinic unconscious. Against, and from this lamentation, Dewsbury casts in hope – and in beckoning a 'cartography of apprehension' – that social scientists and geographers remain open to the event, and likewise that they grin in the face of deterritorialisation. As Dewsbury (this volume: 88) remarks, "[w]hatever the case of thinking about something else, the event of its happening 'exists a little bit everywhere' (Guattari, 2012: 17)". There is a tinge here, perhaps, of Isabelle Stengers' (2011: 153) take on Guattari and the need to operate – with simultaneous ease and truculence – in the comfort of gravitational vortices:

> for those of us who teach and breath the academic air, reclaiming the machinic freedom of cartography, which Guattari's operative constructs require, may well mean learning the signature of the black hole that threatens

> any (academic) relaying, and transforms relayers into sophisticated, spinning babblers: it is the fear of exposing oneself to the accusation of being duped, to compromise oneself with what others may be able to debunk.

A little phrase. A sonata. "Thinking is all this: a taste, a light, a sound, a grin" (Dewsbury this volume: 88). If we can afford to be less infantile, less decadent with cartography (surely we have but no choice?), a time of non-representational gentleness awaits.

References

Deleuze, G. and Guattari, F. (2004) *A Thousand Plateaus: Capitalism and Schizophrenia* (trans. B. Massumi). London: Continuum.
Genosko, G. (2009) *Félix Guattari: A Critical Introduction*. New York, NY: Pluto Press.
Guattari, F. (2008) *The Three Ecologies* (trans. I. Pindar and P. Sutton). London: Continuum.
Guattari, F. (2011) *The Machinic Unconscious: Essays in Schizoanalysis* (trans. T. Adkins). Los Angeles, CA: Semiotext(e).
Guattari, F. (2012) *Schizoanalytic Cartographies* (trans. A. Goffey). London: Bloomsbury.
Guattari, F. (2016) *Lines of Flight: For Another World of Possibilities* (trans. A. Goffey). London: Bloomsbury.
Querrien, A. and Boundas, C. (2016) Anne Querrien, La Borde, Guattari and left movements in France, 1965–81: Interviewed by Constantin Boundas. *Deleuze Studies*, 10(3): 395–416.
Stengers, I. (2011) Relaying a war machine? In: Alliez, E. and Goffey, A. (eds.) *The Guattari Effect*. London: Continuum. pp. 134–155.

1 Through a net darkly

Spatial expression from glossematics to schizoanalysis

Marcus A. Doel and David B. Clarke

We return to the swamp of spots.

(Guattari, 2015: 179)

Pressed for words, as is our fate, suffice to say four things by way of contextualisation, as we plunge into the swamp of signs. First, we will be forging a constellation of geo-graphical (earth-writing, earth-inscribing) terms that give spatial expression to what Félix Guattari and Gilles Deleuze famously called *schizoanalysis* and *geo-philosophy*. For we are "speaking always as geographers", as they say (Deleuze and Guattari, 1983: 83). For our part, we are not so much interested in *the schizo* as a conceptual persona or a mental ecology/pathology (we lean towards anti-humanism, after all), but as a figure of splitting and splintering (from the Greek, *skhizein*, to split), which, incidentally, has a close affinity with the X or χ of deconstruction (the twofold asymmetrical process of reversal and re-inscription chimes with the twofold process of deterritorialisation and reterritorialisation). Second, and to drive home the point that we have just made about speaking always as geographers, this constellation of fissured and fractured terms takes flight from an impasse bequeathed to us by what has come to be known as *structuralism* and *poststructuralism*. After Ferdinand de Saussure, the signifying fraction of the sign (the 'signifier', as he famously dubbed it; the portion or splinter that 'makes sense', so to speak, by giving expression) was set afloat, and its tendency to drift (to disseminate) could only be temporarily arrested by Saussure through the forcible and arbitrary imposition of chains. Some of these chains sought to anchor the signifier onto something meaningful (the 'signified', as Saussure famously dubbed it; the portion or splinter that 'made sense', so to speak, by being expressed), which many have mistaken for solid ground. Most of these chains, however, were used to attach the signifier onto an illimitable network of other signifiers in a vain attempt to pin and peg down its identity through a summation of negative differences, explicated differences, and differentiations (Deleuze, 1994). Such enchainment could be called a 'structure'. After Jacques Lacan and Jacques Derrida, however, the signified, having slipped beneath the signifier, more or less sank without trace, and the signifying chains that once held the signifier captive in its allotted place were loosened to such an extent that the signifier was set adrift. And just as the play of foam amidst breaking waves may allude to the phantom presence of a sinking or sunken ship, so the play of the signifier on the surface of language may allude to the phantom presence of a sinking or sunken signified. There are those who claim to make out the flotsam and jetsam of meaning here and there amidst the torrent of breaking signs, but the wreckage of sense is barely discernible from the nonsense within which it floats. Hereinafter, there is nothing but broken signs stretching before us in all directions: "the universe is mute. Only men speak. Though there is nothing to say" (Coover, 1998: 83). Third, while empty signs may captivate and seduce us, as Jean Baudrillard (1990) aptly put it, the insatiable desire that this enchainment unleashes should not be mistaken for revolutionary desire. Doggedly pursuing the tantalising yet elusive "*objet-machine petit 'a*'" (Guattari, 1984: 115) will never exhaust, let alone overturn, the labyrinthine 'structure' within which such a futile chase takes place (Deleuze, 2004; Derrida, 1981). Finally, for those of us, like Guattari and Deleuze, who yearn for a desiring revolution, it is sobering to be reminded that while Sigmund Freud pronounced that there are three impossible professions – *governing*, *educating*, and *healing* – Lacan added a fourth: *inciting desire* (Boni, 2014; Lacan, 2007; Zupančič, 2006). To put it bluntly, the militant injunction *Revolt!* is indistinguishable from the reactionary injunction *Enjoy!* Accordingly, Lacan scoffed at credulous responses to the solicitation of revolutionary desire in May '68. "If the events of May demonstrated anything at all, they showed that it was precisely that structures had taken to the streets!" (Lacan, quoted in Dosse, 1997: 122). Here as elsewhere, once enthralled and enraptured by its alluring chains, there is no escaping the signifier. All of the exits are blocked. And yet, it was by taking into account a-signifying semiotics that Guattari discovered a way to circumvent this impasse. This seemingly oxymoronic notion occurred to Guattari during a forced reading of Louis Hjelmslev in contradistinction to Lacan, to which we now turn.

The tyranny of the signifier and the resistance of a semiotic war machine

> What, in fact, does Lacan say? He says that the unconscious is structured like a language and that a signifier represents the subject for another signifier. . . . You accede to desire by the signifier and by castration, and the desire to which you accede is an impossible desire.
>
> (Guattari, 1986: 18)

Guattari (1986: 18) added a crucial twist to Lacan's algorithmic and paralytic analysis of the signifier. "I think that Lacan is completely right in terms of the unconscious of the capitalist social field", he concurred, but not necessarily for other social fields. Indeed, Guattari "goes so far as to argue that the unconscious itself has become an institution: 'one finds oneself rigged out with an unconscious the moment one dreams, délires, forgets, or makes a slip of the tongue'" (Goffey,

2016: 41, quoting Guattari, 1995: 10). Capitalism employs regimes of signs – 'sign machines' – that institute a 'structuralized' unconscious, but there is still *something more* than the tyranny of the *sliding signifier*, after Saussure, and the empire of the *algorithmic sign*, after Lacan (Nancy and Lacoue-Labarthe, 1992). And what's more, this excess may take us beyond the horizon of capitalism, the horizon of the signifier, and perhaps even beyond the horizon of structuralism (Guattari, 2011). A forced reading of Hjelmslev was pivotal for enabling Guattari and Deleuze to break through the 'white wall' of the *sliding signifier* and the 'black hole' of *serial erasure* in order to fabricate a "semiotic [war] machine against structural semiology" (Dosse, 2011: 232). One of the crucial aspects of this *semiotic* war machine is that it avoids the detour through either the imaginary or the symbolic registers by remaining in constant touch with the real: "the real is not the impossible, as Lacan thinks, but the artificial" (Guattari, quoted in Goffey, 2016: 46). It is a resolutely materialist and immanent semiotic in which the real is machined or manufactured (*made* to appear, *forced* to appear). "Guattari reincorporates the real into semiotic theory by identifying a kind of a-semiotic encoding that, borrowing Hjelmslev's terminology, bypasses 'substance' in order to couple 'form' directly with 'matter'" (Watson, in Young et al., 2013: 158). Barbed wire, for example, sadistically instructs animals and humans by lacerating their flesh, rather than by forming semiotic substances that signify something.

Now, recall that Saussure inaugurated the rendering of linguistics as two 'heterogeneous realms' or 'floating kingdoms', one composed of an articulated mass of 'signifiers' (sound-images) and the other of an articulated mass of 'signifieds' (concepts). The problem for Saussure and his heirs was understanding how each of these two masses is internally articulated, and how they are articulated with one another; and the solution to this problem was shaped by the recognition that linguistic articulation is not only linear (enchained), but also arbitrary and incorporeal: *arbitrary*, since the slicing up and suturing of the otherwise 'shapeless' and 'confused' masses could always have been otherwise; and *incorporeal*, since language works on combinations and permutations that produce "*a form, not a substance*" (Saussure, 1974: 113). What crystallised out of linguistic articulation were points (e.g. anchoring points), lines (e.g. the bar along which the signifying chain slides), and surfaces (e.g. the array of differential traces); and the drift of post-Saussurean thinking towards 'post-structuralist' thinking was invariably towards a dis-articulation, dis-jointure, and cracking open of these points, lines, and surfaces. "Ex-centric, dis-integrated, dis-located, dis-juncted, deconstructed, dismantled, disassociated, discontinuous, deregulated . . . de-, dis-, ex-. These are the prefixes of today", as Bernard Tschumi (1994: 225) put it. In other words, an "and . . . and . . . and . . ." (Deleuze and Guattari, 1983: 57). Stuttering . . . stammering . . . scission . . . (Deleuze, 1997). And yet, all of this mind-boggling discombobulation remains sandwiched between the two heterogeneous masses, despite them becoming increasingly maladjusted and malformed; both the one that drifts (thereby becoming inexpressive: a mass of signifiers that fail to recall any signifieds) and the one that sinks (thereby becoming unexpressed: a mass of signifieds that have seemingly vanished without trace). It is this waning bi-univocalisation that must be undone; the one-to-one coupling of the two masses, under the *direction* (i.e. domination, determination) of one of the masses. So, rather than a duality of signifier *and* signified, there is a bi-univocality of a signifier *for* a signified or *in lieu* of a signified. Bi-univocality, then, undergirds linearity, hierarchy, and transcendence: one term (the signifier) comes to dominate, eclipse, and foreclose the other term (the signified). Signifiance is what remains once signification is paralysed; once the signified is laid off. Such is the tyranny of the signifier. Hereinafter, signifiers are inclined to have intercourse amongst themselves, to purloin the phraseology that Karl Marx famously employed to characterise the fetishism of commodities, which exist for each other only as exchange-values, rather than as use-values. Indeed, it was Jean Baudrillard's (1981) genius to appreciate that commodities and signs have homologous forms: exchange-value is to use-value as signifier is to signified.

In his correspondence with Deleuze, Guattari (2006: 38) wrote: "I think it's with Hjelmslev . . . that we can find the key to 'clearing' out . . . structuralism". The key attraction of Hjelmslev for Guattari is Hjelmslev's desire to fashion "an immanent algebra of language" (Hjelmslev, 1961: 80), without recourse to a transcendental perspective (language as a means or vehicle for reference, representation, knowledge, communication or suchlike), which Hjelmslev dubbed 'glossematics'. And at the heart of this "arid algebra of language" (Genosko, 1998: 175) is the principle of *reversibility* rather than bi-univocality. "What Deleuze and Guattari borrowed above all from Hjelmslev", says François Dosse (2011: 232), "was the distinction between the absolutely reversible planes of expression and content. . . . The distinction had to do with strata, planes of consistency that destroyed Saussure's binarism. In fact, there was to be only a single plane of consistency unfolding in multiple strata". Brian Massumi (1992) offers the example of a carpenter engaged in woodwork to convey this reversibility. "Seen from the perspective of the dominating tool, the wood is content", he says. "But from the perspective of the forces that went into it, [the wood] is an expression, of the water, sunlight, and carbon dioxide it captured and contains" (Massumi, 1992: 12). In the hands of a carpenter, "each singularity can become a substance of expression: a knot may become the eye of a fish; a grain pattern, the waves of the sea; or something else entirely: the content–expression relation here is one of contingency, not necessity" (Holland, 2011: 17). Furthermore, the carpenter "with hand to tool is an agent of expression, but from another angle he is the content of an institution. . . . A content in one situation is an expression in another" (Massumi, 1992: 12). This is why we have stressed the word *force*: "Content and expression are reversible only in action. A power relation determines which is which" (Massumi, 1992: 12–13). Content and expression are both squeezed out – expressed.

In his *Prolegomena to a Theory of Language*, Hjelmslev (1961) contrasts the traditional (i.e. pre-Saussurean) and modern (i.e. post-Saussurean) conceptions of the sign. In the traditional/transcendental conception, "a sign is first and foremost a sign *for* something"; "the sign is an *expression* that points to a *content* outside the sign itself" (Hjelmslev, 1961: 47). In the modern/immanent conception inaugurated by Saussure, however, "the sign is an entity generated by a connexion between an expression and a content" (Hjelmslev, 1961: 47). On Guattari's (2006) reading, Hjelmslev breaks with bi-univocality (whether in the traditional form of 'referral' or the modern form of 'connexion') by introducing the notion of a '*sign function*' posited between the plane of expression (Saussure's plane of signifiers) and the plane of content (Saussure's plane of signifieds). There is no 'sign' *as such*, only a sign *function*. From a functional perspective, content and expression are simply "functives that contract the function" (Hjelmslev, 1961: 48). They are solely formal and operational terms. "There will always be solidarity between a function and . . . its functives: a function is inconceivable without its terminals, and the terminals are only end points for the function and are thus inconceivable without it" (Hjelmslev, 1961: 48). For if a 'sign' were to lose one or both of its functives (i.e. its terminals of content and expression), or the function that contracts them as functives, then it would cease to be a sign.

> The sign function is in itself a solidarity. Expression and content are solidary – they necessarily presuppose each other. An expression is expression only by virtue of being an expression of a content, and a content is content only by virtue of being a content of an expression.
>
> (Hjelmslev, 1961: 48–49)

This insistence on solidarity acts as a kind of "prophylaxis against signifier fetishism", says Gary Genosko (1998: 179); or rather, a prophylaxis against *signifier dissemination*, perhaps. "Therefore – except by an artificial isolation – there can be no content without an expression, or expressionless content; neither can there be an expression without a content, or content-less expression" (Hjelmslev, 1961: 49). In other words, what Guattari stumbles upon in Hjelmslev is the reassertion of 'mutual presupposition', but no longer in terms of a "double entity" (Saussure, 1974: 65) that is bound together through (arbitrary) association. For Saussure, it is always a matter of wholes and parts, which Lacan – amongst others – will delight in shattering (once bound, forever gagged, so to speak). For Hjelmslev, however, the sign is not a 'double entity', but a single function. Guattari's intuition was that this functional approach was amenable to forging a semiotics of multiplicity and reversibility. Hjelmslev's sign *function* will become, for Guattari, a full-blown sign *machine* (Guattari, 2006, 2011).

By applying a *functional* – rather than an associative, an algorithmic, or indeed an anagrammatic – logic to the sign, there are no longer *two* irreversible planes, realms, or kingdoms (i.e. the mass of more or less well formed signifiers and the mass of more or less well formed signifieds) that are more or less well articulated bi-univocally with one another, but rather there is a *single* functional (machinic) plane with *reversible* (foldable) functives. "The terms *expression plane* and *content plane* and, for that matter, *expression* and *content* are chosen in conformity with established notions and are quite arbitrary", says Hjelmslev (1961: 60), rather disconcertingly. "Their functional definition provides no justification for calling one, and not the other, of these entities *expression*, or one, and not the other, *content*. . . . They are each defined only oppositively and relatively, as mutually opposed functives of one and the same function". In other words, although the sign function operates between two planes, these planes are not given in advance, and in want of connexion, but are an effect of the functional operation itself. The sign function is *differential*. It differentiates (splits) into *distinct planes*. But such a distinction is reversible in the differential. What matters, then, is *difference*. Without this differential the sign function would collapse into redundancy and cease to operate. Consequently, for Hjelmslev, a *mono-planar* structure is symbolic rather than semiotic (e.g. in chess the entities are isomorphic with their interpretations). (For a subtle and profound analysis of the notions of 'sign' and 'symbol' in Freud, Saussure, Hjelmslev, and Lacan, see Arrivé, 1992). Moreover, since the sign function is differential rather than binary it can express any number of planes. Indeed, Hjelmslev (1970: 106) mentions in passing that "we must also reckon with the possibility that a language may turn out . . . to have three or more planes, but it would be absurd to suppose that such a structure is found in any everyday language and we can therefore ignore the possibility in our present discussion". With this act of compression, the multiplicity of sense was flattened out to just meaning and reference, and all the rest was squeezed out. One does not need to turn to 'affect', however, to discover that there is more to sense than meaning and reference. The depths of sense remain to be plumbed even within the twofold compression of everyday language (Deleuze, 1990).

By opening up the possibility of multiple planes that would be irreducible to a mono-plane, Hjelmslev thereby anticipated Deleuze and Guattari's (1988) 'rhizomes', 'assemblages', and 'multiplicities' (all with $n - 1$ dimensions, with the One 'subtracted' from the Multiple, so that immanence takes flight from transcendence). We are reminded of an ambivalent and enigmatic phrase that Jacques Derrida was fond of using, '*plus d'un*', which could be variously translated as 'more of one' or 'more than one', but also 'no more one' or 'no more of one'. All is fissured, cracked, and splintered. Each and every 'One' is shattered through and through. This vacillating, differentiating, and disseminating '*plus d'un*' that cannot be counted upon is the sense in which the operation of the sign function – the sign machine – produces "a single plane of consistency unfolding in multiple strata" (Dosse, 2011: 232).

Now, according to Hjelmslev, if one compares different languages so as to "extract, or subtract, the factor that is common to them and that remains common to all languages" (Hjelmslev, 1961: 50), then one will discover a 'common extract' that he calls *purport* (matter, sense). Purport is the unformed, amorphous mass that is nevertheless susceptible to semiotic formation, which thereby becomes what Hjelmslev calls *substance*. ("His *form* is our *code*. His *substance* is our *flow*", was how Guattari (2006: 202) originally pitched Hjelmslev's terminology to Deleuze.) But in keeping with his emphasis on the reversibility of the sign function, there is a purport of both content *and expression*, as well as a substance of both content *and expression*; just as there is a form of both expression *and content*. In short, both the plane of expression and the plane of content have their regions of formlessness (purport, matter), their array of forms (casts, nets, moulds), and their discrete formations (semiotic substances). Accordingly, the sign function articulates a sextuplet of content-form and expression-form, content-substance and expression-substance, and content-purport and expression-purport (or even an octuplet, if one is inclined to include the a-signifying sub-atomic particles of content and expression that Hjelmslev calls *figuræ*, those elementary morsels below the threshold of sense that constitute the raw material out of which signs will come to be composed).

> [T]he unformed purport extractable from all these linguistic chains is formed differently in each language. Each language lays down its own boundaries within the amorphous 'thought-mass'. . . . It is like one and the same handful of sand that is formed in quite different patterns, or like the cloud in the heavens that changes shape in Hamlet's view from minute to minute. Just as the same sand can be put into different molds, and the same cloud take on ever new shapes, so also the same purport is formed or structured differently in different languages. . . . Purport remains, each time, substance for a new form, and has no possible existence except through being substance for one form or another. We thus recognize in the linguistic *content*, in its process, a specific *form*, the *content-form*, which is independent of, and stands in arbitrary relation to, the *purport*, and forms it into a *content-substance*.
>
> (Hjelmslev, 1961: 51–52)

Content, then, has its own form (*content-form*) as well as its own substance (*content-substance*). So too for expression (*expression-form* and *expression-substance*). Hence the parallelism of Hjelmslev's terminology: *expression-purport* and *content-purport*; *expression-form* and *content-form*; and *expression-substance* and *content-substance*. In his book on Michel Foucault, Deleuze offers a wonderful example of this six-fold typology:

> The content has both a form and a substance: for example, the form is prison and the substance is those who are locked up, the prisoners (who? why? how?). The expression also has a form and a substance: for example the form is penal law and the substance is 'delinquency' in so far as it is the object of statements. Just as penal law as a form of expression defines a field of sayability (the statements of delinquency), so prison as a form of content defines a place of visibility ('panopticism,' that is to say a place where at any moment one can see everything without being seen).
>
> (Deleuze, 1988: 47)

Content-substance and expression-substance exist only by virtue of the content-form and the expression-form, "which appear by the form's being projected on to the purport, just as an open net casts its shadow down on an undivided surface" (Hjelmslev, 1961: 57). "In this metaphor", says Winfried Nöth (1990: 69), "the 'undivided surface' is the purport, the 'open net' is the form, and the 'shadow' is the substance". This is why the sign function gives rise to *incorporeal* events.

> The sign is . . . a sign for a content-substance and a sign for an expression-substance. It is in this sense that the sign can be said to be a sign for

> something. [There is] no justification for calling the sign a sign merely for the content-substance, or (what nobody has thought of, to be sure) merely for the expression-substance. The sign is a two-sided entity, with a Janus-like perspective in two directions. . . : 'outwards' toward the expression-substance and 'inwards' toward the content-substance.
>
> (Hjelmslev, 1961: 58)

Many following in the wake of Saussure and Lacan have mistaken the signifier for merely the expression-substance and expression-form, respectively, to the exclusion of the signified qua content-substance or content-form (i.e. the sign under erasure; the sign rendered inoperative; the sign malfunctioning). Guattari, for his part, used Hjelmslev's functional sign to break out of the double bind of Saussure's structural sign and Lacan's algorithmic sign, and in so doing replenished the wealth of expression that had been crushed and squandered by signifier fetishism.

Guattari's Hjelmslev

> This linguistics of the signifier was to be replaced by a completely different linguistics of flux.
>
> (Dosse, 2011: 231)

For Hjelmslev, then, the twofold distinction between the plane of expression and the plane of content is 'overlain' and 'cut across' by a triple distinction between purport, substance, and form. A sign is not the *double* articulation of signifier and signified, but a *six-fold* articulation of content, expression, form, substance, and purport. "This is a *crossroads machine*", says Guattari (2006: 205). "It's the conjunction of two deterritorialization processes, a disjunctive synthesis for an inscription machine. . . . But what's annoying, as far as I'm concerned, is that there are plane*s* and not *a* plane, a pure plane of consistency, of the filiation of deterritorialized machinic inscriptions". Hjelmslev remains faithful to the 'pseudo-dualism' of content and expression, of signified and signifier, whilst nevertheless betraying this fidelity through reversion.

> Why pseudo? Because . . . it's implied that these terms are *absolutely reversible*. . . . Hjelmslev's axiomatic machinism . . . spills out onto two planes: that of the signified – content – and that of the signifier – expression. Impossible once and for all to call it quits with Papa Saussure. Not to mention that form conserves a taste of eternity and substance a taste of transcendence, the signified keeps one foot in spirit and the signifier keeps one foot in matter. And yet, Hjelmslev, after the *Prolegomena*, did everything he could to mess this all up. It's pure *strata*, he started to say. Substance and form are a pure relation of *manifestation*.
>
> (Guattari, 2006: 204)

For Guattari, then, there are not just two planes of content and expression, but a single plane of consistency with innumerable strata and countless sign functions/sign machines. Each is *plus d'un*. Each is manifold. "One does indeed find folds everywhere", quipped Deleuze (1995: 156). Guattari highlights three 'regimes of signs' or specific formalisations of expression. The first is what he calls '*non-semiotic encodings*' or '*natural encodings*', which function without the need for a semiotic substance; without the need for signs (e.g. the genetic code or photosynthesis). They encode directly, without recourse to an intermediary inscription or script. By way of illustration, Guattari and Deleuze return time and again to the 'machinic' encounter between a wasp and an orchid that deterritorialises each in relation to the other whilst reterritorialising them transversally as a conjoint becoming-other (a *becoming-wasp* of the orchid and a *becoming-orchid* of the wasp; a block of becoming that is *plus d'un*): "the production of a continuum of intensities in a nonparallel and asymmetrical evolution . . . a capturing, a possession, a plus-value, but never a reproduction or an imitation"; "the capture of a fragment of the code, and not the reproduction of an image" (Deleuze and Guattari, 1986: 13–14). Each becomes implicated in the explication of the other. Each is *plus d'un*.

> With machines the question is one of connection or non-connection, without conditions, without any need to render an account to any third party. It is from that that the surplus-value of encoding originates. The situation is like that of the [wasp] which, by being there, became part of the genetic chain of the orchid. The specific event passes directly into the chain of encoding until another machinic event links up with a different temporalization, a different conjunction.
>
> (Guattari, 1984: 125)

The second regime is what Guattari calls '*signifying semiologies*', which are those sign systems that employ semiotically formed substances to make sense. This regime has two kinds of sign system. On the one hand, what he calls '*symbolic semiologies*' or '*presignifying semiologies*', which employ various semiotic substances, each with its own autonomous territory, such as a semiotics of gesture, of hysteria, or of clouds (e.g. Pearson, 1992; Hustvedt, 2011; Hamblyn, 2001, respectively). On the other hand, what he calls '*semiologies of signification*', where all manner of "substances of expression (of sound, sight, touch, and so on) are centred upon a single signifying substance. This is the 'dictatorship of the signifier'" (Guattari, 1984: 75). This is also the vortex of signifying slurry into which everything plunges: the vortex of nonsense around which the 'mania for interpretation' and the 'delirium of capital' revolve (Genosko, 2009, 2012). Semiologies of signification work by "crushing all that wealth of expression, all that opening-out to reality" (Guattari, 1984: 79). The gloopy mush that flows from this "machine of semiotic disempowerment" (Guattari, 2016: 122) bathes the world in an insipid blue-grey, which, as Jean-François Lyotard (1998) reminds us, is the ghostly and ghastly colour of exchange-value, money, and capital. Hereinafter, all signs are signs of signs.

> There is a simple general formula for the signifying regime of the sign (the signifying sign): every sign refers to another sign, and only to another sign, ad infinitum. That is why, at the limit, one can forgo the notion of the sign, for what is retained is not principally the sign's relation to a state of things it designates, or to an entity it signifies, but only the formal relation of sign to sign insofar as it defines a so-called signifying chain. The limitlessness of signifiance replaces the sign. . . . It is this amorphous continuum that for the moment plays the role of the 'signified,' but it continually glides beneath the signifier, for which it serves only as a medium or wall: the specific forms of all contents dissolve in it.
>
> (Deleuze and Guattari, 1988: 112)

Finally, the third regime of signs is what he calls '*a-signifying semiotics*' or '*post-signifying semiotics*', which use the signifying semiologies of the second regime of signs as "an instrument of semiotic de-territorialization, making it possible for the semiotic fluxes to form new connections with the most deterritorialized material fluxes. Such connections operate independently of whether or not they signify anything to somebody" (Guattari, 1984: 75). The avant-garde has taught us that the destiny of every sign is to be led astray and forced to work otherwise. "For nonsignifying language anything will do", declare Deleuze and Guattari (1984: 240); "no flow is privileged in this language, which remains indifferent to its substance or its support". "To put it another way: at this level, absolutely anything goes – any ideology, or even religion will do, (sic) even the most archaic: all that matters is that it be used as the raw material of existence" (Guattari, 1992: 20). Hereinafter, there are nothing but flows and interruptions, implications and explications, and experiments and becomings. And here we finally see the vital importance of Hjelmslev for Guattari and Deleuze:

> a substance is said to be formed when a flow enters into a relationship with another flow, such that the first defines a content and the second, an expression. The deterritorialized flows of content and expression are in a state of conjunction or reciprocal precondition that constitutes figures as the ultimate units of contents and expressions. These figures do not derive from a signifier. . . ; they are nonsigns, or rather nonsignifying signs, points-signs having several dimensions, flows-breaks or schizzes that form images through their coming together in a whole, but that do not maintain any identity when they pass from one whole to another.
>
> (Deleuze and Guattari, 1984: 240–241)

There are, then, always *several* regimes of signs in play (*plus d'un*), the multiplicity of which *resists* any power take-over that would collapse or reduce the severalty into a unity (Deleuze, 2008). By cross-cutting and traversing one another, by taking up, processing, and emitting each other's signs (e.g. 'voting with one's feet'), the regimes could be called an *assemblage*. Most geographers will be familiar with the notion of 'assemblage' (Anderson et al., 2012), which Guattari and Deleuze developed and refined in numerous books. The notion is loosely taken to mean an arrangement of disparate materials that performs some work. It is a structural and diagrammatic notion, but also a functional and machinic notion (Watson, 2009). An assemblage, then, is not simply a performative gathering of heterogeneous materials (*it* operates, *it* performs, *it* produces, *it* transforms, etc.), which has more or less severalty and more or less consistency (*plus d'un*), since *it* is also torn between the work of the *structure* and the work of the *machine* (Guattari, 1984; Lazzarato, 2014; Raunig, 2010). For Guattari, while structure is "the generality characterized by a position of exchange or substitution of particularities", the machine has the "function of detaching a signifier as a representative, as a 'differentiator', as a causal break, different in kind from the structurally established order of things" (Guattari, 1984: 111 and 114, respectively). To put it crudely, while *structures* shut down and close down, always resorting to more of the same (the same prefab possibilities, prefab statements, prefab interpretations, etc.), *machines* break down and fall apart, always leaving themselves open to tinkering and experimentation, and to a break through.

Content and expression are machinic fabrications rather than structural components, and since each has its own forms and substances, there is no need for one of them to *stand in for* the other; rather, they *interweave* with one another, and *intervene* in one another, according to the machinations of the various regimes of signs, abstract diagrams, and assemblages that mobilise them. Paradoxically, content and expression ordinarily stand apart from one another, without any natural affinity or essential symmetry. They must be *forced* together, so that expression *acts on* content, and content *acts on* expression. For "signs are at work in things themselves just as things extend into or are deployed through signs" (Deleuze and Guattari, 1988: 87). In other words, the conjunction of content *and* expression entails dis/connection, dis/assembly, de/construction, pro/duction, trans/duction, and trans/formation. Deleuze and Guattari famously called this work 'desiring-production' – "To desire consists of this: to make cuts, to let certain contrary flows run, to take samplings of the flows, to cut the chains that are wedded to the flows" (Deleuze, in Guattari, 2009: 53). And all of this "means nothing. There is no meaning, no interpretation to be given, no significance" (Deleuze, in Guattari, 2009: 54). In other words, there are just "sign machines' that function, malfunction, and function otherwise (Guattari, 2016: 118). And what they produce is the real – the real as process, as fabrication, as artifice.

Conclusion

> *INTERVENTION: Lacan, is psychoanalysis revolutionary?*
> Now, there's a good question!
>
> (Lacan, 2007: 200)

By forging a trajectory from Louis Hjelmslev to Félix Guattari, we have wrought a constellation of geo-graphical (earth-writing) terms that give spatial expression to what Guattari and Deleuze call *schizoanalysis*. This constellation includes: points, lines, surfaces, planes, series, segments, strata, joints, folds, flotsam and jetsam, slips and slides, assemblages, arrangements, articulations, multiplicities, structures, rhizomes, diagrams, machines, and transversals. One could just as easily call it *geo-philosophy* (Bonta and Protevi, 2004; Deleuze and Guattari, 1994; Gasché, 2014; Woodard, 2013). After Guattari and Deleuze, geography should give a Spinozist expression to *schizoanalytic cartographies*. The 'schizo' is not, then, a *figure of madness* or of *reason unhinged* (isn't it always?), but a *creative energy*: splitting, differentiating, deviating, mutating, evolving, explicating, implicating, etc. The 'schizo' is the gust of fresh air that gets everything moving, the molecular twists and turns that are eternally tearing open and pulling apart the very fabric of space and time: *schizosmosis* and *chaosmosis* (Guattari, 1995, 2013).

> Schizzes have to do with heterogeneous chains, and as their basic unit use detachable segments or mobile stocks resembling building blocks or flying bricks. We must conceive of each brick as having been launched from a distance and as being composed of heterogeneous elements: containing within it not only an inscription with signs from different alphabets, but also various figures, plus one or several straws, and perhaps a corpse. Cutting into the flows (*le prélèvement du flux*) involves detachment of something from a chain.
>
> (Deleuze and Guattari, 1984: 39–40)

Accordingly, "the revolutionary period . . . is when the machine represents social subjectivity for the structure – as opposed to the phase of oppression and stagnation, when the superstructures are imposed as impossible representations of machine effects" (Guattari, 1984: 117). The militant task of schizoanalysis, then, is not to incite desire – *Revolt! Enjoy!* – but to repeat differently – *Deterritorialize! Refabricate!* For "the unconscious doesn't mean anything: because machines don't mean anything. They merely work, produce and break down, because all we're looking for is how something functions in the real" (Deleuze, in Guattari, 2009: 76). When all is said and done, then, we need machines, rather than structures, to take to the streets, and we need the empty squares of human habitation – from the smallest cubbyhole to the largest cosmopolis – to buzz with "schizzes, points-signs, or flows-breaks that collapse the wall of the signifier, pass through, and continue on beyond" (Deleuze and Guattari, 1984: 242). Or again: "the schizzes-flows" are "forces that escape coding, scramble the codes, and flee in all directions"

(Seem, in Deleuze and Guattari, 1984: xxi). Finally, it is invariably assumed that a revolution requires a great many and a huge number, if only to lend weight to the enormous effort that it takes to overturn a 'whole' world – and the inordinate effort that it takes to keep it upturned thereafter. But this is not so. Revolutions hinge on the *plus d'un*. An *and* or a *but* may suffice. This is always how a world shatters: from the hairline cracks that run through the middle of things.

References

Anderson, B., Keanes, M., McFarlane, C. and Swanton, D. (2012) On assemblages and geography. *Dialogues in Human Geography*, 2(2): 171–189.

Arrivé, M. (1992) *Linguistics and Psychoanalysis: Freud, Saussure, Hjelmslev, Lacan and Others* (trans. J. Leader). Amsterdam: John Benjamins Publishing.

Baudrillard, J. (1981) *For a Critique of the Political Economy of the Sign* (trans. C. Levin). St Louis, MO: Telos.

Baudrillard, J. (1990) *Seduction* (trans. B. Singer). Basingstoke: Palgrave Macmillan.

Boni, L. (2014) Formalisation and context: Some elements of a materialist reading of Lacan's four discourses'. In: Parker, I. and Pavón-Cuéllar, D. (eds.) *Lacan, Discourse, Event: New Psychoanalytic Approaches to Textual Indeterminacy*. London: Routledge. pp. 128–139.

Bonta, M. and Protevi, J. (2004) *Deleuze and Geophilosophy: A Guide and Glossary*. Edinburgh: Edinburgh University Press.

Coover, R. (1998) *Ghost Town*. New York, NY: Henry Holt.

Deleuze, G. (1988) *Foucault* (trans. S. Hand). London: Athlone.

Deleuze, G. (1990) *The Logic of Sense* (trans. M. Lester and C. Stivale). London: Athlone.

Deleuze, G. (1994) *Difference and Repetition* (trans. P. Patton). London: Athlone.

Deleuze, G. (1995) *Negotiations, 1972–1990* (trans. M Joughin). New York, NY: Columbia University Press.

Deleuze, G. (1997) *Essays Critical and Clinical* (trans. D.W. Smith and M.A. Greco). Minneapolis, MN: University of Minnesota Press.

Deleuze, G. (2004) How do we recognize structuralism? In: Deleuze, G. *Desert Islands and Other Texts 1953–1974*. New York, NY: Semiotext(e). pp. 170–192.

Deleuze, G. (2008) *Proust and Signs* (trans. R. Howard). London: Continuum.

Deleuze, G. and Guattari, F. (1983) *On the Line* (trans. J. Johnston). New York, NY: Semiotext(e).

Deleuze, G. and Guattari, F. (1984) *Anti-Oedipus: Capitalism and Schizophrenia* (trans. R. Hurley, M. Seem and H.R. Lane). London: Athlone.

Deleuze, G. and Guattari, F. (1986) *Kafka: Towards a Minor Literature* (trans. D. Polan). Minneapolis, MN: University of Minnesota Press.

Deleuze, G. and Guattari, F. (1988) *A Thousand Plateaus: Capitalism and Schizophrenia* (trans. B. Massumi). London: Athlone.

Deleuze, G. and Guattari, F. (1994) *What Is Philosophy?* (trans. G. Burchell and H. Tomlinson). London: Verso.

Derrida, J. (1981) *Dissemination* (trans. B. Johnson). Chicago, IL: Chicago University Press.

Dosse, F. (1997) *History of Structuralism, Volume 2: The Sign Sets, 1967–Present* (trans. D. Glassman). Minneapolis, MN: University of Minnesota Press. pp. 175–190.

Dosse, F. (2011) *Gilles Deleuze and Félix Guattari: Intersecting Lives* (trans. D. Glassman). New York, NY: Columbia University Press.

Gasché, R. (2014) *Geophilosophy: On Gilles Deleuze and Félix Guattari's What Is Philosophy?* Evanston, IL: Northwestern University Press.
Genosko, G. (1998) Guattari's schizoanalytic semiotics: Mixing Hjelmslev and Peirce. In: Kaufman, E. and Heller, K.J. (eds.) *Deleuze and Guattari: New Mappings in Politics, Philosophy, and Culture*. Minneapolis, MN: University of Minnesota Press.
Genosko, G. (2009) *Félix Guattari: A Critical Introduction*. London: Pluto.
Genosko, G. (ed.) (2012) *Félix Guattari in the Age of Semiocapitalism*. Edinburgh: Edinburgh University Press.
Goffey, A. (2016) Guattari and transversality: Institutions, analysis and experimentation. *Radical Philosophy*, 195: 38–47.
Guattari, F. (1984) *Molecular Revolution: Psychiatry and Politics* (trans. R. Sheed). New York, NY: Peregrine.
Guattari, F. (1986) *Soft Subversions* (trans. D.L. Sweet and C. Wiener). Los Angeles, CA: Semiotext(e).
Guattari, F. (1992) Regimes, pathways, subjects. In: Crary, J. and Kwinter, S. (eds.) *Incorporations: Zone 6*. New York, NY: Zone. pp. 16–37.
Guattari, F. (1995) *Chaosmosis: An Ethico-Aestehtic Paradigm* (trans. P. Bains and J. Pefanis). Sydney: Power Publications.
Guattari, F. (2000) *The Three Ecologies* (trans. I. Pinder and P. Sutton). London: Athlone.
Guattari, F. (2006) *The Anti-Œdipus Papers* (trans. K. Gotman). New York, NY: Semiotext(e).
Guattari, F. (2009) *Chaosophy: Texts and Interviews 1972–1977* (trans. D.L. Sweet, J. Becker and T. Adkins). Los Angeles, CA: Semiotext(e).
Guattari, F. (2011) *The Machinic Unconscious: Essays in Schizoanalysis* (trans. T. Adkins). Los Angeles, CA: Semiotext(e).
Guattari, F. (2013) *Schizoanalytic Cartographies* (trans. A. Goffey). London: Bloomsbury.
Guattari, F. (2015) *Psychoanalysis and Transversality: Texts and Interviews 1955–1971* (trans. A. Hodges). Los Angeles, CA: Semiotext(e).
Guattari, F. (2016) *Lines of Flight: For Another World of Possibilities* (trans. A. Goffey). London: Bloomsbury.
Hamblyn, R. (2001) *The Invention of Clouds: How an Amateur Meteorologist Forged the Language of the Skies*. London: Picador.
Hjelmslev, L. (1961) *Prolegomena to a Theory of Language* (trans. F.J. Whitefield). Madison, WI: Wisconsin University Press.
Hjelmslev, L. (1970) *Language: An Introduction* (trans. F.J. Whitefield). Madison, WI: Wisconsin University Press.
Holland, E.W. (2011) *Nomad Citizenship: Free-Market Communism and the Slow-Motion General Strike*. Minneapolis, MN: University of Minnesota Press.
Hustvedt, A. (2011) *Medical Muses: Hysteria in Nineteenth-Century Paris*. London: Bloomsbury.
Lacan, J. (2007) *The Seminar of Jacques Lacan: Book XVII: The Other Side of Psychoanalysis* (trans. R. Grigg). New York, NY: Norton.
Lazzarato, M. (2014) *Signs and Machines: Capitalism and the Production of Subjectivity* (trans. J.D. Jordan). Los Angeles, CA: Semiotext(e).
Lyotard, J-F. (1998) *The Assassination of Experience by Painting – Monory* (trans. R. Bowlby). London: Black Dog.
Massumi, B. (1992) *A User's Guide to Capitalism and Schizophrenia: Deviations from Deleuze and Guattari*. Cambridge, MA: The MIT Press.

Nancy, J-L. and Lacoue-Labarthe, P. (1992) *The Title of the Letter: A Reading of Lacan* (trans. F. Raffoul and D. Pettigrew). Albany, NY: SUNY.

Nöth, W. (1990) *Handbook of Semiotics*. Bloomington, IN: Indiana University Press.

Pearson, R.E. (1992) *Eloquent Gestures: The Transformation of Performance Style in the Griffith and Biograph Films*. Berkeley, CA: California University Press.

Raunig, G. (2010) *A Thousand Machines: A Concise Philosophy of the Machine as Social Movement* (trans. A. Derieg). Los Angeles, CA: Semiotext(e).

Saussure, F. de (1974) *Course in General Linguistics* (trans. W. Baskin). New York, NY: Philosophical Library.

Tschumi, B. (1994) *Architecture and Disjunction*. Cambridge, MA: The MIT Press.

Watson, J. (2009) *Guattari's Diagrammatic Thought: Writing Between Lacan and Deleuze*. London: Continuum.

Woodard, B. (2013) *On an Ungrounded Earth: Towards a New Geophilosophy*. New York, NY: Punctum.

Young, E.B., Genosko, G. and Watson, J. (eds.) (2013) *The Deleuze and Guattari Dictionary*. London: Bloomsbury.

Zupančič, A. (2006) When surplus enjoyment meets surplus value. In: Clemens, J. and Grigg, R. (eds.) *Jacques Lacan and the Other Side of Psychoanalysis: Reflections on Seminar XVII*. Durham, NC: Duke University Press. pp. 133–178.

2 Mapping the unconscious[1]

Manola Antonioli

Make a map, not a tracing

We know the importance of maps and all types of cartography, probably under the initial influence of Fernand Deligny and his *lignes d'erre* (wandering lines), in Deleuze and Guattari's works. The primary role of maps is explained in the introduction to *A Thousand Plateaus*, in which the two authors invite us to "make a map and not a tracing" (Deleuze and Guattari, 1987: 33). My argument in the following pages is that the new version of analysis that Deleuze and Guattari propose under the very enigmatic name of 'schizoanalysis' can also be interpreted as an activity of 'mapping the unconscious', a proposition that I'll try to establish through a reading of *The Machinic Unconscious* and *Schizoanalytic Cartographies*.

In contrast to psychoanalytic traditions inspired by structuralism, such as those found in Freud, Jung, or Reich, Guattari proposes his own version (schizoanalytic) of an unconscious *bricoleur* and *bricolé*, "as something that we drag around with ourselves both in our gestures and daily objects, as well as on TV, that is part of the zeitgeist, and even, and perhaps especially, in our day-to-day problems" (Guattari, 2011: 10).

The 'machinic unconscious' is also a deterritorialised one, because it is not located only in the *interior* world of individuals, but equally in the exteriority of their connections to bodies, territories (existential or spatialised), to school, work, society, etc. The Guattarian unconscious has also a new orientation in time: it is not concerned only with crystallisations of the past, but focuses on the future:

> To think time against the grain, to imagine that what came "after" can modify what was "before" or that changing the past that the root can transform a current state of affairs: what madness! A return to magical thought! It is a pure science fiction, and yet. . .
>
> (Guattari, 2011: 11)

The machinic unconscious cannot be interpreted in a straightforward fashion. It cannot be interpreted using a hermeneutic model, nor through an archaeological model which tries to gain access to its strata and stratifications buried in the past (or it would be an 'archaeology' in a Foucauldian sense, therefore an archaeology of statements, externalised and spatialised from the beginning), nor even by a genealogical or a narrative model. Instead it needs to be interpreted using a semiotic based on a 'collective assemblage of enunciation'[2] always two-sided or multi-faceted, according to the pragmatics of language and signification which Deleuze and Guattari were developing at the same time in *A Thousand Plateaus*, in the plateau '587 B.C.-A.D. 70: On Several Regimes of Signs'. Connections are created by a reciprocal deterritorialisation of natural things and language, and produce assemblages that are always concrete, fixed in time, situated, never modelisable, eternal, or merely formal.

It is for this reason that this kind of unconscious should be mapped, not interpreted: the schizoanalysis will be a form of cartography. In the heterogeneous field of a 'machinic unconscious' it is always necessary to identify the regions fixed in the past, immobilised in strata and segments, fossilised and necrosed, in order to find and situate the changing 'crystals of the possible'. This approach to the unconscious is oriented by two essential assumptions:

1 the *most* deterritorialised assemblage must solve the impasse and untie the stratifications;
2 "deterritorialization in all its forms 'precedes' the existence of the strata and the territories" (Guattari, 2011: 16).

This revision of the unconscious will therefore be oriented towards the constitution of a 'pragmatic schizoanalysis', which never forgets political and micropolitical questions, a pragmatics constructed by the rhizome used as a 'non model', which never has the stability or the iterability of a 'model'. Contrary to each treelike structure and model, rhizomes can connect any point to any other; a rhizome does not necessarily refer to a linguistic sign, but it belongs in a semiotic field that associates heterogeneous elements at each step (biological, political, or economic). In the introduction to *The Machinic Unconscious*, Guattari therefore explicitly chooses to maintain and reinforce the distinction between "*semiology* as a trans-linguistic discipline that examines sign systems in connection with the laws of language (Roland Barthes' perspective); and. . . *semiotics* as a discipline that proposes to study sign systems according to a method which does not depend on linguistics (Charles Sanders Peirce's perspective)" (Guattari, 2011: 22).

A cartography of the unconscious is always a part of a complex semiotics, which can never be reduced to interpretation of language or to the dimension of a signifier. In schizoanalysis, the maps that belong to 'internal' dynamics of the subject and those which belong to 'external' dynamics, are situated in a merely linguistic field and anchored in a generalised semiotics and an extra-linguistical pragmatics. This kind of map would try to show at the same time elevations and bodies of water, administrative boundaries and territorial ones, natural resources and economic resources, natural frontiers and political ones, the ethnic and religious composition of inhabitants and their density, etc. It would result in a slightly crazy adventure for each geographer, in much the same way that schizoanalysis is for each analyst. Instead of a face-to-face relationship between the analyst and the analysand, but instead involved in a collective construction.

This 'pragmatics of rhizome' equally abandons the idea of a 'deep structure': the machinic unconscious is a spatialised one, which must be constructed, explored, and analysed like a map that is "open and connectable in all of its dimensions; it is detachable, reversible, susceptible to constant modification" (Deleuze and Guattari, 1987: 33). In the simultaneous movements of territorialisation and deterritorialisation which move through this kind of unconscious, nothing is irrevocably fixed: a rhizome can give place to a treelike structure but, conversely, "a tree branch or root division may begin to burgeon into a rhizome" (Deleuze and Guattari, 1987: 36).

In contemporary philosophy, we can find an equivalent project to spatialise the unconscious in the first volume of the 'critique of spatial reason' which the German philosopher Peter Sloterdijk develops in his *Spheres* trilogy. But Sloterdijk (radically, deeply, and permanently influenced by Heidegger's thought) privileges the themes of 'staying', which is sedentary and inscripted in a territory, and rather different to the deterritorialised and deterritorialising approach of Guattari. Sloterdijk's trilogy proposes an exploration of spaces under three forms: psychology (*Bubbles*) (Sloterdijk, 2011), politology (*Globes*) (Sloterdijk, 2014), and technology (*Foams*) (Sloterdijk, 2016). The exploration of 'interior space' developed in *Bubbles* leads to a vision of human space with a folded, limited, and participatory structure, which consists in the connection of several interior spaces, where no one can ever occupy a purely phantasmic position of a secret interiority or sovereign exteriority.

That which is called, for want of a better name, the 'interior world' cannot be conceived according to the archaeological model of a deep psychology, but must be interpreted using an architectural model of a series of interior spaces which exclude any closed monad.[3] The 'interior life', therefore, can exist only by incorporations, connections, bendings, resonances and common rhythms, or (in psychoanalytic terms) by successive identifications. This, then, is the reason why the unconscious is accessible neither by decrypting, nor through discovering latent interpretations; there is no 'product' but, rather, a construct which cannot be interpreted, only mapped.

It is not by chance that both Sloterdijk and Guattari are fascinated, each in their own manner, by all forms of creation of spaces and territories: existential territories, but also by forms of aesthetic and cultural production of territories, as well as the production of habitable territories in an architectural context. The convergence points between the spatial perspective of Sloterdijk and that which is explored by Guattari could be found (despite their different theoretical and philosophical references, a few of which are perhaps incompatible) in an interest in the production of 'pre-architectural' spaces and territories (spheres or protective bubbles in Sloterdijk's work, which progressively become foams; and refrains or existential territories in Deleuze and Guattari's philosophy).

In the rhizomatic structure of *The Machinic Unconscious*, we can see a transversal connection between the introduction and part six, 'Reference Points for a Schizoanalysis', where Guattari writes that "the unconscious is constituted by machinic propositions that no semiological or logico-scientific propositions can ever grasp in an exhaustive fashion" (Guattari, 2011: 149), because "concepts must be folded onto realities, not the other way around" (Guattari, 2011: 149). In part six, several pages are devoted to trees and tracings, maps and rhizomes, and (as per the introduction) Guattari speaks like a cartographer.

Before any production of statements and any translation through to language, schizoanalysis presupposes (yet again) the role of geographer, cartographer or even land-surveyor, which consists in creating (for each case and each situation, then without any recourse to a universal and methodologically valid interpretative model) a map of the unconscious, which will include each single time-fixed strata, deterritorialisation lines, and black holes. This map must allow an opening on experimental perspectives for the 'analytical assemblage' or the 'analyst'. Crucially, the analyst need not be an individual, but could also be a team or an institution, that necessarily involves an analysis of economic, political, material, and technological flows.

The schizoanalytical *praxis* is defined this time as a 'transformational *praxis*': all the unconscious processes, blocked as they may be, can in this way find exit points with the conjunction of disparate semiotic elements. Put differently, "machinic consistency is not totalizing but deterritorializing" (Guattari, 2011: 171). A rhizome, by definition, cannot be reduced into a model and therefore can neither be defined by any psychoanalytical topic nor structural model. In this kind of real-world experimentation, maps work as laboratories. Opposed to the idea of structure, the map can open itself to multiple dimensions; it can be torn apart; it can be adapted to any kind of montage. This is a 'pragmatic map' which can be produced by an individual or a group, drawn for the purpose of orientation or conceived like a piece of art, like a political action or like a meditation tool.

The rhizome is altogether different, a *map and not a tracing*. Make a map, not a tracing. The orchid does not reproduce the tracing of the wasp; it forms a map with the wasp, in a rhizome. What distinguishes the map from the tracing is that it is entirely oriented toward an experimentation in contact with the real. The map does not reproduce an unconscious closed in upon itself; it constructs the unconscious (Deleuze and Guattari, 1994: 33).

This kind of map is always context-specific and never presupposes larger competencies: a map which would adequately describe a certain type of territory, might not work in another context. For instance, a map conceived as a simple orientation tool might not work as a piece of art, and vice versa. The pragmatic competence of one cartography in respect to another depends on the fact that it puts into place a more molecular, deterritorialised, machinic, segmentarity. Maps can produce a draft of a possible which cannot become real in an existing context, but they can also produce real machinic mutations. In any situation, we can construct a cartographic or diagrammatic micropolitics which refuses any kind of divine, economic, structural, or hereditary fatality, all kinds of visions of the unconscious like a destiny or a structure.

To give an example of this kind of mapping, we might turn to Guattari's rhizome-map of the phobic surroundings of Little Hans in Freud's work. First of all, we find in this analytical case, heterogeneous elements: places (the family home, parental bed, the street, . . .), becomings (the becoming of sexual body, becoming-guilty, but also an 'accumulation of deterritorializations' from the becoming-imperceptible to the becoming-guilty, to the becoming-sexualised-body, to becoming a social body to becoming-animal), lines of facialisation (the mother's face, the face of Professor Freud, and the 'facialisation transfer') and connections which move from one element to another.

But rhizome-maps are not only tools for an analysis of an individual unconscious, since schizoanalysis is also a tool for a political *praxis*, for molecular revolutions. After the rhizome-map of Little Hans, Guattari describes the rhizome-map of the Leninist divide and the begetting of Stalinism (Guattari, 2013: 176–177). Pragmatic assemblages are machinic and they never depend on universal laws, they are subject to historical mutations which are first and foremost unpredictable and therefore always susceptible to happen again (the 'romantic complex', the '*Front populaire* complex,[4] the 'Resistance complex') without the universalistic quality that psychoanalysis lends or lent to the Oedipus complex. Each segmented, stratified, and molar balance of power can be disrupted at any moment by the uprising of a revolutionary situation, which upsets pre-existing maps, by showing that a new assemblage was eating away, in a subterraneous manner, at a balance of power which formerly seemed immutable, fixed, and eternal.

The mere act of starting to actively draw and construct maps could initiate mutation and deterritorialisation effects: to write one's dreams rather than listening passively to their interpretation, to draw them, to act them out, could also transform the unconscious map. Schizoanalytic cartography is also a dangerous practice, which requires a great amount of prudence and has nothing to do with wild interpretations. It would consist of making at any moment, micropolitical choices which engage the 'analyst' and their responsibility for the processes they take care of. They should opt for accelerating or slowing down an internal transformation of assemblages, in order to facilitate or put the brakes on the creation of a collective assemblage, to explore and experiment with a becoming unconscious, rather than indefinitely tracing complexes or pre-existing interpretations. A cartography, therefore, which never excludes compromises, steps forwards and backwards, revolutions, does not aim to control or overcode unconscious processes but only to assist and accompany them.

What could be particularly interesting is the way in which Guattari (like Deleuze in some of his texts) looks at maps and cartography and anticipates the current evolution of visions and functions of maps by geographers. The map is less and less often perceived as a simple 'representation' of a territory or as an objective and neutral presentation of a group of phenomena (like natural and economical resources, demographics, ethnic groups, etc.). According to the indications given by French philosopher and landscape epistemologist Jean-Marc Besse (2010) in his introduction to an issue of *Carnets du paysage*, we can identify the following transformations:

1 The statement (very machinic, in the technological sense of the word) that concrete spaces and virtual spaces are constantly interpenetrating in our experience (GPS, Google Earth, etc.).

2 The interest that artists show more and more for mapping operations and which reveal the aesthetic, creative, and imaginary dimension of cartographic operations (the importance which Guattari gave to the 'aesthetic paradigm' is well known). As noted by French art historian Gilles A. Tiberghien[5] (2007), cartographic processes show the way in which the imaginary and the imagination work underneath even the most "scientific" activities. Each cartographer (we could think, for example, of Christopher Columbus imagining a new route to India) imagines the world before representing it and in his representation gives an image constructed through conventional relationships with the reality, which has very little to do with *mimésis*: "There is no such thing as a truly cartographic reality, but only multiple ways to understand the world through maps" (Tiberghien, 2007: 6).[6] The map is not only a mimetic tool but is always a constructive system. This is what Deleuze (1994 [1968]) understood as a 'new image of thought', where any concept is never simply an act of representation but also a tool for experimentation and creation.

3 In the context of the 'spatial turn' already identified by Michel Foucault, maps are no longer merely a tool for architecture, for territorial analysis, for geography or geo-politics, but they have become more and more important in epistemology, particularly in the modelisation of types of connections and links allowed for new communication technologies.

4 The progressive abandonment of Euro-centrism and the end of its absolute mastery over the representation of the world order provoked the recognition of native mappings which have renewed social and cultural relationships to maps. The map is no longer merely a tool of power and spatial appropriation but is starting to become a way in which traditional societies can recover a specific way to perceive the space. Maps, more and more distant from their use as reproduction models, can become means of appropriation and of exploration of real and existential territories. Maps of the desert for nomadic people of the Sahara, maps of the ocean sensitive to all empirical variations of navigation, maps of ice formations, maps of cyberspace, maps of smooth or striated spaces:

> Ultimately, the decisive position that cartography holds in territorialisation operations, that is in any kind of actions through which societies give meaning to their environment, has never seemed so evident (Tiberghien, 2007: 8).
>
> We can particularly discover that a road map presupposes a kind of habit, perception and conception of space radically different from a geometrical and mathematical map of modernity which was based on objectivity and transparency. A road map must include, in different ways, the subjective and time-based experience of the territory that it represents, which is never perceived as independent from practices and wanderings which take place within it.

5 Cartographers call into question the idea of a 'neutral', 'objective', or scientific map, and they take into account the implicit discourses that construct

> maps, issues of political, economic, and cultural power which are expressed within the individuals or institutions who produce them.

From this point forward, maps can only be read in the plural. Moreover, transversalities between different cartographic practices (of geographers and urban planners, of artists, of philosophers, and of anthropologists) are more and more frequent. The cartographic act, as Guattari wrote, is never neutral or without consequences but has a performative and creative nature, which can give life to new realities; it includes all the temporary steps of its conception (sketches, preparatory drawings, diagrams). Deleuze and Guattari's schizoanalysis anticipated and foreshadowed this new context of a transversal cartography, with multiple entry points, creative and concerning many theoretic and aesthetic fields.

Cartography of existential territories

In 1989, in *Schizoanalytic Cartographies* (Guattari, 2013), Guattari expanded his cartographic project "to the cartography of existential Territories, implying sensible, cognitive, affective, aesthetic, etc. Universes, for clearly delimited areas and periods of time" (2013: 3). Elsewhere, he describes this as "the cartographies of concrete processes of subjectification" (2013: 5). The analytic cartographies which are the subject of the first chapter are structured around 'assemblages of enunciation'. Using assemblages of enunciation can allow us, according to Guattari, to finally escape from the concept of 'unconscious', so as not to reduce the facts of subjectivity to intra-subjective instances and inter-subjective relations. The schizoanalytic cartographies must exceed the problematic of the individualised subject in order to think by heterogeneous ensembles, that are indifferently material and semiotic, individual and collective, machinic and subjective and whose components are interacting with radically heterogeneous domains.

Schizoanalytic subjectivity is established at the intersection of Flows of signs and machinic Flows, at the junction of the facts of sense, of material and social facts, and, above all, in the wake of transformations resulting from the different modalities of Assemblage (Guattari, 2013: 20).

Through this new cartography, Guattari aims to reach the separation of the problematic of consciousness and unconscious from that of the representation of the Ego and the unity of the person, working with the help of the category of deterritorialisation:

> The idea of a totalizing, even totalitarian consciousness ("I am the master of myself and of the universe") participates in a founding myth of capitalist subjectivity. In fact, there are only diversified processes of conscientialization, resulting from the deterritorialization of existential Territories that are themselves multiple and tangled.
>
> (Guattari, 2013, p. 26)

Guattari introduces four 'functors of deterritorialization' which allow him to trace these dynamic maps: Flows (material and signaletic), existential Territories, Phyla (abstract and machinic), and incorporeal Universes.

The Plane where these dynamic cartographies of the subjectivity are drawn is a 'chaotic Plane of immanence'. Chaos is a central figure in Guattari's thought,[7] where it is not interpreted in a static manner, by refusing theories that represent it in the form of a mixture, of holes, caverns, or dust. Engaged in a process of continuous chaosmosis, chaos only exists as a process of 'chaotizing' where it never maintains its consistency or a stable configuration. It escapes to the logic of discursive sets, but it generates as much disorder as it does complex virtual compositions, that appear and disappear at infinite speed: "Chaos thus becomes the primary matter of virtuality, the inexhaustible reserve of an infinite determinability" (Guattari, 2013: 103).

Guattari maps the chaotic Plane of immanence through two types of relations: relations of reference and relations of consistency. The reference functions in the mode of holding things together, in a kind of existence where there are no interactions but only relationships of co-existence, trans-existence, transversality, without any subject to hold together the world. If reference is rather a spatial entity, consistency is of a more temporal order: the connective processes are fragile and finite, made of "contingent arrangements at heterogeneous levels" (Guattari, 2013: 106). Behind its apparent stability, the consistency of existence is continuously cracking and its cracks are the condition of possibility for becomings; they constantly allow the passage of a regime of connective passivity into one of processual connectivity. In the chaotic Plane of immanence, "there isn't any brute form of being, rooted there, once and for all, independent of the Assemblages that apprehend it, so as to undergo its effects or inflect his trajectory of destiny. Being is the modulation of consistency, the rhythm of putting together and dismantling [montage et démontage]" (Guattari, 2013: 107).

Therefore, two types of ontological states populate the Plane of immanence: "chaotic multiplicities composing and decomposing complex arrangements at infinite speeds; [and] existential filters selecting relatively homogeneous sets of arrangements characterized by the iteration of local and localizing decelerations" (Guattari, 2013: 108–109). Filters act as 'hooks' for chaotic multiplicities and in the Plane of immanence we can always find two dynamics (of crossing and uncrossing) between the dimensions of reference and consistency: filters confer a relative stability to multiplicities, whereas multiplicities create hyper-complexity in the Assemblages. But crossed and uncrossed regimes envelop each other constantly. The supremacy of crossing makes us enter the domain of the possible; the supremacy of uncrossing keeps us in the domain of the virtual. This vision of the Plane of immanence and Agencements is clearly inspired by quantum physics, where the same quantum of energy can take forms that are simultaneously wave *and* particle, discontinuous *and* continuous, separable *and* inseparable. Schizoanalysis can thus map the component of unconscious Assemblages in apparently contradictory forms (for example, a corporeal Territory with a slow consistency, and deterritorialised Universes associated with this Territory but with a rapid consistency):

> What does our problem come down to? To trying to construct ordinary consistencies and temporalities from infinitely slow speeds of separability and infinitely rapid speeds of continuity.
>
> (Guattari, 2013: 161)

In these kinds of maps and Assembages, there is never a linear temporality, with only one dimension: times are composite, and they can be recomposed on the basis of new instrumentations and experiments. Among other elements that constitute these cartographies we can find heterogenesis. Heterogenesis produces deterritorialisations by the encounter and the knowledge of the other and through the other; the 'other' is not only another subject, an identifiable person, but also impersonal affects, animal, machinic, and cosmic becomings, in a process of a generalised transference.

Everything derives from the fact that the ultimate reference point is cracked, that absolutely consistent Alterity (the big Other) does not exist. It is from this relativity of alterity that a differential enunciation through enunciative fragments, scattered to the four corners of the cosmos, originates. It speaks in the margin; a self-organising process can be set off where nothing precise could be expected. (Guattari, 2013: 189)

By explicitly choosing to operate according to a sensitive and aesthetic paradigm rather than following a scientific model, schizoanalytic cartographies open up to at least three directions for mapping the unconscious:

1 They call into question the essentially linguistic dimension of Freudian and Lacanian psychoanalysis, which cleaves to a model of the interiorised and individualised dimension of the unconscious, this chapter has argued that this ought to be replaced by open subjectification processes that integrate heterogeneous and machinic components.
2 This approach might best be understood as a dynamic cartographic project, and is one which could be able to represent spatialised components of reference, but also dynamic component of time on a chaotic plane of immanence on which it could be possible to represent complex processes (with their stops and starts, with their bifurcations and becomings).
3 From a philosophical point of view, a 'chaosmotic ontology', where Being is a modulation, a rhythm of 'montage et démontage'.

We can add that many instances of creative mapping have been undertaken in contemporary French geography (sometimes under the wake of Ferdinand Deligny's experimentations). As early as 1993, Françoise Péron has been asking young children and teenagers living on islands to draw 'mental maps' of their displacements within the island and toward the continent (Péron, 1993). These maps imply both: the displacement from home to school and back, which is most often a walk, and the mobility toward the main land, which is a boat trip. They were of such interest that coastal planners used them. In a less political way, many maps have also been drawn inspired by Debord's 'derive'. Sometimes these mappings do not take the appearence of a drawing with a scale and an orientation, but are also simply texts (Gwiazdzinski, 2017). The maps of the unconscious developed by Guattari echo a processual transformation of the vision of maps and plans in the field of the geography and of the 'psycho-geography', which try to decipher the unconscious of the territory.

Notes

1 A shorter version of this paper was published in French (see Antonioli, 2012).
2 According to Deleuze and Guattari, language is always a 'collective assemblage of enunciation', a non-subjective production. The source of utterance is social, physical, technical, political, and language is not *within* the speaker (the 'subject'), but always *outside*: "On a first, horizontal, axis, an assemblage comprises two segments, one of content, the other of expression. On the one hand it is a *machinic assemblage* of bodies, of actions and passions, an intermingling of bodies reacting to one another; on the other hand it is a *collective assemblage of enunciation*, of acts and statements, of incorporeal transformations attributed to bodies" (Deleuze and Guattari, 1994: 98).
3 In *The Monadology*, Gottfried Leibniz (1992[1714]) sketches a metaphysics of substances, or *monads*. Monad is a 'simple' substance, which means that it has no parts and is therefore indivisible.
4 The Popular Front in France (*Front populaire*) was an alliance of left-wing movements, including the French Communist Party (PCF), the French Section of the Workers' International (SFIO), and the Radical and Socialist Party, during the interwar period (founded in 1936 and dissolved in 1938).
5 I am referring here and in the following remarks about cartography to Tiberghien's *Finis terrae*.
6 My translation.
7 For more information on the theory of chaos in Deleuze and Guattari's work, I refer the reader to Félix Guattari's (1995[1992]) *Chaosmosis* and Deleuze and Guattari's (1994[1991]) *What Is Philosophy?*

References

Antonioli, M. (2012) Cartographier l'inconscient. *Chimères*, 76: 91–100.

Besse, J.M. (2010) Editorial. *Carnets du paysage, Cartographies*, 20: 5–9.

Deleuze, G. (1994 [1968]) *Difference and Repetition* (trans. P. Patton). London: Continuum.

Deleuze, G. and Guattari, F. (1987[1980]) *A Thousand Plateaus* (trans. B. Massumi). Minneapolis, MN: University of Minnesota Press.

Deleuze, G. and Guattari, F. (1994[1991]) *What Is Philosophy?* (trans. H. Tomlison and G. Burchell). New York, NY: Columbia University Press.

Guattari, F. (1995[1992]) *Chaosmosis* (trans. P. Bains and J. Pefanis). Sydney: Power Institute.

Guattari, F. (2011[1979]) *The Machinic Unconscious* (trans. T. Adkins). Los Angeles, CA: Semiotext(e).

Guattari, F. (2013[1989]) *Schizoanalytic Cartographies* (trans. A. Goffey). London: Bloomsbury.

Gwiazdzinski, L. (2017) Nouvelles explorations urbaines. In: Caritoux, N. and Villard, F. (eds.) *Nouvelles Psychogéographies*. Milan: Mimésis Éditions. pp. 177–196.

Leibniz, G. (1992 [1714]) *Discourse on Metaphysics and Other Essays* (trans. D. Garber and R. Ariew). Indianapolis, IN: Hackett Publishing Company.

Peron, F. (1993) *Des îles et des homes: L'insularité aujourd'hui*. Rennes: Ouest France.

Sloterdijk, P. (2011[1998]) *Bubbles: Spheres Volume I*. Microspherology (trans. W. Hoban). Los Angeles, CA: Semiotext(e).

Sloterdijk, P. (2014[1999]) *Globes: Spheres Volume II*. Macrospherology (trans. W. Hoban). Los Angeles, CA: Semiotext(e).

Sloterdijk, P. (2016[2004]) *Foams: Spheres Volume III* (trans. W. Hoban). Los Angeles, CA: Semiotext(e).

Tiberghien, G.A. (2007) *Finis Terrae*. Paris: Bayard.

3 Guattari's incorporeal materialism

From individuation to aesthetics (and back again)

Tom Roberts

Introduction

> It is not easy to extricate oneself from the politics of signification and interpretation.
> (Guattari, 1984: 88)

Guattari's philosophy is nothing if it is not a machine: his concepts acquire meaning and significance only in as much as they continue to function, plugging into those material forces that suffuse and animate the present in order to extract something new. And like all machines, it requires a minimum amount of energy to jump-start it into action on each new occasion. The strange beauty of Guattari's enigmatic style lies precisely in its propensity to deflect interpretation, time and time again, addressing its 'readers' as experimental mechanics who have been tasked with maintaining an unfamiliar machine. For this reason, the significance of Guattari's philosophy resides in the relational disturbances that its machinic functioning demands; which is to say, it is up to the reader to get the Guattarian machine up and running.

It is perhaps in this regard that Gerlach and Jellis (2015a) celebrate Guattari's philosophy as an inherently "impractical" undertaking, where impracticality is understood not in terms of a withdrawal from the messy materialities of practice *per se*, but rather in terms of a capacity for short-circuiting established structures of signification and meaning. The impracticality of Guattari's thought stems, therefore, from the requirement that its users assume the role of experimental machinists, forever cognizant of the fact that a machine can never simply be applied in the manner of a tool. As Gerlach and Jellis explain, "no thinker provides ready-made 'tools' to dissect and understand a situation; it is not a matter of simply applying Guattari's ideas, unchanged" (2015b: 179). Not a tool to be applied for obvious practical gain, then, but a machine whose impracticality testifies to the incessant demand placed upon its users to think again, and, in doing so, to think anew.

I have, in my own Guattarian encounter, experienced the intense and often bewildering dynamism of this impracticality in relation to a particular refrain that recurs throughout Guattari's philosophy, namely, that of the incorporeal. This chapter is an attempt to draw out the significance of the incorporeal for Guattari's materialist account of subjectivity and its process of production. More specifically, I argue that it is not possible to grasp the distinctly aesthetic tenor of Guattari's ethical problematic without first understanding the role played by incorporeal individuations. The main difficulty posed by the incorporeal is that it appears, at first blush, to signify a realm that altogether transcends materiality. Indeed, a cursory glance through an English dictionary is enough to confirm that, by convention, the word incorporeal refers to a supposedly immaterial reality: it is that which is "of, pertaining to, or characteristic of immaterial beings" (Oxford English Dictionary Online, 2016). We might be forgiven, then, for wanting to associate such an incorporeal reality with that of the spirit, the soul, or the transcendental subject. Yet, it seems clear to me that this tendency to bifurcate reality into, on the one hand, the objective materiality of non-thinking nature, and, on the other, the immateriality of thinking subjects, is precisely what Guattari is trying to challenge by accounting for subjectivity in terms of processes of machinic production.

Turning to Guattari's conceptual machines in greater detail, *Schizoanalytic Cartographies* (2013) leaves us in no doubt that it is the Kantian figure of the transcendental subject, cut-off from the concrete flux of materiality, that prevents us from apprehending subjectivity in terms of radically immanent processes. In order to grasp subjectivity from the perspective of its production, Guattari endeavours "to minimize the use of notions like those of subjectivity, consciousness, significance . . . as transcendental entities that are impermeable to concrete situations" (2013: 23). In place of subjectivity, Guattari will often use the term "assemblages of enunciation" (2013: 18) as a means of short-circuiting those Kantian relays that posit the existence of a transcendental subject as *the* foundation for any experience whatsoever. Guattari's conceptual machines, Jellis notes, are cartographic in precisely this sense, namely, that they enable their users to 'map out' and explore the production of subjectivity through "relational configurations which change state and status as a function of particular assemblages" (Jellis, 2014: np.).

Contra the Kantian lineage, Guattari is not at all interested in maintaining a relation of mutual exclusivity between subjects and objects. Rather, his aim is to approach the problem of subjectivity – that is, of the creative potentiality of thought as an experience that is felt – without importing a transcendental entity to act as its *raison d'être*. There is, therefore, no subject for Guattari, at least not in the sense of an individual capable of transcending its relations, because subjectivity is thoroughly inseparable from the material process that constitutes its conditions of possibility. Or, as Guattari (2013: 23) articulates:

> The most abstract, radically incorporeal, references mesh with the real; they cross the most contingent of Flows and Territories. On account of this, nothing protects them against historical alteration and cosmogenetic mutations.

What we are faced with, then, is a radically materialist approach to subjectivity, one that, whilst refusing the metaphysical distinction between subjects and objects, strives nonetheless to articulate the possibility of incorporeal consistencies in terms of processes that remain, nonetheless, intensely material. Guattari's philosophy is certainly a materialism, but – and herein lies the twisted roots of its dynamic impracticality – it is a materialism that eschews the certainty of a solid foundation through its cartographic style.

In a move that resonates in many ways with the speculative, process philosophy of A.N. Whitehead, Guattari manages to transform the relationship between materiality and the incorporeal in such a way that it no longer falls back upon the impasse of a metaphysical bifurcation (Whitehead, 2004). Like Whitehead before him, Guattari shows us that the only way to actively respond to this apparent paradox is to invent new concepts, and in doing so to experiment with a materialism that does not presume to know what matter can do, prior to the events of mattering that compose experience.[1] It is, in Isabelle Stengers' words, "to accept the risk, to try the adventure, to explore what the rejection of a bifurcation of nature obliges us to think" (2011: 40). Impractical, indeed! The risk that I wish to harness, alongside Guattari, is that we simply cannot understand the ontological texture of, say, the deterritorialised constellations of Jazz music, if we restrict our materialism to the coordinates of corporeal life (1995: 93–94). Similarly, the drama of a Van Gogh is not a thing that we can locate in space or time, for its existential vibrancy expresses a different kind of consistency than that which characterises its material-energetic supports (Guattari, 2011a: 52).

It bears repeating, however, that Guattari categorically refuses to attribute these incorporeal consistencies to the psychological emanations of a transcendental subject. It is here that we find the core of Guattari's cartographic materialism, for it is the imperative of materialist philosophy to locate these consistencies not within the mind of a human subject but rather upon a plane that remains immanent to nature – or better, whose immanence simply *is* a plane of nature. I would now like to draw out two particular implications of Guattari's thought to take forward in this chapter. First and foremost, an incorporeal materialism necessitates a somewhat unusual approach to the problem of individuation, for it posits the reality of material consistencies that remain irreducible to the physical coordinates of space, time, and energy (2013: 52). It requires, in other words, a theory of individuation that is no longer tethered to the finite corporeality of individuated forms. Second, by situating these individuations on a plane of nature that precedes the subject, Guattari is thus able to highlight the constitutive role of these processes in the production of subjectivity, in such a way that to experiment with incorporeal consistencies becomes an ethico-aesthetic imperative of life (Hynes, 2013). I now turn to unpack these two strands and their interrelations in more detail, beginning with the problem of incorporeal individuation.

Incorporeal individuation

> There is a mode of individuation very different from that of a person, subject, thing, or substance. We reserve the name haecceity for it. A season, a winter, a

> summer, an hour, a date have a perfect individuality lacking nothing, even though this individuality is different from that of a thing or subject.
>
> (Deleuze and Guattari, 2004: 288)

How best to approach this strange problem of an incorporeal materialism, of individuations that forever exceed the finite coordinates of physically individuated entities? One avenue that I have found to be particularly fruitful in exploring these questions has been through the concept of 'haecceity', which Guattari develops alongside Gilles Deleuze in *A Thousand Plateaus* (2004). Haecceity is an important concept for thinking through the incorporeal because, as the opening extract suggests, it allows us to grapple with the idea of an individuality that is very different from that of an finite entity, whether object or subject. For this reason, haecceity is a concept that is very much at odds with the figure of the transcendental subject, which requires a world that remains thoroughly amenable to the representational claims of scientific epistemologies – if not in practice then at least in principle, for the transcendental subject that forms the basis of the Kantian lineage is inseparable from the humanist fantasy of achieving an a-historical, disembodied gaze.

I ponder these thoughts one summer's evening whilst strolling out across some nearby fields, the sky broken here and there by the branches of trees stretching upwards towards the ever intensifying skies. What would it mean, I wonder, to write the world not as a subject concerned with representing objects, but to amplify a sense of consistency that is all the more material in its irreducibility to the finite coordinates of individuated forms? The infernal cogs of the Guattarian machine begin to shudder, slowly but surely, into motion. I recognise the tree that now stands before me through the finite outline of its form, the texture of its bark, the translucent green of its leaves and the shade that it affords me on a summer's afternoon. To the extent that it is recognised, the tree remains unremarkable, fading into the background of experience. Yet, there is a moment in which *this* instance, of which the tree before me is a part, seems to exceed the familiar boundaries of its 'tree-ness'. Such an experience could remain fleeting, for instance, in the almost-unconscious register of rhythmic consistency as the leaves dance enticingly above me on the breeze. But it could also be catastrophic: an imperceptible threshold is crossed, and I feel myself swept away by a becoming-other (Guattari, 1995: 93).

Throughout his writings, Guattari urges us to consider these moments not simply as events of psychological trauma but as profoundly ontological movements, expressive of a cosmic fusion that interpolates subjectivity into the processual weave of non-thinking forces (1995: 120). As a consequence, it no longer makes sense to reduce reality to the finite coordinates of individuated objects. I do not get any closer to the reality of the swaying-glowing-greenness, for example, by enquiring as to what the tree is: to ask this question is to already approach the world solely in its relation to a transcendental subject. For Guattari, it is no longer a matter of parsing the world into, on the one hand, an objective reality composed of individual identities (trees, chairs, atoms, and substances) and, on the other, human experience as mediated by representations. The reality of the tree is not reducible to an individuated form but is also expressed through intensive forces that are themselves irreducible to the form of an individual. Thus, I recognise a tree as an object, but at a particular time of day, when sun and breeze enter into a particular combination, I am overwhelmed by an intensity of greening that renders the distinction between subject and object obsolete. The world hangs together, acquiring an intensive consistency before there is a subject capable of recognising and representing the form of an object.[2]

In the event of greening, the tree that I habitually recognise as a discrete object gives way to a singular composition of intensities. Deleuze and Guattari reserve the concept of haecceity for these kinds of compositions, which, whilst admitting neither object nor subject, "have a perfect individuality lacking nothing" (2004: 288). As Francois Zourabichvili (2012: 127) explains, the concept of haecceity that Deleuze and Guattari borrow from the philosophical writings of Duns Scotus provides a means of affirming the reality of an 'evental' individuation:

> Haecceity refers to an evental individuality, and is opposed to the received idea according to which there can be no individuality without form (whether in existence, as a body or a person, or in art, as in the work).

Contrary to what we might expect, however, the evental nature of haecceities does not confine them to the instantaneous or the momentary. As individuations without form, haecceities cannot be located according to finite spatio-temporal coordinates. When Deleuze and Guattari speak of an hour as a haecceity they are not referring to the determinable instant at which the bell begins to toll. Rather, the hour as haecceity expresses an existential consistency that gathers and articulates incorporeal intensities without circumscribing the finitude of a form. Unlike an actual state of affairs that can be determined empirically in space and time, a haecceity has no beginning or end as such: it is "a simultaneous too-late and too-early, something that is both going to happen and has just happened" (Deleuze and Guattari, 2004: 289). In short, a haecceity's individuality is expressed not through the extensive differences that determine individual forms but through the intensive force of difference as such: it is an individuation whose sole consistency lies in its capacity to differentiate itself in the mode of "a passage or a change" (Zourabichvili, 2012: 128).

Deleuze and Guattari's concept of haecceity thus directs our attention towards processes of individuation that do not proceed according to the development of a particular form: atmospheric variations, air temperatures, an intensity of light, an event of greening. These individuations take place on a plane of consistency that is composed not of forms (subjects and objects) but consists of intensive differences. "There are only relations of movement and of rest, speed and slowness between unformed elements, or at least elements that are relatively unformed, molecules and particles of all kinds" (Deleuze and Guattari, 2004: 293–294). The body transfixed beneath the trembling green of the early evening feels itself being addressed not in the form of an individuated subject but as a composition of relatively unformed elements, that is, of singular intensities. But what is this body, in its unformed elements, being addressed by? It is certainly not the tree-as-object, for the plane of consistency admits neither the identity of a subject nor the form of object. Rather, it is in the register of unformed elements that individuations by haecceity are able to take hold.

In the event of greening – suspended beneath the wavering leaves – a moment takes hold. Not an instant in chronological time, but a haecceity whose consistency lies in the zone of proximity produced between degrees of speed and slowness. Individuated as haecceity, this moment attains a power that exceeds the spatialised temporalities constitutive of corporeal finitude and the phenomenological closure of the subject's world; "the moment, instead of being passively enshrined between past and future, becomes germinative, it develops ontological coordinates" (Guattari, 2011a: 53). The moment is thus a *movement* without origin and with no direction. It is not directed towards a goal, nor is it concerned with the development or evolution of a particular form. There is no progress on the plane of consistency, no evolution when it comes to haecceities. To this end, the moment remains indeterminable within the context of measurable time; it is eternal, not because it lasts forever but because its consistency is irreducible to the finite coordinates of individuated forms. It is "a zone of indetermination, of indiscernibility, as if things, beasts, and persons . . . endlessly reach that point that immediately precedes their natural differentiation" (Deleuze and Guattari, 1994: 173).

We learn nothing from the concept of haecceity, however, unless we realise that we do not exist without first being caught in the impersonal consistencies of these incorporeal becomings. By over-dramatising my arboreal encounter, this account risks reducing haecceity to a moment of crisis, as if it could only be understood as a disruption of continuity. Yet, for Deleuze and Guattari, our lives simply *are* haecceities, nothing but haecceities. The finite continuities that we attribute to individuated subjects overlooks the sense in which our lives are also the winds and atmospheric variations that move us, the hours that holds us (perhaps for much longer than an hour), the intensities of light that set new thoughts aglow. As Deleuze and Guattari (2004: 289) note:

> You are longitude and latitude, a set of speeds and slownesses between unformed particles, a set of nonsubjectified affects. You have the individuality of a day, a season, a year, a life (regardless of its duration) – a climate, a wind, a fog, a swarm, a pack (regardless of its regularity). Or at least you can have it, you can reach it.

Whilst it might be tempting to read Deleuze and Guattari's valorisation of haecceities as an outright refusal of the finite and the discursive, for Guattari, however, it is only by transversally interfacing *between* these heterogeneous domains of individuation that mutations in the production of subjectivity become possible.[3] It is to this inherently cartographic possibility that I now turn.

Ethico-aesthetics

> Subjectivity finds itself threatened by paralysis. It loses the taste for difference, the unpredictable, and for the singular event. TV game shows, the star system in sport, variety shows, political life, work on subjectivity like neuroleptic drugs which guard against anxiety at the price of infantilization and de-responsibilization.
>
> (Guattari, 2015: 98)

One of the most striking aspects of Guattari's incorporeal materialism lies in its capacity to transform the field of ethical problematics into an aesthetic concern for felt intensity and the production of new modes of experience capable of interfacing between heterogeneous domains. By framing his concern for the production of subjectivity in explicitly aesthetic terms, Guattari hoped to distance himself from a psychoanalytic tradition that, for many years, had reduced the conditions of subjective experience to the signifying semiologies of a universal structure, accessible only through the representational frameworks of linguistic interpretation (Guattari, 2009: 173–174). Yet, Guattari's cartographic alternative remains thoroughly irreducible to the heroic narratives of self-help, self-discovery, and self-improvement that reverberate through the mass-mediatised circuitries characteristic of 21st-century materiality. Indeed, Guattari clearly understood this phenomenon as heralding the emergence of an Integrated World Capitalism directly invested in the reproduction of capitalistic subjectivity through the proliferation of its own universalising semiotic regimes (Guattari, 2000: 48).

This shift from structural psychoanalytic approaches to an aesthetics of experience disrupts any attempt to reduce subjectivity to the representational frameworks of linguistic signification. Indeed, when reading Guattari's philosophy I am often left with the disorienting and somewhat dissatisfying sensation that, despite his professed concern for the production of subjectivity, the subject itself is nowhere to be seen. It is as if Guattari's commitment to the mapping-out of felt intensities in excess of signification precludes any attempt at representing what the subject of this experience might look like. This is, I believe, a deliberate and thoroughly ingenious gesture on Guattari's part; it is a short-circuit that has been engineered into the very workings of his conceptual machinery, reminding the user that the production of subjectivity does not have an individuated identity as its goal. Guattari's aesthetic problematic, concerned as it is with exploring new trajectories of feeling, cannot be reduced to an ethics of self-fashioning, as if the ultimate aim were to attain at an identity capable of neutralising those pulsations of desire that makes us strangers to ourselves. Rather, the task that Guattari sets before us is to approach subjectivity as a process of production in its own right. The goal is not the production 'of' a subject, whether individual or collective, but a becoming conscious of production as an experience of felt intensity, in and of itself.

Herein lies the source of the dissatisfaction, then, for Guattari's approach to the production of subjectivity bears all the hallmarks of an infinite task. It is an adventure rather than a journey, a becoming rather than an evolution, a cartographic impulsion rather than a structural model. Indeed, it is by mistakenly prioritising the dream of satisfaction over processes of production that capitalistic subjectivity finds itself, in Guattari's words, "threatened by paralysis" (2015: 98). This paralysis culminates in the fantasy of a transcendental subject, a Platonic figure that encapsulates the desire to "return to a pure awareness of the sense of self" in which there are no existential itches left to scratch (Guattari, 1984: 80). The lure of self-identity, of the possibility of isolating one's true being beneath the humdrum of life, is a neuroleptic that substitutes the anaesthesia of an imagined certainty for the perpetual anxiety of humanity's improvised cartographies. It is for this reason that we should take care not to interpret Guattari's concern for the "de-responsibilization" endemic to contemporary culture as a reversion to the certitudes of either self-knowledge or self-mastery. On the contrary, what is at stake here is the possibility of fostering a different sense of responsibility, one that is all the more profound – indeed, all the more cosmic – in its capacity to affirm the inseparability of the product from the process of production (Guattari, 1995: 107).

How, then, are we to make sense of Guattari's ethico-aesthetics and the strangely cosmic tenor of the responsibility that it demands? The key lies, I believe, in the incorporeal individuation of haecceities, because it is in the relation between the territorialised materiality of individuated bodies and the deterritorialised consistencies traversing them that subjectivity is produced. This relationship, between the Territory of corporeal perception and the Universes of incorporeal individuation, is, however, by no means a dialectical one. In fact, *Schizoanalytic Cartographies* (2013: 52) sees Guattari approach the production of subjectivity through the processual interrelation of four heterogeneous domains:

> (1) Flows of matter and energy; (2) the abstract machinic Phyla that preside over objective laws and changes; (3) existential Territories, considered from the angle of their self enjoyment (their 'for itself') and, finally, (4) incorporeal Universes, which escape from the energetic, legal, evolutionary and existential coordinates of the three preceding domains.

These four domains are each implicated in the production or 'modelling' (Guattari, 2013: 18) of subjectivity: Flows of matter or energy refer to the actuality of material relations that model experience through the finite coordinates of physical systems; Phyla are the formalised ideas or (techno)logics capable of generating subjective mutations through the concretisation of new combinations of material potentials; Territories express embodiment, modelling experience through the corporeal grasping of a spatilialised milieu or world; finally, Universes involve individuations that model experience through the incorporeality of intensive movements (see Figure 3.1).

It is essential not to intuit any sort of hierarchical arrangement to these domains, which operate independently of each other even as they coalesce in the production of human experience. There is no foundational domain that could serve as the support for an 'authentic' form of subjective experience. Rather, subjectivity is produced through a process of assemblage that problematises and enlarges

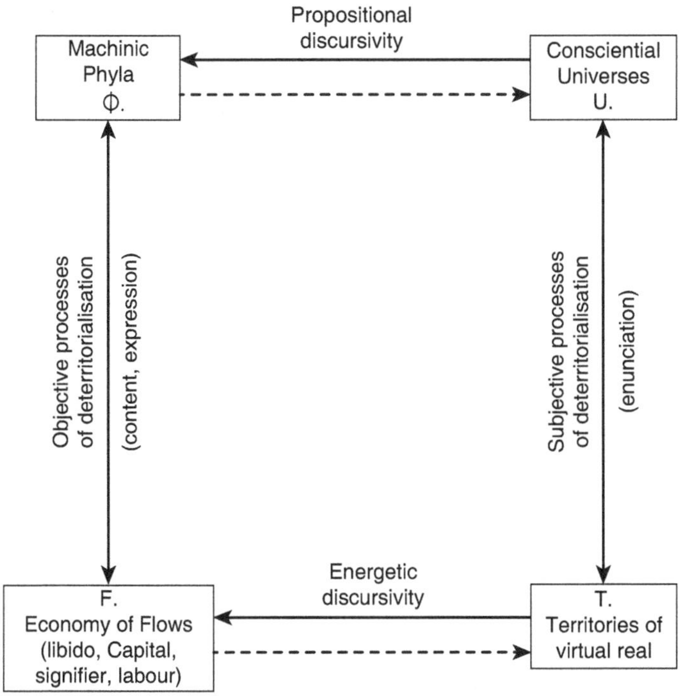

Figure 3.1 Mapping the transversal relations between Guattari's heterogeneous modelling domains (Guattari, 2013: 27)

experience by mapping-out new interfaces between the modelling processes of each domain.[4] Thus, subjectivity always requires a Territory, a certain place to call its home, yet this Territory must remain open to processes of recomposition – that is, to transversal processes of deterritorialisation – if it is to avoid the existential stagnation of a fixed identity. Although we cannot dwell without a Territory, this domain is unable to produce new modes of subjectivity independently of its interrelations with the remaining three; or, as Guattari (2013: 27) explains, the domains "will only be able to sustain their own configuration through the relations that they entertain with each other".

Of particular importance here is Guattari's distinction between, on the one hand, the *actuality* of material Flows and abstract Phyla, and, on the other, the *virtuality* of existential Territories and incorporeal Universes. In other words, whilst Flows and Phyla are expressed discursively through the finite coordinates of space, time, and energy, existential Territories and the infinite Universes that traverse them are non-discursive and so cannot be located or known in the same way. This distinction between the two kinds of domain is significant because it situates the Territory between the actuality of material-energetic Flows and the incorporeal consistencies of virtual Universes. The heterogenetic "grasping" (2013: 77) that constitutes the Territory as an ontological domain emerges through a kind of processual relay, "straddling" (2013: 52) the domains of Flows and Universes to produce the existential consistency of a world. Moreover, in relaying between these adjacent domains the Territory also finds itself caught between the territorialised (Flows) and the radically deterritorialised (Universes), with the effect that its own territorialisations are never fully actualised in the manner of material-energetic Flows.

There is, as we might expect, a hint of the unresolvable to Guattari's four-fold scheme, which forces readers to experiment with different kinds of transitions in a way that always seems to leave one domain out of reach. And indeed, perhaps the beauty of this strange creation, this infernal machine, is that it simply doesn't resolve into the kinds of bifurcations that have come to anesthetise contemporary narratives regarding subjectivity. There are neither subjects nor objects: there are only assemblages of enunciation relaying mutations between the four heterogeneous domains. However, what this immanent cartography does enable us to grasp is the fundamentally constitutive role of incorporeal Universes in the production of subjectivity, and, moreover, the implication of deterritorialised movements at the heart of all existential Territories. As Brian Holmes (2009: np, emphasis in original) explains:

> The territory is the **virtually real**, the material place where subjectivity emerges. But it cannot be understood, and indeed, no word can be spoken, without the deterritorialization that precedes it and gives it form.

What Holmes makes clear, therefore, is that existential Territories are necessary but not sufficient when it comes to the production of subjectivity, for there can be no movement towards the actual domain of material-energetic Flows or the Phyla of formalised combinatorics without affirming the virtuality inherent within the Territory itself; that is to say, the immanence of the outside to the inside, the deterritorialised Universe to the finitude of the Territory, the chaotic to the complex.

It is at this point that we can return to the question of Guattari's ethico-aesthetics, because it is through the conflation of the Territory with fully actualised coordinates that the neuroses of self-identity are established and maintained. Yet, as I have suggested, Guattari's analysis hinges on the contrary assertion that the Territory remains virtual and that it is made manifest not through discursive categories but through a kind of embodied, pre-cognitive grasping. The point I wish to emphasise, however, is that it is not enough to shift the ethics of self-knowledge from the discursive to the non-discursive registers, for to do so would be to ignore the inherence of deterritorialising movements within the very grasping of the Territory itself. It would be a severe misreading, then, to confuse Guattari's philosophy with a phenomenology grounded in embodied experience, the reason being that such approaches tend to romanticise the search for a territory through the narrative of a 'return' to the authenticity of the body's lived experience. Phenomenology's perpetual reversion to the corporeality of the lived falls drastically short to the extent that it cannot think the incorporeal individuations that traverse and agitate the Territory from within.

Conclusion

> 'I is an other', a multiplicity of others incarnated at the crossroads of components of partial enunciation, everywhere exceeding individuated identity.
>
> (Guattari, 2011b: 21)

Here, then, is the rub: Guattari's ethics is not concerned with 'getting back to' the lived body, as if a Cartesian neglect for all things corporeal could be blamed for the cult of the self with which we have become all too familiar (Colebrook, 2014: 85). It seems to me that Guattari is offering us a much less conservative, and indeed a much more challenging ethics in this regard, one whose aim is to nurture a sense of responsibility not for the protection or maintenance of a Territory (a form of Being) but towards the incorporeal vectors that animate and inflect it (forces of becoming). Necessary as they may be, the problem with Territories is that they can thicken like concrete, hardening into rigid structures in such a way that any movement between domains becomes inhibited. These structures give a sense of security and comfort to our lives, no doubt. The problem, however, is that this security is achieved at the cost of relieving us of all responsibility when it comes to experimenting with the transversal tendencies of felt intensity. Thus, Guattari's ethical project is about freeing our experiential and imaginative potentials from the stultifying Territory of the individuated self, loosening the infantilising cult of personality with the hope that we might take greater responsibility for subjectivity's cosmic capacities.

In my reading, Guattari's ethics is an ethico-*aesthetics* precisely because of its refusal to found itself on the individuation of an actualisable form, whether subject or object. The use of the term 'aesthetics' thus signals towards the importance of incorporeal individuations when it comes to understanding subjectivity as a process of production. The value of these haecceities lies not in any sense of belonging: to this extent, their constellations remain ontologically irreducible to the domain of existential Territories and the subjective modelling processes that they entail. Rather, the value of haecceities resides in the incorporeal consistencies that they are capable of making felt, and in making felt perhaps also making thought. I feel the need to stress, finally, that Guattari's ethics was never a matter of abandoning Territory altogether, of foregoing all sense of existential belonging in order to plunge into deterritorialised chaos – as if such a humanist gesture were possible, let alone desirable! As Deleuze and Guattari (1994: 201) remark in their final collaborative act: "We require just a little order to protect us from chaos". The point, then, is that the experience of thinking beings invariably requires a Territory of some form, for this experience cannot exist without the inertial stupidity of its habits, opinions, and assumptions. Yet we should be under no illusion, for as Guattari reminds us, the degree of order that this experience requires is *always* less than that of an individuated identity.

Notes

1 The resonances linking Guattari's thought to Whitehead's process philosophy require a more thorough investigation than can be attempted here. Massumi has recently suggested, however, that Guattari's concern for an aesthetics of material processes might be brought into fruitful conversation with a Whiteheadian concept of 'feeling', which emphasises the primacy of felt transition in the becoming of any actually existing entity, whether human or otherwise (Alliez and Massumi, 2014: 16–17).

2 Here, I would suggest that Guattari is much closer to Whitehead's process philosophy than he is to the kinds of 'object-oriented ontology' that have recently gained traction within the social sciences and humanities through the work of Graham Harman (2010, 2016), Timothy Morton (2011, 2013), and Ian Bogost (2012). Like Whitehead before him, Guattari emphasised the processual nature of existence, in such a way that the most concrete actuality is not simply the thing itself but the affect or felt intensity upon which its endurance depends (Guattari, 2013: 203–206).

3 The production of subjectivity is 'transversal' to the extent that it proceeds through the establishment of dynamic interfaces linking the finitude of corporeal existence with the infinite field of incorporeal individuations that ceaselessly traverses it. Moreover, these machinic interfaces do not function on the basis of signification, for the goal of transversal interventions is "the opening up of new fields of experience" beyond pre-established structures of meaning (Alliez and Goffey, 2011: 7).

4 Guattari describes this transversal assemblage of subjectivity across the distinct domains as a process of 'meta-modelling': "Not as general model, but as an instrument for deciphering modelling systems in diverse domains, a meta-model, in other words" (Guattari, 2013: 17).

References

Alliez, E. and Goffey, A. (2011) Introduction. In: Alliez, E. and Goffey, A. (eds.) *The Guattari Effect*. London: Continuum. pp. 1–14.

Alliez, E. and Massumi, B. (2014) Performing the ethico-aesthetic paradigm. *Performance Research: A Journal of the Performing Arts*, 19(3): 15–26.

Bogost, I. (2012) *Alien Phenomenology, or, What It's Like to Be a Thing*. Minneapolis, MN: University of Minnesota Press.

Colebrook, C. (2014) *Death of the PostHuman: Essays on Extinction, Vol. 1*. Ann Arbor, MI: Open Humanities Press.

Deleuze, G. and Guattari, F. (1994) *What Is Philosophy?* London: Verso.

Deleuze, G. and Guattari, F. (2004) *A Thousand Plateaus: Capitalism and Schizophrenia*. London: Continuum.

Gerlach, J. and Jellis, T. (2015a) Guattari: Impractical philosophy. *Dialogues in Human Geography*, 5(2): 131–148.

Gerlach, J. and Jellis, T. (2015b) A lovesick note to the impractical. *Dialogues in Human Geography*, 5(2): 177–182.

Guattari, F. (1984) *Molecular Revolution: Psychiatry and Politics*. Prescott, AZ: Peregrine.

Guattari, F. (1995) *Chaosmosis: An Ethico-Aesthetic Paradigm*. Bloomington, IN: Indiana University Press.

Guattari, F. (2000) *The Three Ecologies*. London: The Athlone Press.

Guattari, F. (2009) *Chaosophy: Texts and Interviews 1972–1977*. Los Angeles, CA: Semiotext(e).

Guattari, F. (2011a) On contemporary art: Interview with Oliver Zahm, April 1992. In: Alliez, E. and Goffey, A. (eds.) *The Guattari Effect*. London: Continuum. pp. 40–53.

Guattari, F. (2011b) Schizo Chaosmosis. In: Alliez, E. and Goffey, A. (eds.) *The Guattari Effect*. London: Continuum. pp. 17–24.
Guattari, F. (2013) *Schizoanalytic Cartographies*. London: Bloomsbury.
Guattari, F. (2015) *Machinic Eros: Writings on Japan*. Minneapolis, MN: Univocal Press.
Harman, G. (2010) *Towards Speculative Realism: Essays and Lectures*. Ropely: Zone Books.
Harman, G. (2016) *Immaterialism: Objects and Social Theory*. Cambridge: Polity Press.
Holmes, B. (2009) *Guattari's Schizoanalytic Cartographies, or the Pathic Core at the Heart of Cybernetics*. Available online at: http://miriamgrossi.paginas.ufsc.br/files/2013/02/25090965-Guattari%E2%80%99s-Schizoanalytic-Cartographies.pdf
Hynes, M. (2013) The ethico-aesthetics of life: Guattari and the problem of bioethics. *Environment and Planning A*, 45(8): 1929–1943.
Jellis, T. (2014) Schizoanalytic cartographies, by Félix Guattari. *Society and Space*. Available online at: http://societyandspace.org/2014/07/19/schizoanalytic-cartographies-by-felix-guattari-reviewed-by-thomas-jellis/#
Morton, T. (2011) Here comes everything: The promise of object-oriented ontology. *Qui Parle: Critical Humanities and Social Sciences*, 19(2): 163–190.
Morton, T. (2013) *Hyperobjects: Philosophy and Ecology After the End of the World*. Minneapolis, MN: University of Minnesota Press.
Oxford English Dictionary Online (2016). Available online at: http://www.oed.com/
Stengers, I. (2011) *Thinking with Whitehead: A Free and Wild Creation of Concepts*. Cambridge, MA: Harvard University Press.
Whitehead, A.N. (2004) *The Concept of Nature*. New York, NY: Prometheus.
Zourabichvili, F. (2012) *Deleuze: A Philosophy of the Event; Together with the Vocabulary of Deleuze*. Edinburgh: Edinburgh University Press.

4 Metamodelising the territory

On Teddy Cruz's diagrammatic urbanism

Christoph Brunner

Introduction

Diagrams play a crucial role in architectural, artistic, and scientific practices. Diagrams often define a technique for visually expressing the relations between entities or movements as part of a complex system. They sketch out, provide overviews, and express the relational nature of complex sets of information. Beyond the representation of such complexities, diagrams can also emphasise process and movement, in addition to connecting entities. In the following, I want to conceive of the diagram as a "relation-specific" tool for speculative thought and practice following Derek McCormack's (2010) elaborations on Brian Massumi's work. Through the term relation-specific, rather than site-specific, I understand diagrams not as spatializing representation but rather as sets of relations enabling a dynamic understanding of complexity and movement through their coupling. In other words, from a relation-specific point of view, diagrams provide a vital material investigation of the movements between a visual expression of complex systems *and* their potential for activating movement at the same time. Through the work of architect Teddy Cruz and his *Estudio Teddy Cruz* (ETC),[1] I will outline a conception of the diagram as both a visual technique attentive to relations, and as a concept emphasising and generating the movement across and among these relations. The particularity of diagrams, if used as technique of relation, is their double function as visual tool of confinement and a means for expressing movement beyond confinement. Especially in architecture and the arts, diagrams are used to open up a dynamic dimension in processes of formation.[2]

ETC's architectural procedures and practices make extensive use of diagrams, as it folds social, economic, and material dimensions into each other to develop propositions for what Cruz calls a *transborder* urbanism "beyond the property line". The diagrams often include structural elements, common to visualisations of specific states of affairs, and these are combined with a wild range of colours, photographs, lines, and arrows, weaving highly complex visual expressions suggestive of, for example, Situationist International's work on psychogeography or Dadaist collage art. However, their purpose is purely pragmatic, seeking to communicate problems of urban development in migrant communities to a public audience, while paying specific attention to the power of aesthetic refinement.

ETC's aesthetic approach fuses with political and activist concerns while avoiding the subsumption of one under the other.[3]

Through the diagrams an aesthetic dimension operating on the level of perception charges political empowerment and activism. ETC's techniques focus on how to find modes of political practice based on emergent relational qualities in experience, across different modes of existence and ways of acting and thinking. The studio provides inventive *techniques for activation* that address socio-political problems pertaining to issues of trans-border migration, exploitation, and community activism. These diagrams operate as techniques for aesthetically opening up a political issue towards a process of empowerment and direct engagement seeking change. In the first instance I will outline ETC's procedural architectural approach. A second movement addresses how its practice resonates with Félix Guattari's developments on metamodelisation, as an attempt to think the emergent quality of practices in activist research.

Procedures over products – towards a diagrammatic urbanism

ETC's practice is a hybrid mix between architectural thinking and artistic ways of communicating its ideas or, in Cruz's words, "between research and practice".[4] From an architectural point of view the studio's practice is defined by a unique relation between urban re-development projects, mostly for Hispanic migrant communities in the USA, fused with conceptual developments in community organisation and activism, and a particular interest in flows of humans and goods across national border zones worldwide. Born and raised in Guatemala and trained as an architect, Cruz teaches in the Visual Arts department at University of California San Diego, where he has worked for almost 20 years on urban development along and across the San Diego-Tijuana border zone.[5]

While architectural propositions are often the initial motive for his work, Cruz states, "we can not only design fantastic buildings, but also configure social, political, and economic agendas that can yield particular architectures and special configurations" (Cruz and Sokol, 2008: np). His unconventional architectural approach becomes apparent when looking at the former website of ETC, which is now Estudio Teddy Cruz + Fonna Forman. The works presented were not necessarily "architectural" in the conventional sense of built structures.[6] The site contained only videos showing ETC's projects and design research models, ranging from design propositions often developed along specific devices (like a light-weight scaffolding), to documenting workshops held along the San Diego-Tijuana border, or a diagram-movie explaining the negotiation processes for urban re-development between communities, political stakeholders, and industry representatives. Only one project, *Casa Familiar*, is a housing development, explored under the title of "the performance of a small parcel".[7] Symptomatic of all of the project presentations is the use of bright colours and collaged photo-material. Often the technique of colouring is not just used to separate different domains or signifiers, such as social groups in a specific suburb, but it is a highly refined design strategy as part of a more general social and political practice at the heart of ETC.

By design strategy, we can understand a relational and participatory approach towards architecture which attempts to insert specific techniques or tools into the existing urban fabric. Accordingly, ETC's approach is different from architectural endeavours that pursue the construction of a new building as the minimum point of departure. The particularity of ETC resides in the aesthetic strategies it chooses to underline its practice, which is as much carried by an architectural approach towards urbanism as research strategies from design, social work, art, and cultural theory. One could say that its approach is site-specific, basing its mode of conceptual, social, political, and material engagement on the problem at stake. On the website, the videos or slide shows interlaced visual elements with processual aspects such as animated graphs and short accompanying texts explaining the foundational concepts and stakes for each project while emphasising its overall dynamism. Similarly, the projects bear titles which include a sense of movement and process: "the informal: not its image but its procedures", "60 linear meters of transborder conflict", or "a micropolicy for the neighbourhood", to mention just a few. While, on first glance, these titles seem like suggestive aphorisms, they refer directly to the procedural level carrying each project. The entire website itself emphasises processes over products and avoids confined representations, such as the building sketches and models common in architectural practice.

For Cruz, one cannot treat a particular public or private territory without investigating the underlying power relations such as political or economic stakeholders, existing values, social needs, and flows of movement at the heart of local communities.[8] Cruz calls this process of instigating new relations between these factors "trans-border urbanism" where design aims at changing conditions for social and architectural processes to emerge. ETC's practice focuses on *conditions of emergence* and their *transversal* (or trans-border) qualities, rather than assuming pre-set and confined structures.[9] Such a thinking extends accustomed habits of architectural development by focusing on "complexity, hybridity, and improvisation" providing a *temporal urbanism of insurgency* (Heath, 2009: 94–98). Accordingly, ETC has developed a body of work that focuses on procedures, operations, and retooling architectural (and artistic) means of intervention and representation. One can conceive of such a procedural practice as a "populist architecture of hypothesis approached as a community-wide collaborative initiative" (Gins and Arakawa, 2002: 61).[10] It is the intensification of heterogeneous elements through aesthetic means, such as the visualisations, power point presentations, and actual diagrams, that renders ECT's practice what might be called a *diagrammatic urbanism*.

The notion of the diagram refers in a first impulse to Michel Foucault's analysis of disciplinary power exhibited by the infamous example of the Bentham Panopticon. Foucault describes disciplinary power as "a diagram of a mechanism of power reduced to its ideal form" and as "a figure of political technology that may and must be detached from any specific use" (1995: 205). With his conception of the diagram he provides a crucial element immanent to ETC's work: the analysis of underlying power structures permeating heterogeneous domains of a politically charged territory operating through the mutual insertion of forces into each other. These rather programmatic points focusing on a political understanding of power nonetheless hint at an ethico-aesthetic expansion of the analytical scope, as developed in the works of Félix Guattari. For Guattari the question of power and its relational character differs strongly from a mere unveiling of already existent power relations as structures. Power as force, and here he refers to Foucault, is always productive, an "existential production" which he calls the "diagrammatic function" (1996: 178). While Guattari, similar to ETC, acknowledges the "power formations" held together by "relations of forces", he emphasises that these relations are of a primordial constituent quality that is dynamic and relation-specific. In a first instance, ETC's approach starts with an analysis of existing power structures. However, instead of schematically outlining these structures – like a mere mapping of the relevant stakeholders – the visual techniques deployed by the studio shift the main focus towards its capacities of enabling new potentialities for thought and action by mobilising the sensous field. Most crucially, this happens through aesthetic techniques likes animated videos, collages, colours, and diagrams that supplant different and often heterogeneous elements (i.e. social relations combined with material flows) into a complex relational and transversal cartography.

Similar to Gilles Deleuze's adaptation of Foucualt's concept of the diagram, ETC's main practice resides in multiplying diagrammatic functions, if we consider these functions as the very capacity of heterogeneous forces to interlace (such as social, political, or material forces) and generate new expressions. In this vein, Deleuze writes: "a diagram is a map, or rather several superimposing maps. And from one diagram to the next, new maps are drawn. Thus there is no diagram that does not also include, besides the points which it connects up, certain relatively free or unbound points, points of creativity, change and resistance" (1988: 44). The diagram takes on a double function in Cruz's practice. On the one hand it is an analytical tool for understanding power relations. On the other hand, it becomes a creative technique that functions through aesthetic forces of (visual) diagramming. Throughout ETC's practice, diagrams in their analytic and expressive (visual) function actively shape and "modulate" how complex relations affectively interlace and how to enable modes of resisting dominant power relations. Modulation, in this context, can be understood in proximity to Guattari's conception of meta-modelling as a technique similar to ETC's diagrammatic practice undermining the power relations of dominant representations. Such representations also concern the approach of how to conduct research and how to communicate insights. At the same time, Guattari and Gilles Deleuze resonate in their assertion that modulation becomes also a primary technique of late capitalist uses of information fragmenting former relations in order to render them compatible (in a modular fashion) with the "mechanics of social domination" (Guattari, 1995: 105; Deleuze, 1995: 177–182). For Guattari, meta-modelling, in a nutshell, becomes a way of deciphering the modelling of systems, for instance the overall approach of a method, and ensuring "transversal passages between different problems of different kinds" as can be found in the heterogeneous visual assemblages of ECT's practice (Guattari, 2012: 17). Modulation as core element of the process of modelling then becomes a technique not only adapted by operational modes of capitalist value extraction but also in terms of opening up new relations while anchoring them in the concrete concerns (problems) at stake like communal housing projects or trans-border migration.[11]

To give an example, the housing project in collaboration with NGO Casa Familiar, a community centre for mostly Hispanic migrant communities in San Diego's suburb San Ysidro, entails a complex set of procedures enabling its final realisation. Due to the informal social and economic structures of local migrant communities, a modular and dense building complex was devised, including different spaces for smaller and extended families to co-inhabit varying and modular units, community spaces for micro-entrepreneurial exchange such as weekend markets, as well as studio-flats for artists to live in and provide services to the community (see Bratton, 2004). To effectuate the building of the centre, a change of municipal zoning laws of San Ysidro had to be achieved. In other words, architectural procedures as means to address representational politics were needed – or, different diagrammatic functions and their maps had to be interlaced.

There was no way for Cruz to move forward on the project other than getting elected to the city hall's urban planning board (or so he contends). Once elected, Cruz had to convince the board to modify its single-unit-per-parcel policy towards a multi-unit architecture. In other words, he had to shift from the usual suburban family home (which ETC calls "McMansions") to a denser urban multi-unit architecture. This political and representational procedure took many years before reaching its goal, and Cruz's withdrawal from the board after achieving his goal is part of the procedure. Overall, ETC conceives of the architectural procedure of Casa Familiar as a process from "official conforming" to "informal non-conforming". And what Cruz calls "urban pedagogy" encompasses all conceptions of such extended architectural procedures as ways to "enable new institutional protocols by producing new interfaces with publics and unorthodox cross-institutional collaborations, rethinking the very meaning of infrastructure, housing and density, and mediating top-down development and bottom-up social organization" (Cruz, 2011: 111). What ETC terms "institutional protocols" might be called diagrams in the sense that they express the weaving of heterogeneous domains (such as infrastructure or housing density) with practices such as cross-institutional collaborations.

The "ecology of practices" (Stengers, 2005) composing the architectural procedure called *Casa Familiar* are felt most concretely through the visual diagrams accompanying and visually communicating the project and its genesis. In ETC's work, diagrams become an active operational tool of complexification, communication, and transformation of architectural, as well as social, and research practices. A procedure, as outlined in Cruz's work, is a folding of heterogeneous acts, materialities, and practices through "techniques of relation" (Manning and Massumi, 2014: 91–92, 101), including diagrams as aesthetic tools and concepts as lures for thought. ETC's procedures attempt to rethink architectural practice in the way it engages with social and cultural concerns, its use of communicating beyond the confined entity of built structures, and deployment of concepts as mostly abstract explanations. ETC transforms the power of conceptual invention into a procedural approach attentive to social and political issues, and their material and economic entanglements. The creative potential of concepts thus instigates a speculative movement of thought that co-evolves with social, political, and material movements. In such a diagrammatic and procedural unfolding, these practices constitute new *ecologies of relation*.[12] Through their interlacing, the conceptual, social, and political do not pre-exist one another in ETC's practice but mutually shape each other according to their shared ecological field of emergence. In other words, the ground from which the heterogeneous forces composing such ecologies emerge might be considered as *transversal geographies*, in this case most prominently exhibited by the San Diego-Tijuana border zone.

What Cruz calls the procedure of an "urbanism of insurgency" involves a rethinking of the social dimension in relation to contemporary urbanism and possible architectural interventions. In general, his focus on procedures, flows, and movements underlines a reworking of scale, density, and confinement towards a mobile conceptualisation of urban activity. Many of his projects address the material circulations of debris and labour power across borders. A major concern for ETC lies in detecting these flows and movements, investigating their temporal dimensions, and identifying spatial urban "fragments waiting to be activated through synergistic development, their potential residing in a certain contingency rather than cure-all master planning" (Bratton, 2004: 120). ETC emphasises a strong political agenda, not only in terms of re-negotiating local and often conflictual situations. It also critically approaches architecture as spatial structuring in relation to contemporary forms of global neoliberal capitalism, and the way it affects the organisation of (social) space (Cruz, 2008, 2009, 2011). At this point it becomes clear that the diagrammatic and procedural techniques deployed by ETC underline the relevance of aesthetic dimensions of such political processes. In this sense, the studio's work resonates strongly with what Félix Guattari terms "ethico-aesthetics", where the way political processes yield effects cannot be separated from the way they are felt and experienced (Guattari, 1995: 127). In activating a sensitivity for the complexities of political intricacies at stake in ETC's work through the sensuous power of diagrams, the studio opens up an experiential dimension that is both felt and thought at the same time. Diagrammatic urbanism comprises what Brian Massumi calls "speculative pragmatism" as part of an affective politics (McKim and Massumi 2009: n.p.). Put differently, the diagrammatic functions immanent to ETC's work open up an affective and aesthetic politics capable of mutually including pragmatic as much as speculative elements, where the speculative is what activates new modes of thinking and feeling while grounding them in pragmatic techniques producing concrete effects (such as a change in zoning laws).

ETC's projects work through redefinitions of the social by extending its scope. On the one hand, it analyses the social conditions for emergence, such as confinements, movements, economics, relations, modes of production, and the ethics implied in them. On the other hand, by focusing on the procedural, the notion of the social extends to "more than human" dimensions (Manning, 2013: 228, n.2), generating concepts for thinking relations across various domains. In this type of diagrammatic urbanism, the social includes both human and more than human forces, such as material flows and power relations, affects and intensities. From an architectural point of view, Cruz re-imagines the urban along such "vectors of force", which are not based on individual actors but *ecologies of relation* (Cruz and Tate, 2010: 82). As a result, the procedural and movement-based practice of such an urbanism requires new ways of conceptualising and aesthetically expressing these transversal geographies and their activities.

On metamodelisation and Translocality

The most common critique of Cruz's practice is the lack of actually building things: many of ETC's projects exist on paper, in PowerPoint presentations or beautiful visual expressions, but not in "reality". Such critiques, however, underestimate the diagrammatic activation of forces yielding effects in political practice *and* their ability to alter ways of thinking about contemporary forms of urbanism, that is, how to conceptualise with them. In opposition to such critiques, this section asks how a shift in the methods and models of contemporary architecture need to be deployed to arrive at a diagrammatic urbanism. The proposition outlined through Cruz's practice not only accounts for movement and dynamic flows in urban environments but also the invention of new modes of analysis and expression.

A diagrammatic urbanism attentive to movement and flows abandons disciplinary boundaries for the sake of emergent, immanent, and procedural practices. The question might be, then: is it possible to devise such procedures independent of their context? Or, in a more architectural manner, is there an underlying model to be deployed and applied in varying environments? A diagrammatic practice is neither entirely local nor global, which Cruz hints at in his project *Political Equator*. Tracing a line around the globe marking zones of conflict – called "urbanities of labour and surveillance" like Tijuana-San Diego, Palestine-Israel, Ceuta-Melilla, Indian Kashmir/Pakistani Kashmir, and China – this work demonstrates flows between the "Functioning Core" (North) and the "Non-Integrating Gap" (South). While remaining slightly schematic on first glance, the crude divides are supplanted by very specific and case-sensitive (or relation-sensitive) movements that ETC traces. The phenomenon of the re-enforcement of strong boundaries and borders seems even more pressing under contemporary political regimes and their conceptions of border control. The same accounts for the control of migration and the different translocal flows of humans, goods, and information that have been recently described as part and parcel of a general logistics (see Mezzadara and Neilson, 2013). Arriving from the south, migrants are seeking work, while the functioning core outsources its production zones to the south. While there are global processes, which can be assimilated between these border zones, Cruz emphasises that his practice consists of "series of minor modifications" and not building a "humanist utopia" (Cruz and Tate, 2010: 81).

In Cruz's work, the site-specific tendency transgresses and translates into what might be called a *relation-specific* approach through the ethico-aesthetic technique of a diagrammatic urbanism. Conceiving of borders as a tool for thinking – thus making this endeavour an ecology of practices – allows us "to speculate on how we can address all the cities and all the territories around the world" (Cruz, 2008: np.). A relation-specific approach requires one to thinking the urban as a translocal structure of converging and diverging forces. Accordingly, the local never transforms into the global, or vice versa; this is an insufficient binary. On the contrary, a relation-specific and diagrammatic approach enables us to embrace the heterogeneous and open flow of social, material, and political forces and to identify techniques and tools for addressing them – both locally as a situation and in resonance with a more globally encompassing context. While we perceive new modes of capitalist integration on a global scale through what Guattari calls an "Integrated World Capitalism" (preceding some of the north-south concerns later raised by ETC), the main point in ETC's practice is that the ecologies of relation perceived as geo-political and site-specific concerns require an extension towards more transversal, ethico-aesthetic dimensions and their speculative and pragmatic tendencies (Guattari and Negri, 2010). This means generating practices of radical inclusion rather than exclusion. At the same time, inclusion is never arbitrary but itself defines a complex process of interrelating diagrams. For Cruz's the notion of the *informal* emphasises such a tendency:

> I see the informal not as a noun but as a verb, which detonates traditional notions of site specificity and context into a more complex system of hidden socio-economic exchanges. [. . .] I see the informal as the site of a new interpretation of community, citizenship and praxis, where emergent urban configurations produced out of social emergency suggest the performative role of individuals constructing their own spaces.
>
> (Cruz, 2010: np.)

In this case, the conceptual exploration of the informal instigates architectural procedures from specific situations and provides the potential of relating this process to other situations. Such forms of situated emergence and trans-situational relaying become clearest if understood as the operational capacities of a diagrammatic urbanism interlacing forces and power relations both in their felt effects but also in their inventive potential for new modes of thinking and making. In other words, diagrams need to be actualised always anew in specific situations where they shift the entire set of relations through their expressive capacities. With each actualisation, Deleuze writes, a diagram integrates itself into other diagrammatic processes: "a collection of progressive integrations that are initially local and then become or tend to become global, aligning, homogenizing, and summarizing relations between forces" (1988: 37). Drawing attention to these processes we can see how the interlacing of the visual, conceptual, and material fold with their very own diagrams into the social, political, and architectural circulations in the San-Diego-Tijuana border zone – and how they might transgress the local boundaries into larger (global) concerns.

Moving from site-specificity to relation-specificity produces an "interdisciplinary collaboration" focused on "operative dimensions" targeted at "exchanging procedures" (Cruz and Tate, 2010: 87). This exchange of procedures might be better understood as a process of relaying and modulation, which shifts ecologically when inserted into a new situation. Developing specific procedures tailored to the concerns at stake while generating new cartographies or maps – as Deleuze and Guattari call them (1987: 203) – for relation thus explodes the conventional conception of the model and moves towards what Guattari termed *metamodelisation* or *meta-modelling*. Similar to Cruz's conception of the informal as a verb and not a noun, one might envision the process of modelling as an alternative to the rather confined notion of the model. Guattari's urge for developing practices beyond disciplinary boundaries leads him to invent new strategies of *emergent modelling* rather than relying on pre-existing models for application.[13]

Cruz's emphasis on temporal urbanism and his conception of interdisciplinary research resonate strongly with Guattari's idea. In both works the key point revolves around defining and generating interfaces of resonance between heterogeneous approaches of analysis and their materials while accounting for their inventive relational capacities. For Guattari metamodelisation

> has concerned something that does not found itself as an overcoding of existing modelizations, but more as a procedure of 'automodelization,' which appropriates all or part of existing models in order to construct its own cartographies, its own reference points, and thus its own analytic approach.
> (Guattari, 1996: 122)

Auto-modelisation and auto-affirmation are key concepts in Guattari's account of metamodelisation. Diagrammatic urbanism and Guattari's practice of metamodelling both aim at making new dimensions of potential felt in expression through the prism of ethico-aesthetic forces – and thus transforming the very realm of experience. They both emphasise "complexification [and] processual enrichment" (Guattari, 1995: 61). Operating both conceptually and in the visual format of the diagram, ETC has developed a "pragmatic cartography" that opposes reductive representations of dominant signifiers such as political stakeholders while opening up new dimensions of thought through aesthetic techniques (1995: 60).

ECT's diagrammatic urbanism finds connections with existing systems of thought and procedures, activates new dimensions like material or human flows, and generates an auto-modelisation, providing each element with a procedural (diagrammatic) agency. However, such a process can never become universal (i.e. global) in its aims, which is what models often aspire to – and herein lies a distinct difference to the automatised and genralised operation of capitalist integration through modulation. On the contrary, as Guattari writes, "each modelization is always grounded and reaffirmed in a singular situation" (Guattari, 2000: 13). In other words, relation-specifity grounds the emergence of a new practice and the invention of novel procedures; it generates effects. Without these effects, nothing in the overall discourse would change. By finding new modes of working between artistic experimentation and social engagement, ETC diagrammatically metamodelises the emergence of new techniques and procedures of an insurgent architectural and social practice – with both global and local repercussions.

Far from being a method or the cross-linking of different approaches, *metamodelisation thinks modelling as modulation*. Modulation names the process of a phase-shift of an unfolding process, a change of state, a new relational capacity.[14] In addition, Guattari's insistence on self-referential enunciation and auto-affirmation underlines the autonomy of affective forces (Massumi, 2002: 35) inherent in each process of metamodelisation (Guattari, 1995: 60, 106). Auto-affirmation means that each modelling deployed consists primarily in resonance with its immediate milieu and is not a mere accumulation of parts. Auto-affirmation emphasises process and immanent change. Thus, metamodelisation as "operative diagramming" moves beyond the initially interesting but finally hollow statement that the whole is more than the number of its parts. Metamodelisation attempts to develop techniques based on extensive and excessive potential operating actively across an ecology of relation.

Metamodelisation is a way of accounting for diagrammatic movement as part of ethico-aesthetic activist practices. In relation to San Diego-Tijuana, it would seem evident that local phenomena refer to global causes such as poverty, exploitation, or capitalism. As a consequence, one might claim that each of the local effects has its very specific ecology of relation producing the singularity of the underlying state of affairs. Metamodelisation, on the other hand, concerns the emergent collectivity of practices in their ecological formation and endurance, their heterogeneous, heterochronous, and differential qualities. The process of meta-modelling requires tracing different historical, social, and economic factors, but also their operational potential, in terms of what they might become and how they may change along the way.

The auto-modelisation immanent to diagrammatic practices emphasises the self-perpetuating activity of each process and plugs it into a more collective constellation of transversal becoming across different geographies/cartographies. In the case of ECT's work, diagrams function as the relaying device of an architectural metamodelisation of procedural practice and its relation to social, political, and material matters of concern. The diagram as force field, that is, as an emergent and dynamic cartography in the process of relating heterogeneous elements while constantly shifting – contracted into a visual form – allows for making the co-composition of the situated and the transversal felt as part of the same experience. If the diagram were only conceptually conceived without being felt, it would lose all its rigour, becoming a mere symbol devoid of any movement. The force toward expression defines the necessary passage for an affective contagion beyond mediation. ECT's practice would not gain such interest, nor a refined degree of complexity, if he did not choose the appropriate discursive and non-discursive *functions* to yield felt effects. The varying degrees of intensity and matter mobilising in this practice traverse personal political involvement, the highly specialised craft of architectural and urban development, the active participation of social community workers, and the migrant inhabitants themselves in their everyday desires and activities. To find appropriate *forms* of communicating complexity without reduction affords more than creative representations or communicative skills. His practice asks concretely what new modes of a "distribution of the sensible" (Ranciére, 2004) yield an affective aesthetic politics undermining dominant forms current power relations.

As an immediate process of formation, diagrams enable a perceptual process that opens a complex concern to a wider audience while being specific to the relational ecology at stake. As part of a procedural practice, ETC's diagrams become a field for collective attunement, while simultaneously enabling singularisation. In their openness and precision diagrams invite individuals to resonate with and relate to a context in their singular way. From here, a mode of subjectivity emerges that is relationally composed and not situated in the subject of the perceiver. In addition, the relation between diagrammatic urbanism, metamodelisation, and subjectivity is crucial for its operational value. As Guattari writes, such a relationship "produces its own existence across processes of singularization, because it engenders itself as *existential territory*" (1996: 125). By existential territory, Guattari refers to lines of subjectivation coursing through a collectively sensed ecology of relation, thus grounding sensation in experience and rooting it in bodily, spatio-temporal continuities. This non-identitarian element of subjectivity is one essential aspect of a temporally extensive (that is, endured) and lived practice of existence. Similarly, ECT's architectural procedures would not have any effects if they could not find ways of inserting themselves into the micro-social desires and needs populating a specific field of relations. In other words, "Metamodels are not just abstractions because they require the putting into place of the organizational and institutional means for their collective realization" (Genosko, 2003: 138).

The pragmatic aspect of the process of metamodelisation should not distract from its speculative openness. The diagrammatic functions exposed throughout ETC's work provide the ground for a production of subjectivity that moves across an entire ecology of relation (a field of forces) in a relation-specific manner, without predetermining the effects of this process. However, for such a field-experience to take effect, it requires a marker or attractor taking hold of the encounter and generating its very own rhythm. The primary experiential encounter requires endurance while working itself into memory and feeling. The sought complexity expressed in diagrammatic practices extends not only through sensation, but also through auto-modelisation processes of subjectivation in abstraction: "What distinguishes metamodelization from modelization is the way it uses terms to develop possible openings onto the virtual and onto creative processuality" (Guattari, 1995: 31). Opening onto creative processuality does not mean adapting another creative model to be deployed in one's own practice or thought. Rather, it foregrounds attentiveness to movements populating a tensed field, thus enabling potential transformations or transversal techniques singularising into new expressions. Guattari's insistence on resingularisation and a 'virtual ecology' define the poles of diagrammatic expressions which tend toward future activations of a felt intensity.

From here we might further ask, how can such transversal, procedural, and diagrammatic practices of metamodelisation offer novel approaches of dealing with increasingly complex and charged geographies of conflict, crisis and oppression on a global scale? One of the propositions foregrounded here resides in investigating the ethico-aesthetic dimensions of resistant potentiality to dominant forms of discursive and sensible power relations. Investigating transversal geographies along their diagrammatic functions as fields of forces with specific capacities for affectively interlacing into ecologies of relation, we might approach new modes of metamodelisation that are both sensuously affecting and effective, shifting dominant power relations locally while modulating in resonance with other geographies of struggle. Metamodelisation, then, might offer new perspectives on translocal forms of activism through an aesthetic/affective relaying of practices.

Notes

1 Emphasising the collaborative and collective nature of the work, I will use *Estudio Teddy Cruz* throughout the chapter. In case of citations of Cruz himself, I will deploy his name referring to his personal utterance.

2 Amongst the many explorations of the diagram between art and philosophy the German essay collection "Materialität der Diagramme: Kunst und Theorie" (The Materiality of Diagrams: Art and Theory) by Susanne Leeb (2012) is closest to the relation-specific approach at stake in this article. Another, yet more research-creation oriented publication would be "Drawing a Hypothesis: Figures of Thought" by Nikolaus Gangsterer (2011).

3 In a way, Cruz's popular but aesthetically finely crafted diagrams echo Walter Benjamin's developments on the relation between aesthetics and politics. For Benjamin, art cannot be subsumed under politics. Aesthetics, if subsumed under politics – as an "aestheticizing political life" – leads to war. Politicising aesthetics, on the other hand, enables a wider range of individuals (for Benjamin, "the masses") to collectively create a new political ground for society, which Benjamin attributes to Communism (2008: 41–42).

4 Cruz provides a distinct definition of his vision of research, while the notion of practice seems to remain rather underdeveloped as a concept. He writes, "Many architects talk about research and practice. I'm trying to problematize that relationship as well. This not only means researching issues for the sake or researching, but also that architects must enter into certain institutions to actually see the way that they are operating" (Cruz, 2008: np.).

5 The notion of San Diego-Tijuana received prominent exposure as part of the art-event inSite_05 including exhibitions, interventions, walks, and artist workshops including, Teddy Cruz in San Diego-Tijuana. See Osvaldo and Conwell, 2006.

6 See www.estudioteddycruz.com. For all further references to works of Estudio Teddy Cruz it is recommended to use the website in order to experience the full breadth of the aesthetic effects.

7 If not indicated otherwise, ETC's expressions and short citations are derived from the short video-clips at www.estudioteddycruz.com.

8 Cruz differentiates his practice quite strongly from contemporary architectural phenomena, which he clearly criticises: "It has been unsettling to witness that some of the most 'cutting edge' practices of architecture rush unconditionally to China and the

Arab Emirates to build their dream castles, reducing themselves to mere caricatures of change and camouflaging gentrification with a massive hyper aesthetic and formalist project" (Cruz, 2010).

9 It should be mentioned that Guattari himself engaged in urban re-development projects. One called FGERI (*Fédération des groupes d'études et de recherches institutionnelles*) dealt in parts with the intersection of architecture and psychiatry (see the special number of the journal *Recherches* on "Architecture, programmation, et psychiatrie" (Guattari, 1967)). The other, CERFI (*Centre d'Études, de Recherches et de Formation Institutionnelles*) expressed one of Guattari's life-long concerns, that of institutions, attempting to rethink institutions and institutional practices from a radically transversal point of view. Through several commissions CERFI was asked to conduct in-depth study on the *"système économique urbain"* paid by the Ministère de l'Equipement et du Logement and two projects to initiate urban transformation in two Banlieues, Alma-Gare in Roubaix and Petit Séminare in Marseille. Anne Querrien describes both of these attempts as major failures (2012: 98–113).

10 Architects and poets Arakawa and Gins promote architectural procedures as one of their main concepts for rethinking the relation between architecture and life.

11 The more technical side of modulation in relation to a very singular theory of information beyond the mere reduction to bits in binary logics of digital media technologies have been explored in the works of Gilbert Simondon (1989: 130) and recently explored by Yuk Hui (2015).

12 "Ecologies of relation" is a key concept I develop in my PhD-dissertation entitled *Ecologeis of Relation:Collectivity in Art and Media*. Available at: https://spectrum.library.concordia.ca/979665/1/Brunner_PhD_S2015.pdf

13 Guattari himself successively developed a conception of transdisciplinarity through the notion of meta-modelling towards the end of his life. His thoughts in this direction are most concretely stated in Guattari (1992).

14 One of the most succinct explorations of the amplificatory power of modulation can be found in Gilbert Simondon's article "Forme, Information, Potentiels" (2005, 531–551).

References

Benjamin, W. (2008) *The Work of Art in the Age of Its Technological Reproducibility, and Other Writings on Media*. Cambridge, MA: Belknap Press of Harvard University Press.

Bratton, D. (2004) Practice profile – Estudio Teddy Cruz. *AD – Architectural Design*, 74(1): 117–124.Cruz, T. (2008) Interview with Teddy Cruz. *Artkrush – International Art Online*, 93.

Cruz, T. (2009) A city made of waste. *The Nation*, 16 February.

Cruz, T. (2010) Mapping non-conformity: Post-bubble urban strategies. *Emisférica*, 7(1). Available online: http://hemisphericinstitute.org/hemi/en/e-misferica-71/cruz

Cruz, T. (2011) Latin American meander in reach of a new civic imagination. *AD – Architectural Design*, 81(3): 110–117.

Cruz, T. and Sokol, D. (2008) Repositioning practice: Teddy Cruz. *Architectural Record*, 16 October. Available online: www.architecturalrecord.com/articles/6429-repositioning-practice-teddy-cruz

Cruz, T. and Tate, J. (2010) Design ops: A conversation between Teddy Cruz and Jonathan Tate. In: Choi, E. and Trotter, M. (eds.) *Architecture at the Edge of Everything Else*. Cambridge, MA: The MIT Press.

Deleuze, G. (1988) *Foucault* (trans. S. Hand). Minneapolis, MN: University of Minnesota Press.

Deleuze, G. (1995) *Negotiations* (trans. M. Joughin). New York, NY: Columbia University Press.

Deleuze, G. and Guattari, F. (1987) *A Thousand Plateaus: Capitalism and Schizophrenia II* (trans. B. Massumi). Minneapolis, MN: University of Minnesota Press.

Foucault, M. (1995) *Discipline and Punish: The Birth of the Prison*. New York, NY: Vintage Books.
Gangsterer, N. (2011) *Drawing a Hypothesis: Figures of Thought*. Berlin: De Gruyther.
Genosko, G. (2003) Félix Guattari: Towards a transdisciplinary metamethodology. *Angelaki: Journal for Theoretical Humanities*, 8(1): 129–140.
Gins, M. and Arakawa. (2002) *Architectural Body*. Tuscaloosa, AL: University of Alabama Press.
Guattari, F. (ed.) (1967) Programmation, architecture et psychiatrie. *Recherches*, 6.
Guattari, F. (1992) Fondements éthico-politiques de l'interdisciplinarité. In: Portella, E. (ed.) *Entre savoirs. L'interdisciplinarité en acte: enjeux, obstacles, résultats*. Paris: UNESCO. pp. 101–109.
Guattari, F. (1995) *Chaosmosis: An Ethico-aesthetic Paradigm* (trans. P. Bains and J. Pefanis). Indianapolis, IN: Indiana University Press.
Guattari, F. (1996) *The Guattari Reader* (ed. G. Genosko). Oxford: Blackwell Publishers.
Guattari, F. (2000) Vertige d'immanence. *Chimeres*, 38: 12–30.
Guattari, F. (2012) *Schizoanalytic Cartographies* (trans. A. Goffey). London: Bloomsbury.
Guattari, F. and Negri, A. (2010) *New Spaces of Liberty, New Lines of Alliance* (trans. M. Ryan, J. Becker, A. Bove and N. Le Blanc). Brooklyn, NY: Autonomedia.
Heath, K.W. (2009) *Vernacular Architecture and Regional Design*. Oxford: Architectural Press.
Hui, Y. (2015) Modulation after control. *New Formations*, 84–85: 74–91.
Leeb, S. (ed.) (2012) *Materialität der Diagramme: Kunst und Theorie* (Materiality of Diagrams: Art and Theory). Berlin: b_books.
Manning, E. (2013) *Always More Than One*. Durham, NC: Duke University Press.
Manning, E. and Massumi, B. (2014) *Thought in the Act: Passages in the Ecology of Experience*. Minneapolis, MN: University of Minnesota Press.
Massumi, B. (2002) *Parables for the Virtual: Movement, Affect, Sensation*. Durham, NC: Duke University Press.
McCormack, D. (2010) Thinking in transition: The affirmative refrain of experience/experiment. In: Anderson, B. and Harrison, P. (eds.) *Taking-Place: Non-Representational Theories and Geography*. Farnham: Ashgate. pp. 201–220.
McKim, J. and Massumi, B. (2009) Of microperception and micropolitics – an interview with Brian Massumi. *Inflexions: A Journal for Research-Creation*, 3.
Mezzadara, S. and Neilson, B. (2013) *Border as Method, or, the Multiplication of Labour*. Durham, NC: Duke University Press.
Querrien, A. (2012) Von der Architektur für die Psychiatrie zur Ökologie der Stadt. Ein Ensemble von Aktionsforschungen inspiriert durch Félix Guattari [From architecture for psychiatry to ecologies of the city: An ensemble of action research inspired by Félix Guattari]. In: Lorey, I., Nigro, R. and Raunig, G. (eds.) *Inventionen II*. Zurich, Berlin: Diaphanes. pp. 98–113.
Rancière, J. (2004) *The Politics of Aesthetics: The Distribution of the Sensible*. London: Continuum.
Sánchez, O. and Conwell, D. (eds.) (2006) *InSite_05/ [Situational] Public> Público [situacional] InSite_05/ Art Practices in the Public Domain San Diego-Tijuana*. San Diego: Installation Gallery.
Simondon, G. (1989) *Du mode d'existence des objets techniques*. Paris: Aubier.
Simondon, G. (2005) *L'individuation à la lumière des notions de forme et d'information*. Grenoble: Millon.
Stengers, I. (2005) Introductory notes on an ecology of practices. *Cultural Studies Review*, 11(1): 183–196.

5 Schizoanalytic cartographies[1]

Anne Querrien

As a friend of Félix Guattari, I can testify to the very special way he had of constructing a geography for the problems he was addressing. I have followed his work in various fields: the construction of an alternative to asylum psychiatry, the conception of 'collective equipment' and state intervention in everyday life, the clinic of artistic and institutional creativity in his analytical work at La Borde and with many intellectuals and artists. In all these fields he sought to expand the space in question in order to understand the problem posed at the outset from a different perspective, to discover the lines of flight, and to propose partial resolutions that are – if only provisionally – acceptable to those actors involved. This is with the clear awareness that this is a postponement; that geography does not serve to stop history, but to provide it with space for decision as events unfold.

Félix Guattari's geo-politics is tethered to a real space: an ensemble of points that might be affected by the problem posed, and mobilised in new relations to move it forward, or rather to break with the morass in which it repeats itself (Guattari, 1984). In the background of his thought, struggles are always present – by networks of friends, by militant actions. The multiple geographies of such worldly struggles provide opportunities for differentiated interventions according to the position and commitment of various actors. Geography is not understood merely as a science at the service of the ruling class or colonial power, which the militant geographer should denounce. It is also a matter of discovering, through the crystallisation of plural geographies, the points of view from which micropolitics are possible. This more lively [*active*] geography is not only an alliance of human and physical geography but, above all, a transformation in how territories are perceived.

Among these transformations, it is deterritorialisation that plays an essential role. This works throughout the earth and the universe, and is a sensitivity to the force of material flows that takes people to new horizons. There can no longer be stable representations; the earth and territories are carried away by movements we need to understand in order to build alliances, and with which we must compose. The ecosophic perspective is created with other coordinates of value than those allowing geography to accompany the exploitation of the earth.

I now want to examine three areas to develop these somewhat abstract statements: the alternative(s) to psychiatry, the apparatuses of power, and the 'clinic of creativity' (the schizoanalytical work for intellectual and artistic creation).

The alternative to asylum psychiatry

In the mid-1960s psychiatry in France found itself in a peculiar situation: some private clinics and several hospitals developed a new practice called institutional psychotherapy. This had the economic advantage of most patients not being hospitalised for very long. However, this practice did not correspond to the description of the normal functioning of the reimbursement of care through Social Security. The La Borde clinic where Félix Guattari worked with Jean Oury is one of the institutions that propelled this movement. It constantly struggled with the health department to have its own approach recognised, and to receive a sufficient subsidy in order to operate. This subsidy depended on the hierarchy of qualified staff employed by the clinic (doctors and nurses), while the clinic actually involved numerous 'instructors' [*moniteurs*] for cultural workshops and various activities. Above all, the clinic did away with the hierarchy of care and upkeep [*le ménage*] in order to be more efficient and to generate more opportunities for caring contact between staff and residents. The executives of the social security absolutely refused to take into account that the clinic functioned in this way. And yet it could take care of around 2,000 patients with only 120 beds, which was exactly what the World Health Organization had been asking for since the early 1960s.

The geography of French psychiatry was, therefore, completely frozen in large hospitals (of more than 2,000 beds) far from the cities. These hospitals were built in order to permanently exclude the mentally ill, who would be interned there until their death. And as life expectancy was increasing, it was noted that there was a lack of psychiatric hospitals. There was, then, a need to extend this geography, increasing the number of sites of detention. The government, which sought opportunities for the steel industry, launched an appeal for psychiatric hospitals in industrial buildings! But the birth of psychoanalysis in the interwar period and the scandal of the premature deaths of thousands of mentally ill people during the war due to lack of food, created a new sensibility and it was not too difficult for Jean Oury and Félix Guattari to collect hundreds of signatures against this invitation to tender to the steel industry. In so doing, they publically raised the question of the need for a new geography of psychiatric care.

Doctors, architects, and hospital administrators were sent to La Borde to investigate and take note of the new way of doing things. Gradually, the large psychiatric hospitals in the countryside were invited to open offices in the general hospitals in cities for times of crises and to provide specialised consultations, centres, and workshops. This did not happen without strong resistance: the administration continued to measure in terms of beds the set of differentiated institutional activities. By trying to bring psychiatry into the workings of the general hospital, the administration ignored completely psychiatry's specificity.

What was available in the fight for an open psychiatry, aiming to return patients to their civil and political life? If we focus on each innovative institution practising at its own level the policy of openness, the balance of power with the administration was weak. These innovative institutions represented a tiny minority of the profession that, admittedly, was a leading group and in line with international expertise. But administrative inertia was always stronger, especially because each institution – taken on its own – was under the control of its medical director. So as innovative as he or she might be, he or she was incapable of contributing to a new geography of psychiatric care needed so that patients can move relatively freely between places of differing therapeutic intensity and in connection with other networks.

It was at this point that Guattari seized the opportunity offered by the doctors investigating on behalf of the ministry, who had come to La Borde to develop alliances and to augment the space for reflection. A working group was formed around the investigation with psychiatrists, architects, students and several people from other, nebulous, working groups created at an earlier point by Guattari under the name of FGERI.[2] The interventions by psychiatrists only repeated the findings of the institutional psychotherapy group but the architects were able to bring lessons from a different practice. For them, the State was fighting the wrong battle. It responded to the requests from the unions for extending existing equipment – a purely quantitative demand – yet could rarely meet these due to budgetary reasons. The state ignored the innovative professional practices, often developed in the private sector because of regulatory constraints, which could be implemented in the public sector with the right means if these constraints could be lifted. What was affirmed in these transversal meetings of professions was a formula that would make it possible to transform the approach to collective equipment [*équipements collectifs*] (Guattari, 1967; Fourquet, 1976; Rostain et al., 1976).[3]

Shortly afterwards, Félix Guattari summarised this transformation as one which would need to introduce an institutional developer of collective equipment: a collective assemblage of enunciation that would take into account the particularities of the territory and the therapeutic practices – in particular psychoanalysis and the use of psychiatric drugs. This developer would, through trial and error, search for the best means of care in a particular space, and establish the best configuration of equipment at that time. For Guattari, then, this was about a perpetual questioning of contingent elements that meant his research team was always in a state of desire and permanent adaptation.

We must keep in mind the geography of the institution and never consider that this is settled. We must apprehend the institution in all its transversality, which is to say in both its hierarchical vertical lines and in its proximate horizontal connections – but also in terms of the micro-events which compose the institution and which resonate from the top to the bottom (or bottom to top). By transversality, Guattari expressed his opposition to the widespread ideas that it is the will of the medical director that dictates the configuration of the clinic and makes it belong to institutional psychotherapy. This view supposes that an institution is subject to the will of its director and does not exist in a milieu, a set of socio-economic strata, which largely determine the behaviour of the staff. Those at La Borde learned this the hard way. The rules for taking holiday leave, which were connected to the status of the staff, meant that one weekend there was no longer a sufficient number of qualified staff such that the clinic should have closed. It was therefore necessary to innovate and invent other rules of rotation and ways of measuring time both at work, and on holiday. Of course, the clinic could have shut down – sending the sick out onto the street, to the emergency room, or back to their families – in order to keep to the rules, but also to run much greater risks. Guattari's pragmatism reconfigured the social geography of the clinic in such a way that it was able to overcome this crisis.

This event was an opportunity to create a 'grid' [*grille*], linking people and tasks, anchored in a geography of the clinic in which the whole staff circulates in a relatively egalitarian way – transversal compared to the hierarchy of the statutes. In this communitarian space of the clinic, then, some of the residents would go out to fetch members of staff and visitors arriving at the nearby station. However, under the influence of Fernand Deligny, who spent some time there, the clinic began to think of itself as close to the city of Blois, where housing and jobs for the residents could be found for those who felt able to do so. This was a shared accommodation, an appendix to the clinic, but a reconfiguration of the space to be experimented with since the proposal attracted some residents – a space between the city and the clinic, close to the lifestyle of the staff who cared for them at the clinic (Rostain et al., 1976).

This new geography of hospitalisation, with its urban satellites – with the rise of social and financial assistance for the disabled – is not unique to France. In Italy, for example, the movement of leftist psychiatrists, *Psychiatria democratica*, was much more radical and focused more on taking into account the suffering of the mentally ill – rather than on the development of effective care as in France. Guattari nevertheless tried to link the different psychiatric practices in a network of intellectual exchange and resistance against the technocratic tendency of wanting to limit the budgets allocated to psychiatry by any means possible. The Alternative Psychiatry Network, which first met in Brussels in 1974, traced a new geography that ranged from England and experiences of anti-psychiatry to the urban policy of social intervention in Italy, not to mention the voices of psychiatric nurses in France or the practice of family therapy by social workers in Belgium. For Guattari, all these experiments, with their contradictory theoretical frameworks, constituted the same research plan for which it would be necessary to develop a 'metamodelization' that could serve as a line of flight, a perspective for reflection – similar to the beginnings of institutional psychotherapy. But in the context of hardening opposition to budgetary control, these exchanges tended to be ungrounded – disconnected from the practices that needed to be transformed or made anew.

This hardening took something of a 'security turn' [*une tournure sécuritaire*] under the presidency of Nicolas Sarkozy, long after the death of Félix Guattari. In a context marked by a willingness to imprison reoffenders or to keep dangerous suspects imprisoned, psychiatric hospitals were ordered to take security measures and return to their role of confinement – protecting society from the 'mad', once again, rather than treating the latter. A movement against this reactionary policy was established by 39 psychiatrists, mainly senior doctors, claiming to belong to the institutional psychotherapy movement. Doctors, psychologists, and nurses all joined this movement, denouncing the reduced budget allocated to psychiatry.

The movement's transversality was sought by coordinating the calls of different intellectual groups. As it was not tethered to real spaces of practice, the movement had no real impact on institutional life, and caregivers consolidated their radically critical position, dissociated from their practices. Friendly groupings took place and statements were drawn up, yet no new geography emerged.

Now it seems to me that we could draw from Guattari's experience – or from those movements which marked his last years – some propositions on what might constitute a militant geography. It is, first, a geography which is adjacent to a specific territory, which progressively constitutes itself by shaping this territory, by experimenting with modes of relations that can be created between elements bound by ideology and which have an economic impact, earn money, and gain new means of acting in that context. Basically, it is a matter of transforming the logic of enterprise through transversality and to give it the possibility of becoming a collective assemblage of enunciation. This cannot be done without concern for the geography in which this experiment/experience takes place, without negotiating the use of resources with others present in this territory, nor without an alliance with all the stakeholders.

In 1967, Guattari set up a research centre in the social sciences, CERFI.[4] In response to requests from the administration concerned to understand May 1968 and produce more adequate collective equipment, we worked in a very open way for around five years, with groups, general assemblies, and frequent institutional innovations, in order to maintain a collective assemblage of enunciation. It was a force for invention and sometimes of intervention. The general meeting every Tuesday was a key time to regroup; it was a space for expressing all kinds of minor demands and propositions. This fixed space was the sounding board for subjects we wanted to explore: feminism, the homosexual movement, child care, continuing education, sewing, cookery, support for Palestinians, work in new towns and poor neighbourhoods. It was a place of memory and turning back to specialist groups. These groups were forever divisible, reconfigured by the geography of the problems mentioned, and became pertinent through the acknowledgement of such problems and the discussions which accompanied in a sort of movement of general assemblies.

The way in which Guattari used space was notable for this balance between a central point and a multiplicity of peripheries – between Paris and La Borde, between overexposure in the general assemblies and the hidden office work. This is an image of transversality as something that we travel, that we regularly (re-) make, that we do not notice, that does not pre-exist, that must be made by a wandering gaze – at once sharp at the centre and blurred at the edges, a gaze which cannot be still.

Apparatuses of power

The fundamental question for Guattari is: 'what is the optional subject in this situation in which we find ourselves?' How are the multiple determinations that we are subjected to, governed by forces one needs to recognise, how these determinations have gaps from which a new route can appear? How can there be an assemblage of enunciation, an acknowledgement of these possibilities? How can we not give in to the lesson of powerlessness, to which Althusserian Marxism was reduced to at the time, chanting over-determination, and justifying inertia, the pre-eminence of the status quo, respect for hierarchies, and in doing so, making psychoanalysis an apprenticeship in adaptation?

This status quo is regulated in the case of collective equipment following a long historical evolution (Murard et al., 1973; Querrien, 1976). Whenever the idea of a new service is born, it is in a particular place, which created the conditions for bringing about what it proposes. If those concerned were right, the same innovation will occur in a number of other places with different coordinates. Conflicts of method will exist, and in a very centralised country like France, it will be the State that asks for one method to become standardised. Each semiotic specific to a particular field of activity is in this way ground down in order to produce a mythically unified geography – even though the legacy of history and the economic conditions, maintain palpable differences to the populations concerned. This tension of territory from the centre animates a movement of deterritorialisation in the search for better living conditions and better services. The action of the State to take control of its population, paradoxically, sets it in motion. It is not massive but at the margin, with all the more force that this movement takes the tangent of the normal inertia. Migration seeks social promotion and draws a new geography of territorial values. The republic – far from being one, or egalitarian – becomes a patchwork of differences with changeable borders. The recurrent denunciation of inequalities reinforces and stigmatises those who are its victims. As long as a call is made to the central power to erase differences, to define a new status quo and take into account the places most in need, we can only hope for a lower average performance and an exacerbation of differences. The advantages in terms of the reputation of the services are cumulative: they benefit new methods reinforcing their reputation, and so on. Thus, the top-rated schools are those established the longest, to which good teachers and good, committed students converge (Querrien, 1976).

Does the current proliferation of collective equipment lead to an irreversible alienation of the economy of desire? Is everybody obeying the different semiotics which model them and make them function? Or else transversally to these services does a collective assemblage of enunciation not weave itself, undoing the repressive aspects? But what do we mean by transversally? In administrative practice, the data collection – both through statistics and fieldwork, of detailed information of the breach of the norm – is the object of an examination understood as a transversal, i.e. the average. Yet the transversality that interests Guattari is of a different nature: it is the tension between exceptional performance through creativity or obedience – in any case, out of the ordinary – and the performance which is just as exceptional for not meeting the given standard of a time or place. This transversality defines a space with multiple dimensions of work on the transformation of the institution of the relevant service. This transversality applies to the collective assemblage of enunciation of a given institution; it does not apply to all the national occurrences of said institution, but by the sum of all these transversalities – of all these programmes of work and desire – we would find our national average. There is a possible space for speech between national guarantors of the functioning of these institutions and, in particular, their budget guarantors, and the local teams as soon as they define their own workplace. The management of the clinic at La Borde was subject to constant negotiation since its inception.

The concerns of central power for urban collective equipment appeared with the rise of industrial capitalism. Machines allowed production to deterritorialise and to develop in new settlements, linked to resources and not to the presence of a trained workforce. Frenzied exploitation was incompatible with productivity. Children's health had to be preserved and the school took on a new meaning: isolation from work. This isolation was used to teach the rudiments, in other words, obedient behaviour: silence, clean notebooks, ordered movements. The content of the teaching was strictly limited and the attempts by monitorial [*mutuelles*] schools to teach the capacity to teach others – and to further learning – was strongly repressed. The vocabulary mobilised by academia – as a whole and in various disciplines – combined to make language the principal instrument for subjugation to economic and political power. Each public service contributes to this semiotic subjugation through its own history and coordinates, more or less in line with families, and conformed – morally and ethically – to the average and ideal national 'subject'. This domination by language, by the signifier, far exceeds the framework of waged labour (which Marxism has studied), and seeps through all the pores of society, into all the relations framed by good manners. But this subjugation also sees failures, resistances, which result in extreme social inequalities for those concerned – but which can also be read as expressions of other unconscious, minoritarian formations. The work of normalising (or standardising) behaviour by the State and capitalism seeks to produce productive individuals, who are obedient and limited by the social positions they occupy. Failure of this standardisation is articulated by the health and education institutions in terms of individual disabilities even though this is a collective phenomenon.

A geographical analysis shows this failure is directly correlated with the concentration of unskilled labour in the given territories. Families are blamed who are not able to support the schoolwork of their children. But many studies have shown that the anticipation of failure by caregivers and educators dictates a conduct likely to produce it. There would, therefore, be an unavoidable social destiny that would overshadow certain parts, while others could benefit from institutions that, in theory, were made for everybody. In contrast, Guattari's proposal is to construct social machines that disrupt this inevitability [*inéluctabilité*]. To do this, we must build on what exists, that is to say, a few teachers, educators, or caregivers ready to try something else. We must transform the individual geographical disability into a force of differentiation and conjugation, in the production of an alternative. It is not a question of building an abstract, corporate, transversality that is outside of the situations in which young people are themselves involved. But we must give ourselves the means to collect, and articulate, the desires in instances that support the will to change. For example, different analytical groups were set up in the framework of FGERI. Change is to be produced smoothly [*en douceur*] along the lines of flight drawn by the new values to be implemented. The circulation – of information, of examples, of questions – through specific media is essential for this and can be promoted through the development of new tools for communication and social networks. Félix Guattari (2015) often wrote about the need for a post-media era in which communications tools would be in the hands of grassroots groups and allow for the construction of new social machines.

A *transversal* approach to all the forces involved in a situation creates opportunities for action by weak forces and marginal actors, especially when it comes to debating the national collective equipment by actors from peripheral or minoritarian territories. While everything that is natural, traditional, and self-evident recalls order, the construction of an experimental situation can express the economy of desire. By placing the machines of signs of different domains in a transversal relation, new relations can be discovered and made to vary. As Deleuze (2004) said, beings and things are caught up in relations that, by the repetition of tiny differences, continue to vary, according to infinite possibilities.

Collective equipment can intimidate inhabitants with its architectural facades, but also by the signifying faces of their agents, reproduced by paintings or photos. A certain face and bodily disposition signifies a particular time, as shown in paintings in museums or collections of photos. This constitutes a 'super-equipment' [*superéquipement*], a social ideal of the individual, which has reached all homes due to media, and which proposes to each one a benchmark in normality. Everybody feels that they are becoming a puppet in an increasingly complex system of machines; everybody feels at fault, in a state of waiting. The face expresses the division of the sexes and social hierarchisation, even if small differences can still be felt. Speaking then becomes the act of instituting the discourse of power, which takes place through you, and which is opposed to the one facing you. We are responsible for what we say, and regain control of the slightest of digressions. The mass media reinforce this system through the promotion of models' ersatz faces, and through a policy of systematic miniaturisation of the effects of power. Multiplicities are transformed into black holes, into accusing eyes: the backbone of white, male, adult, and heterosexual power (see Deleuze and Guattari, 2004).

Faciality is not the big 'other' as in Lacan, but the multiples small differences that singularise a face for itself and for others. It is on the basis of these small differences that a micropolitics of desire can be elaborated. It is from these lines of flight that 'look at us' and not through the central black hole which blinds and distresses us. In 1968 a faciality with long hair shook the world and began to speak and write differently. But the takeover of power was made possible by an 'ethnicisation' of the face (the revolutionary face was assumed as French by the militants themselves, and a nationalistic desire was mixed with the revolutionary one): it is from here or not from here, which synthesises all forms of binary.

Collective equipment is the abstract machine of capital, which controls the whole of the social field through binary values such as Signifying/Nonsense, Useful/useless, Beautiful/ugly. The 'schizoanalysis' (a term coined by Guattari to describe an analysis that does not propose to annihilate these contradictions) establishes, instead, continuous variations and multiple disruptions. Is it just a problem of analysis, of description, or does it also include production and history? When a group stops putting statements in particular boxes, when we make them proliferate, new thresholds of crystallisation operate, and the situations finds itself requalified.

Guattari – with Deleuze – used the figure of the rhizome in opposition to that of the tree to characterise the multidirectional and lively form of the relations of desire (Deleuze and Guattari, 2004). The rhizomatic assemblage transmits virally the choice made in the direction of liberation. On the other hand, the State which is, by definition, an agent of stabilisation of the socius, is organised like a tree, firmly anchored in ancient origins which are law, not only written law but ethnic traditions. The collective assemblage of enunciation, the rhizome, is formed gradually from small deviations to this law, from the suffering that is expressed in obedience. The collective assemblage of enunciation undermines these stabilities by producing new territories, new – offset – geographies. At the same time, Guattari argued that it is not enough to liquidate everything and to contribute to the production of a nihilist black hole. This new assemblage must be built with prudence and discernment, with a careful attention to the vectors of change. And it is in the consciousness that this movement of change will bring about other changes in dominant forms, in parades, and conservation actions. It is a national geography to be challenged in priority.

The collective assemblage of enunciation is constituted by gatherings [*prélèvements*] of relevant traits which combine to produce real shifting of situations. This assemblage is not necessarily the same as a group; it is often spoken by a person in which the forces of the moment intersect. The swing between two territories (which as we noted already, is characteristic of Guattari) predisposes this opening to the constitutive diversity of the new transversal geography. It is here that deterritorialisation intervenes, a movement out of the territorial anchoring, a disaffiliation, a setting in motion both intimate and material. The assemblage will become machinic if it combines with others in the production of a process of collective transformation. Different deterritorialisations are at work at the same time in flows of materials, humans, and signs, and are no in phase with one another, hence certain combinations can result in a reterritorialisation, a resumption of the process in reverse. The deterritorialisation felt in the forms and the functions and forms of successive machines, expressed by the notion of 'machinic phylum', does not coincide with the constitution of existential territories in which humans and animals live.

The three main systems of flow constituting a society are money, work, and language – three lines of organisation of populations and an inscription of their desires, their productions, and their exchanges. The living milieu, which affects everyone and everything [*l'existant*], reproduces itself according to its lines, which also support the invention of new configurations, in an incessant battle that in *A Thousand Plateaus* (Deleuze and Guattari, 2004) is characterised as that between a nomadic war machine against the sedentary State apparatus. People in analysis or in psychiatry are those who, in various ways, confront the state institution. The analyst's attention is then drawn to the areas where the order of signs does not represent the order of things, where the reign of the signifier is called into question by unexpected events. Normal, or what Guattari called 'molar', is not perceptible except as a certain discomfort provoked by multiple, tiny shifts that fail to fit the norm (what Guattari (1977) describes as the molecular level). This is particularly evident in the case of the sexed body assigned to a binary division, which models its subjectivity on very restrictive figures – even though it is in fact sensitive to a multiplicity of different elementary sensations. The liberation of women has been, as much as access to a series of new rights, the possibility of living one's body as it is or as one wishes to transform it. It is a struggle in which they were joined by all the minorities of desire.

The rhizome through which Guattari proposes to connect elements of all domains – arts, sciences, and social struggles – is not a system of representation but a gradual constitution of a collective assemblage of enunciation. The experimentation respects all matters of expression. But, on the other hand, the powerlessness produced by the blockage of the signifier can reduce matters of expression to immobility. Subjectivity functions like an interaction and accumulation of differences. The semiological substance of individual assemblages of enunciation is deployed on a surface of representation divided into the substance of expression and content – the sides of which are held in a biunivocal relation. The machines of power produce their own forms of expression, which obey respect this law of the geographical average that we mentioned earlier. The normal form of expression in whichever domain is imposed throughout the national territory by coercive measures outside of the matter of expression itself, but that fixes it. For example, musical writing was kept on a leash from the Middle Ages to the 18th century by a religious power which subjected it to mathematical arrangements.

The slogan of the pragmatic micropolitics that Guattari proposed would be 'do it'. The problem for all of us is what we have to do, in our eyes, for our own bodies. But this advice would be in vain if everybody did not begin by giving him/herself a cartography (Guattari, 2013) of nonsignifying coordinates, points of passage towards the deterritorialisation and the exit from the system of signifiers, a cartography of feelings and desires, of his/her inscription in an existential territory capable of supporting his/her deterritorialisation from his/her place of origin. We must also map out places of power, places where other social machines are armed. Above all, we must follow the emergence of new machinic arrangements that 'minoritize' the influence of the signifier, via various experiments, bringing forth new values.

Every human being is faced with the solitude of the human condition, and is obliged to bear it by relying on the stratifications of the society to which they belong, which are expressed themselves through well known refrains [*ritournelles*] in the work of Proust (Guattari, 2011). Like Swann, everyone is also sensitive to different small sentences, expressing a new world, offering us a new path of deterritorialisation. Capitalism tends to saturate time with technologies making certain refrains invasive, making them almost mandatory collective equipment, ubiquitous jingles. Kafka emphasises presciently that the collapse of specific, territorialised, refrains risks immobilising us, like the mice of *The Burrow*

(Kafka, 2017), only much more subdued due to the anguish it brings (Deleuze and Guattari, 1986). Contemporary music since Schumann seeks to ward off the black hole of madness by increasingly fugitive, increasingly deterritorialised, childhood refrains, thanks to multiple creations of new assemblages.

The clinic of artistic and institutional creativity

Despite the complexity of the writing, *Schizoanalytic Cartographies* (Guattari, 2013) is addressed to everyone. It is about providing landmarks [*rèperes*] – of whatever scale (world, group, individual) – on the major components of the production of subjectivity. First, to take note of the reliance of subjectivity on the multiplicity of machines that haunt everyday life and that transform possibilities for better and for worse. The nature of human activity is to add a machine to the already existing machines to bring others into the operation that was imagined. These machines do not need to be complicated, nor metallic: the breast, heavy with milk, forms a machine with the mouth of the feeding baby. A machine is what cuts a flow and feeds on it to produce another effect. But technical machines will inscribe themselves in the phylum which escapes from the material flows through deterritorialisation, while social machines will stratify and weigh down [*lesteront*] the socius. As we can observe today, the greater the deterritorialisation of material flows, the more the social body is reluctant to enter into phase with and propose reterritorialising on limited, or even fascistic, existential territories.

The flow is of energy, information, material, water, electrons. . . . Its value in the production of subjectivity only appears when it deterritorialises – leaving its initial form – to invest in existential territories where it is transformed, converted into new systems of action or invested in machinic productions which cut and subjugate it to new lines (Guattari, 2011). The sense of belonging to the great material and human flows, and the adjacent character of subjectivity that can be developed from them, is a foundation of Guattari's therapeutic framework. The body is not closed in on itself, it flees from every part; it has no fixed point of sustenance. It is in welcoming this flowing [*mise en flux*] of the body that it becomes capable of listening to the world and of playing its score there. The material flows are the first expression of this constitutive otherness of human experience – what Lacan calls the real.

Unlike flows that are ubiquitous and not located, we live in existential territories with qualities recognised by all those who participate in it, which correspond to defined spaces and times, moments of human history characterised by a certain place and a certain zeitgeist [*air du temps*]. There can be overlap between existential territories – passages from one to another – and transformations of subjectivities. Contrary to those who describe these existential territories in terms of identity, Guattari considered that these are heterogeneous assemblages, marked by features of singularisation which dissonate [*dissonnent*] with the environment, like the small musical phrase of Vinteuil in Proust. Something signals a break with the monotony and the ambient homogeneity and will attract, aggregate, and bifurcate pre-existing trajectories'; it will transversalise the situation and transform itself into a rallying sign for a new existential territory.

An extreme example is given by the work of Fernand Deligny on autism (Deligny, 1975; Deligny and Joseph, 1975, 1976). Janmari was a young autistic child, who was entrusted by his mother at the age of eight to Deligny because he terrorised institutions by almost endlessly spinning around [*tournant sur lui-même*] and yelling. Nothing else had been done except tie Janmari to a bed so that he would stop moving. Deligny lived at La Borde at that time but considered that it was too striated, too loaded with various requests, to be able to welcome Janmari and enable him to flourish. Deligny therefore went with him to the Cévennes, where he formed a welcoming network for autistic children with young militants from the Paris suburbs who were tempted to return to the countryside. Despite their good intentions, these young militants could not cope with Janmari's behaviour [*la gestuelle*]. And so Deligny suggested that they map the comings and goings of autistic children in the reception area. By superimposing the drawings made every day over a long period, he realised that the aimless movements, which he asked them to pay particular attention to, were related to the presence of objects or people in the same place at a more remote time. These movements, what Deligny (2015) called wander lines [*lignes d'erre*],[5] revealed the particular temporalities of the autistic children, and a universe just as a relational as that of so-called 'normal' people. The circles that marked Janmari's unbearable spinning movements gradually disappeared from the maps; yet it was not because Janmari stopped spinning around but because the cartographer ceased to consider it important. Mapping was pivotal in treating anxiety. To give another instance, the adults wondered what Janmari got up to when he was out of their sight, and they worried that he might escape or have an accident. But the careful observation of Janmari's movements showed that he stayed within a large ellipse; the focal points were the two places where the adults lived. The young autistic becomes attached to what Deligny called a 'common body' (or what Guattari would call an existential Territory); in Janmari's case, he became attached to Deligny who he saw as his guardian (Deligny, 2007). In the drawings of the adult Janmari, the unclosed circle and the waves are the two figures repeated endlessly, figures that made his own body before he could write (Jonquet, 2013). Janmari's body was in phase with the flows of subterranean water that he knew how to find when searching for a source.

In *Schizoanalytic Cartographies*, Guattari (2013) proposes a representation of the tensions to which our bodies and minds are subjected – and in relation to which we must situate ourselves, by defining the matter with a particular option that emerges for each of us. Guattari's schema is four-dimensional, and each dimension is itself a multiplicity. The material flows are animated by a great force of deterritorialisation captured in the machines, used to support existential territories, but destabilising for the animal and reproductive dimension of human constructions. The existential territories within which this reproduction is organised have a strong tendency to use the springs of similarity, of imitation. And it is only because there is a proliferation [*pullulement*] of different existential territories – due to the renewal of generations and the cumulative development of the machinic phylum – that a sort of blurring [*flou*] seizes the existential territories developed by multi-memberships and reconfigurations. As we have already noted with the example of Proust, that which indicates the existence of an existential territory is not at all an average and identifying feature (as the machines of power would have it), but a feature of singularity, an indicator of becoming. The new emergent existential territory thus enters into conflict, either directly or indifferently, with the existential territories already stratified by the machines of power. It is a difficult period of innovation but – thanks to deterritorialisation and the flow of history – the relations of strength will move gradually, and the previously emergent existential territory will be recuperated by the dominant reterritorialisations. The fate reserved for Proust is exemplary from this point of view.

In Guattari's four-headed schema we see material Flows (the real for Lacan) and existential Territories (the imaginary for Lacan). The third pole called incorporeal Universes is the symbolic for Lacan. Incorporeal universes are all the human cultural productions that deterritorialise territories on which they have been produced, taking a universal value, and inscribing themselves in the productions of humanity as such. A kind of pantheon of values in which it is necessary to inscribe oneself to escape the repetition of territories, in which it is necessary to inscribe the values of others (the mad, children, the homeless, in general) to participate in the human history. It is very important to know your position in the flow of deterritorialisation between existential territories and incorporeal universes in order not to stagnate in a smug satisfaction of your revolutionary actions, soon to regress into similitude and fascism. There is a question of mental hygiene for the therapist: it requires a progressive inscription between territories and universes. Someone too concerned with incorporeal universes, an intellectual without manual practice, risks not having the strong relation to earth and history [*prise de terre*] that a suitable existential territory confers. Both inscriptions are required. And we can assume that the revolutionary becoming of 1968 quickly fell back into a sad reterritorialisation not having developed the artistic dimension, apart from some graffiti and especially in the poster workshops of the Fine Arts, an artistic machinic dimension and collective which has been reduced to the sympathetic but sad promotion of some very gifted youths.

The fundamental difference between Guattari's schema and Lacan's is that it is four-dimensional; Guattari adds the dimension of the machinic phylum, the historical or local production of desiring-machines. This machinic phylum is carried away by the deterritorialisation of flows of all kinds – phenomena largely unconscious (but increasingly studied by the sciences) that belong to the incorporeal universes and which are able to imagine the capture of these flows; machines, to be precise. And the word 'imagine' is used here on purpose: it is within the context of existential territories that we can imagine the capture of flows by machines, and the production of objects transforming life and coming to reterritorialise in territories for the production of a new cycle. So this is not simply an additional dimension to Lacan's Trinitarian schema, it is a completely new cyclic operation; this makes it possible to understand why humanity does not always have to repeat Oedipus or Adam and Eve; why humanity has a history in which the analyst, the artist, everyone, has to play his or her own part, in an open attention to the flows, universes, and phyla which constitute the world and which take us with it.

Guattari offers us a four-dimensional map, each dimension itself a multiplicity or a set of flows, and invites us to locate ourselves in the middle, to constitute our own milieu in this fluid universe, to hold the subjective and creative together. Guattari's thought is prescient of the first developments of the Internet and of the emergence of the surfer, the navigator, as a new figure of social behaviour.

The production of creative subjectivity, the production of a processual auto-referentiality, that is to say, of its own 'wander line', deterritorialises itself from the subjugations produced by the territorialised power of property and discipline and by knowledge de- and reterritorialised by capital, while relying on the historical contributions from which they are made. Analytical work (coupled with artistic or institutional work) reconfigures the inherited situation and redraws a new existential territory from which to lead innovation and research. Guattari believed that the proliferation of information processing machines creates favourable conditions for the development of processual auto-referentiality (Guattari, 1995), an auto-referentiality which some like Francisco Varela say is the property of computer networks.

In the age of planetary computing, the problem is the collective human capacity to reappropriate the statistical and modelling devices [*dispositifs*] produced by capital, devices which tend towards averaging and a collapse of the differences from which desire operates. At the same time, without desire, without exploitation of the remaining differences, capitalism cannot progress. There is, therefore, a space of tension at the heart of the system within which to operate thanks to: the coordinates given by analysis, the affirmation of a fundamental right to singularity, and the prudent practice of this singularisation on the schizoanalytic map. We might think, for example, of the autistic children who Deligny lived with and of the research that he undertook on the human out of language, contradicting the scientific claim that it is language that characterises the human and distinguishes it from the animal. Erin Manning and Brian Massumi (2014), following Deligny's work, and in a cohabitation with American autistics, discovered that it was not so much language as speech that their friends were lacking; they were capable of writing fantastic, poetic texts from their different perceptions, all without speaking.

For Guattari, this singularity of the existential trajectory expresses itself in the register of opinion, as a right to dissensus, which would allow everyone to develop his or her own self-referential path, which would perhaps "transform this planet, which is lived as a hell by four-fifths of its population today, into a universe of creative enchantments" (Guattari, 2013: 13). The experience of La Borde, like that of Deligny, or that of Erin Manning and Brian Massumi, puts greater emphasis on cohabitation, living together; attending to others with whom we create, day by day, an existential territory, from which we establish a new incorporeal universe capable of crossing material flows to produce new machines. How astonishing, then, to see the constant creation of new musical instruments, and new ways of using existing instruments!

What can the schizoanalytic cartographies bring to geographers? The symposium on transversal geographies has shown a change in the objects of analysis for geography – an attention to the homeless and to migrants, an attempt to see the

Earth from a different point of view other than that of power. Beyond this transversal movement on the social scale, we might imagine new existential territories for geographers that would make them share the lives of different subjects – from other disciplines (at the least) and from other social positions – and to develop new collective assemblages of enunciation of what geography is and can be. They could invent new geographies, and discover new lands and new ways of inhabiting the world. These new geographies would give us a way to see the world differently.

Notes

1 Translated by Thomas Jellis.
2 FGERI stands for *Fédérations des groups d'études et de recherches institutionelles*, or the Federation of Study Groups and Institutional Research. [Translator's Note]
3 See Goffey (2015), who remarks on how the term does not translate easily into English. He also notes that "The idea of collective equipment is notably absent from A Thousand Plateaus, which is, in some respects, regrettable" (2015: 129). [TN]
4 CERFI stands for *Centre d'études, de recherche et de formation institutionnelles*, or Centre for Institutional Study, Research, and Training. [TN]
5 *Lignes d'erre* "might be translated as 'wander lines', 'errant lines' or 'lines of drift'". Deligny's (2015) translators, Drew Burk and Catherine Porter, prefer the first option. [TN]

References

Deleuze, G. (2004) *Difference and Repetition* (trans. P. Patton). London: Continuum.
Deleuze, G. and Guattari, F. (1986) *Kafka: Toward a Minor Literature* (trans. D. Polan). Minneapolis, MN: University of Minnesota Press.
Deleuze, G. and Guattari, F. (2004) *A Thousand Plateaus* (trans. B. Massumi). London: Continuum.
Deligny, F. (ed.) (1975) Cahiers de l'immuable 1: Voix et voir. *Recherches*, 18.
Deligny, F. (2007) *Oeuvres*. Paris: Éditions L'Arachnéen.
Deligny, F. (2015) *The Arachnean and Other Texts* (trans. D.S. Burk and C. Porter). Minneapolis, MN: Univocal Press.
Deligny, F. and Joseph, I. (eds.) (1975) Cahiers de l'immuable 2: Dérives. *Recherches*, 20.
Deligny, F. and Joseph, I. (eds.) (1976) Cahiers de l'immuable 3: Au défaut du langage. *Recherches*, 24.
Fourquet, F. (1976) *Généalogie des équipments collectifs: Historie des service collectifs dans la comptabilité nationale*. Fontenay-sous-Bois: CERFI.
Goffey, A. (2015) Introduction to Guattari on transdisciplinarity. *Theory, Culture & Society*, 32(5–6): 125–130.
Guattari, F. (ed.) (1967) Programmation, architecture et psychiatrie. *Recherches*, 6.
Guattari, F. (1977) *La révolution moleculaire*. Paris: Recherches-Encre.
Guattari, F. (1984) Causality, subjectivity and history. In: Guattari, F. (ed.) *Molecular Revolution: Psychiatry and Politics*. Harmandsworth: Penguin Books. pp. 175–207.
Guattari, F. (1995) *Chaomosis: An Ethico-Aesthetic Paradigm* (trans. P. Bains and J. Pefanis). Sydney: Power Publications.
Guattari, F. (2011) *The Machinic Unconscious: Essays in Schizoanalysis* (trans. T. Adkins). Los Angeles, CA: Semiotext(e).

Guattari, F. (2013) *Schizoanalytic Cartographies* (trans. A. Goffey). London: Bloomsbury.

Guattari, F. (2015) *Lines of Flight: For Another World of Possibilities* (trans. A. Goffey). London: Bloomsbury.

Jonquet, J.M. (ed.) (2013) *Journal de Janmari*. Paris: Éditions L'Arachnéen.

Kafka, F. (2017) *The Burrow: Posthumously Published Short Fiction* (trans. M. Hofman). London: Penguin Books.

Manning, E. and Massumi, B. (2014) *Thought in the Act: Passages in the Ecology of Experience*. Minneapolis, MN: University of Minnesota Press.

Murard, L., Fourquet, F., Lévy, F., Querrien, A. et Vernet-Stragiotti, M.T. (1973) Les équipements du pouvoir: Généalogie du capital 1. *Recherches*, 13.

Querrien, A. (ed.) (1976) L'ensaignement: 1. L'école primaire. *Recherches*, 23.

Rostain, M., Dardy, C., Grass, G., Murard, N. and Préli, G. (eds.) (1976) Histoires de La Borde: 10 ans de psychothérapie institutionnelle à Cour-Cheverny 1953–1963. *Recherches*, 21.

6 Refrains of lost time

Collapse, refrain, abstract

JD Dewsbury

Introduction: modest beginnings, grand designs

Let us start with something as simple as thinking about something else; or, more precisely, becoming aware of thinking about something else. Perhaps this is akin to waking up from a dream, or when your body is inflected towards another course: a taste molecularises a difference not because of its representation of like or dislike but in the chemical connections that change things; a glint of light makes you turn away with a snapshot image burnt into your visual thinking; sounds secreted by instruments affect the ease of your being; words you just happened to say seem to create a surge of excitement as a grin, except it is not the sound nor meaning of the words themselves but the grinning that results that seems at some point to be, well, the point of it all. Thinking is all this: a taste, a light, a sound, a grin. Whatever the case of thinking about something else, the event of its happening "exists a little bit everywhere" (Guattari, 2012 17): it could come first in "a little musical phrase" or "in a mathematical formula, a face, or something found on the shelves of a junkshop" (Guattari, 2016: 234); Guattari "always stayed as close as possible to his everyday experience" (Polack, 2011: 61). What is the focus and the perspective, then, that comes to the fore in this micro-profound, imperceptibly-not-yet alteration of your life? What is the machinic unconscious that operates for you to 'get out of your head', so to speak, and therefore away from the course of thinking you were previously on? Where do you go then to find some form of intrinsic referencing and self-organisation so that the singularising event of this instant makes sense, such that you find yourself moving on in this way and not that? And when do you realise that that has happened, if you ever do? And above all, what can we say about these singularising events? All the uncertainties, or invisibilities, housed within these questions are the unconscious. In this chapter I want to show how thinking these singularising events of the unconscious is central to the aim and ethos of Guattari. Before going on I want to put in mind the exemplary instance of a singularising event that is going to frame this chapter:

> The material quality of sounds secreted by instruments/the presence of one of these invisible realities in which he had ceased to believe.
>
> (Proust, 2003: 211–214; quoted in Guattari, 2011: 243–244)

Proust's *In Search of Lost Time* is, famously, a book in which the orderings of time structuring prosaic experience give way to a productively confused sense of before, after, then, now, first, second, once, and again. Thus, to say that the words "the material quality of sounds secreted by instruments/ the presence of one of those invisible realities in which he had ceased to believe" denote the 'first' reference by Proust's Narrator to the little phrase of music from Vinteuil's sonata is only part true. In any case, when Guattari takes this same little phrase of music as inspiration for his thinking on the concept of abstract machinism in the years 1968 to 1980, between the publications of *Anti-Oedipus* and *A Thousand Plateaus*, he is not concerned with representing or unpacking this singularising event to give it meaning through our current traditions of meaning making and value giving (for example, Proustian scholarship and representational politics of one kind or another). Rather, Guattari asks what *could* be said of this singularising event if we were to invent new cartographies of apprehension, which maximised the potential of the difference that the event possesses. My reading of Guattari is one which sees him offer a kind of science, and therein a kind of method, for social scientific thinking. This might seem provocative given Guattari's critique of the scientific paradigm but in his mode of experiment and diagrammatisation Guattari draws out the transversal effect of science's enunciative power for different ends. For example, Guattari uses the art of exemplification, here of Proust (indeed, as I do of Guattari), not for the construction of a model that would be general but as that which expresses the singular (see Massumi, 2002: 19). As a science Guattari's method offers a syntactical conceptual grammar for making associations about what is happening 'now'. Let me be clear, matters of identity or representation do not yet matter, what matters instead are the potential differences that underwrite the life of the events through which we become, matters of as-yet unfelt allegiance to music, to aspects of other people, to qualities in the physics of places, and to the trajectories of becoming that we are slipping into.

In relation to going on, as we have always gone on in the social scientific tradition of concept and method, the difference Guattari makes is modest and vital. Whilst the questions that Guattari's work asks are modest, in that they could apply to everyone, they ground a grand form of questioning (after Guattari, 2012; see Jellis, 2014) which proffers a vocation for a new discipline. The vocational importance of Guattari's work is twofold; namely, first, in moving away from the traditional use of language as the only means of expression for the unconscious and, second, in refusing to structure our understanding of the unconscious upon signifying horizons (Negri, 2017: 168). Both these aspects of his work are crucial, in that they relate the unconscious more to its social, economic, and political production, providing a means of analysing the processes whereby desire, subjectivity, and individuality operate as always secondary outcomes. These qualities are profoundly significant, critiquing as they do the way the academy studies the most visible, "the most territorialized components of human behaviour" (Guattari, 2016: 225), at the expense of the latent potentials of the machinic unconscious, its molecular revolutions and new desiring-machines for living. Grin and rip up the ease of what living was like before you said those words, although, of course, those words were said as products of so many desiring-machines that any sense of yourself, before or after their enunciation, was always up for grabs. We need to experiment with the ways by which we can get hold of how that 'up-for-grabs-difference' of our becoming happens *as* it happens, and the space for that is what we could call the unconscious. This, taken alongside what Guattari calls the abstract machine – "transformational matter . . . composed of the crystals of the possible which catalyse connections, destratifications and reterritorialisations both in the living and inanimate world" – presents us with machinic unconscious. Literature, theatre, the arts are the laboratories best suited to these experiments in machinic unconscious, possessing as they do a unique and irreplaceable ecological functioning that immediately and intimately perform, and make matters of expression of, the transversal communication orchestrating the unconscious and thus production of the relationship between the mental and social ecology that gives us our sense of subjectivity. As such, this chapter is motivated by the sense that the social sciences, too, can occupy this experimental role. At the very least, a social science that was closed to this machinic unconscious would surely be an impoverished one. In seeking to apprehend its latent potentials, it is imperative that we start on the arts' terms and bring them towards our assumptions, and that is what I think Guattari does when he encounters work like that of Proust. The conceptual invention that we find in Guattari – inventing concepts that are confusing, complex, and problematic – is precisely due to the point of thinking differently between different institutional spaces of knowledge production. These are not static institutional frameworks either as for Guattari the boundaries are fluid.

So, in questioning Guattari, and forming questions *with* Guattari, this chapter proceeds through three conceptual positions that are in their nature dynamic: thought as semiotic collapse, thought as refrains of assemblage, and thought as abstract machine. The conceptual sectioning of the argument focuses us, as a starting point, towards Guattari's work as a whole, and then more precisely towards Guattari's *Refrains for Lost Time* (2011: 231–332). In this text, Guattari presents us with potential orientations to help us answer the question of how to make a specific cosmological fold come into focus in the aftermath of thinking. While concentrated upon a literary space, this project is very much situated in, and speaks to, our more literally present physical and psychological contemporary condition where thinking itself has become objectified and targeted by capitalist semiotisation. It is then less the phrase itself as presented in the fiction of the text and more the abstract machinism it presents for coordinating new associations that singularise everything in a different way that matters. Such singularisations exist but for an instant and yet, nonetheless, potentially for every time henceforth. Don't expect anything more of the present, Guattari seems to be saying, but do expect so much more of the future.[1]

As a social scientist, the broader question orienting the book, "Why Guattari?", poses something distinct, especially something about the distinctness of Guattari's work from that of Gilles Deleuze and Deleuze-Guattari. As for Deleuze, philosophical writing involved the production of concepts for the actualisation and capture of sensation, though this is, for Guattari, specifically always from and towards actual modes of living. These are modes of living that operate through individual bodies, but not as individuals to be ramped up collectively as society, but rather as something to be understood through multiple subjective productions, equally social, cultural, ecological, economic, and political. These subjectivities are never from the outset discrete, nor simply a hybrid of any of our axiomatic categorisations. They are not entirely ours, nor entirely human, but constantly blurring and bleeding perspectives one into the other. In many ways, Guattari's thought responds to the 'situation' of this situation of how to understand the production of so many subjective trajectories, such that we are always changing and always plural. He does this through his development of practical techniques, which capture the shifts that are constantly occurring within us, techniques that are only as effective, as Genosko (2002) illustrates so well, as the collective project that sustains them, the collective assemblage in the midst of which they are placed. These collective assemblages are assemblages of enunciation and, as such, are precisely singularising events. It is in Proust, I will suggest, that Guattari sees a way of exemplifying the complexity of our emergence, as a project rather than as seemingly consistent entity, and yet with some consistency against the multiplicity and chaos of so many disintegrating perspectives.

Conceptual beginnings: semiotic collapse

In thinking with Guattari, what I appreciate is his radical separation of literary and scientific fields, as a way of bringing them into an alliance that recasts them in new ways. A warning, then, as to the tone of this chapter: like, Guattari, I am not writing here as for "an external exam and 'expert' interventions", but towards a "collective taking charge in determinate micropolitical domains" (Guattari, 2016: 36). There is then a genealogy of ideas and a history of thought I am not going to attend to. Untethered from those institutional frameworks, the micropolitical field in this chapter is a new cartography for singularising events, which commits to moving beyond traditional signification points. Indeed, as a singularising event, 21st century academia, specifically the social sciences, could do equally well if they took up transferences of enunciation (Genosko, 2015) and/or new claims for thought (Lapoujade, 2017).

As such, the art of the Proustian language and writing presents itself as something intrinsically entwined with transferring and augmenting the processes of semiotisation available to us with new claims for evaluation. Whilst starting off as objects of study, Proust's mental objects (those secreted sounds, that grin, the glint of light, etc.) dissolve as the practitioner-philosopher Guattari puts them to work. Meaning here does not exist; it is not a discursive wrap upon a visceral agent giving definition. Rather, what comes to be presented are the sensorial foldings of experience, flesh, and material space that are agential in their own right. For me, this is what Guattari is consistently targeting: these *sensorial foldings of experience*. To orient you: your body is a fold, a fold of memories, molecules, fluids, and electricity, which snag, pull, and tear at any inner/outer sense of you being just you. These folds shunt you in this direction, overwhelm you in that.

There is a science here, which is operated through a microscopic psychoanalytical machine bringing into view this non-object: there are no things of definition, more assemblages of enunciation. Yet, given the anti-psychiatry of Guattari, this machine might also be understood as an *art of this science*. This art, Guattari's art, is that which *probes the mutation of perceptive components*, breaking them down, so as to diagrammatically map the phenomena of the magnification, displacement, overlapping, acceleration, or deceleration of sensorial coordinates (see Gerlach and Jellis, 2015). Again, these are not our sensorial coordinates; we are coordinated by them. That is why diagrams were so important for Guattari: we need a way of giving expression, making sense of, orientating ourselves, to the ways in which we become after the world and not before it. Guattari's twofold critique of language and signification is rendered here as a critique of the distanciation typical of classical social scientific methods and writing (see Glowczewski, 2015).

If 'thinking about something' is more akin to matters of expression which may actualise a new aspect of our subjectivity tomorrow, several years later or never, then it is necessary that these expressions do not reflect our life back to us. Expressions are not reflections acting like mirror images that situate us in the world, in and towards the self. Rather, they act as inflections of our being, choreographing our becoming, whether we like it or not. Like it or not at what point do you evaluate what is to like and what's not to like? You just somehow like the beat, the phrase, the association, the grin. Therefore, these inflections are understood, for Proust, and for Guattari, and now for us, *musically*. It is this music effect that in *In Search of Lost Time* is the quintessential instance of what Guattari calls semiotic collapse:

> This music seemed to me more true than all known books. Sometimes I thought that the reason was that the things we feel in life are not experienced in the form of ideas; and so their translation into literature, an intellectual process may give an account of them, explain them, analyze them, but cannot recreate them as music does, its sounds seeming to take on the inflection of our being.
>
> (Proust; quoted in Guattari, 2011: 232)

Now, for Guattari, what Proust does with this 'music effect' is to site it within the narrative of the Narrator, as the moment within which all the vague, fuzzy, evanescent qualities of the sensations that have been assailing him, takes him "to the end of his impression" for the first time, and for a fraction of a second. This impression is the matter of expression. It touches you in its nascent state, which is to say that it has already touched you. To what extent do you already have the proclivity to be touched in this way – touched, displaced, disturbed, distributed – by this event, this fold of life and matter, bearing in mind that it does not emerge from a discursive analysis such as human language can sustain? Proust ruminates upon the music effect, this latent antecedence, via his interest in musicologists: a note taking as to how musicians take phrases, analogues, from previous composers.

So, you are touched by something that is not-yet. This is not any inadequacy on your part as to your ability to give this expression some form and meaning. The way expression touches you is never as a force or something that pertains to you in a mode of recognition. It is a rupture, it is first and always a deterritorialisation. Guattari is urging us to find a way of giving such rupturing forces their due, without reducing them to a part in a past narrative that forecloses the molecular revolutions that they contain for the future; we have to see the cosmos in ways that are more than reactive (Sharpe, 2014). For Guattari, this is the key analytic he addresses in his engagement with Proust: that literature, specifically Proust's, offers a space for the science of the individual. Literary expression exists as a laboratory space for examining and evidencing the production of the individual. Just as the matter of expression is not for you, this science is not for the individual as such. Guattari sees in Proust the idea that "one cannot consider human subjectivity as something empty and undifferentiated that would be filled and animated from the outside" (2011: 234).

In presenting this science of the individual, as expressed through the Proustian landscape, it is as if a matter of expression projects the body, like a penumbra that follows the body around, constituting the fabric of associative potential as subsequent life unfolds. Like light, sound, or taste, these expressions are, for Proust, notions without equivalence, which are nonetheless discernible advents in the cosmos, like a possession that diversifies your inner life even if you have no sense of where that diversification is going. It possesses you and you become through it. It is in this regard that we can understand these matters of expression as asubjective and a-signifying. They are asubjective, having had their own life aside from whether they come to the fore in yours. And they are a-signifying, in that they change you, and that change cannot yet have the orientation to signify anything. You can create new diagrams afterwards to signify, if you want. Indeed, this is why we see so many diagrams in Guattari's work, diagrams that should not be read as universal road maps but as potentially transversal blueprints for producing new routings. The success of this cartography rests precisely in being confusing. The challenge of its 'use' is in the novelty that its unfamiliarity presents: a new-found complexity as to how the cosmos and our place in it comes about. We should not be surprised by this, we should just be surprised.[2]

As notions without equivalence, these matters of expression are semiotic converters and ethological indicators – a room; a place; a time; an object; an emerging entrance or exit; a societal or institutional role doing that something; an associative flow of desire; all towards an abstract machinism for creation (all these indicators respectively echoing the nine assemblages Guattari presents 2011: 243–306). This series undertakes a kind of labour of conversion, which turns existence upside down (after Guattari, 2016: 230). What holds things together (if indeed they need holding together) is neither of individuality, nor anything's individuality, nor any homeostatic conceptualisation of collectivity to offset solipsism. Rather, what does the holding, and gives consistency, are significant redundancies that such singularising events spotlight within our systems of signification.

Withdrawing exemplification not giving examples: refrains (nine assemblages)

> We must not lose sight of the fact that the little phrase will never be self-identical, that it will never lead to the same politics within the various assemblages where we will rediscover it and that, from one field to the next, it will be brought to prioritise and exploit the different aspects of its matters of expression.
>
> (Guattari, 2011: 239–240)

In the second section of "Refrains for Lost Time" in *Machinic Unconscious*, Guattari studies nine of the "little phrases" that constitute the different assemblages of enunciation in Proust's *In Search of Lost Time*. Elsewhere Guattari describes these phrases as a study of the oscillation between the refrain (and its deterritorialising function) and faciality (its reterritorialising function), which represents the labour of the conversion of the subject (Guattari, 2016: 230). Among other things, these studies of the nine assemblages are methodological descriptions of the circumstances and context of the scenes in which the 'little phrase' comes into being. Now these are useful in giving a traditional orientation to the actuality of how such a world exists, not in just fiction but also in everyday life. A stress on the quixotic nature of what is presented in lieu of, and in excess of, what is observed echoes Guattari's broader critique of the simplification of our understanding of human behaviour. There is something else, equally or even more interesting, presented in these studies; namely "the matters of expression", and their "most significant micro political result", "which they set to work" (Guattari, 2011: 241). Twisted out of the deadening scaffolds of language that is both untraditional and disorientating, potential meanings emerge in ways that are productive, machinic, and novel. There is as yet no consciousness of 'something'; the unconscious is not a country to be discovered, it is one or several territories being produced, and it takes 'a year' (2011: 240) or 'several years' (2011: 260) or 'regular returns' (2011: 253) to germinate. Of the nine assemblages presented I want to draw out details from just two, that of "*a salon, the previous year: the proto-phase or system of fac-simile*" (2011: 243–246) and "*an unforeseen exit at the Marquise de Saint-Euverte's home: the reactivation of the refrain and the resorption of the black hole*" (2011: 260–276).

This first assemblage of enunciation "a salon, the previous year" has Swann, the protagonist of *The Search of Lost Time*, in the presence of the little phrase of music for the first time. Or, rather, we should say in the presence of the little phrase of music this time, since a sense of primacy is redundant when all coordinates of association are as-yet undetermined. This first assemblage of enunciation signals that something is coming into the world: a new set of relations pivoted through a human body, or through a series of sounds, or as a moment in literature forecasting new diagnoses of meaning that are still unravelling. It is in celebrating the moment of unravelling that the point of the facsimile becomes apparent. Any conclusions made at this stage could not be placed on the event of music itself "but merely upon certain equivalents, substituted (for his mind's convenience) for the mysterious entity of which he had become aware" (Proust, quoted in Guattari, ibid.). What is important here is that the process of becoming aware is necessarily without definition and as such this assemblage, like all the others to come, acts as "the instrument of production of a different reality, a machine catalysing new semiotic components" (Guattari, 2016: 227). If there is a brutality to the overload of new conceptual vocabulary and co-ordination in Guattari, it is because we have to challenge the "neurotic rituals" of so many "sociological inertias" (ibid.) that suck the intensity of life out of things, constraining thinking as a tired representation of a humanist world.

The secreted sounds in which the 'little phrase of music' are first presented act as kind of signature of a Guattarian social scientific enterprise. An enterprise deserving of the name Guattarian will initiate with deterritorialisation, so as to constitute itself from a-signifying redundancies and, finally, sustain the announcement of a rhizomatic opening, which challenges "a whole micropolitics of conformity to the dominant realities" (Guattari, 2011: 244). These demands of working with Guattari describe the very "vehicle of abstract machinism" that is the little phrase of music, and that is crucially also the very project of Guattari's thought. If you cannot work with such abstraction, don't quote Guattari to bolster your argument: you are compromising both yourself and Guattari. In this first assemblage, it is an abstract machinism that advocates "one of those invisible realities" of force, of desire, as that which dictates the life that is being produced because of its advent. No scientific paradigm of evidence, no sociological justification of appropriateness, and no artistic evaluation of exemplification is welcome here. The ethico-aesthetic paradigm that Guattari counterpoints to these is precisely about experimenting with modes of thinking more sensitive to life's creation (Hynes, 2013). The point in all of this is to criticise existing inherited models of interpretation to emphasise events "as the bearers of the potential for automodelization that yield new maps of subjectivation" (Genosko, 2015). Guattari pushes this many times: don't binarise the assemblage nor aborify the problem (after Guattari, 2011: 149). We cannot have clearly defined referents as there are only registers of co-existence and crystallisations of intensity that parry between deterritorialising and reterritorialising tendencies (Guattari, 1995: 30).

To detail the productive aspect of these oscillating tendencies, the other assemblage I want to look at is the fifth: *an unforeseen exit at the Marquise de Saint-Euverte's home: the reactivation of the refrain and the resorption of the black hole* (Guattari, 2011: 260). Much time has passed in the narrative of *In Search of Lost Time*, as in Guattari's mobilisation of this time passing, and the matters of expression emanating from the first enunciation of the little phrase of music have since proliferated like the processes of mitosis in cellular production – "becoming-child, becoming-wasp, becoming-orchid, becoming-landscape . . ." (Guattari, 2011: 263). There is no appeal to an authentic real, to a nostalgia for the original instance of the little phrase of music, as all the terms involved have irrevocably changed. Demeaning the satisfaction of having a 'fragile anchorage in external reality', and thus critiquing a simple empiricism, Guattari draws out the production of another reality which sees Swann led "towards the paths of creation" (268) where "other interests will be awakened in him" (269). Here it is not a backward looking symbolic mechanism seeking representational meaning in what is happening, but rather a forward looking abstract machinism, which is using what is happening to change the means by which we make things mean. At the heart of this assemblage is a significant and summary argument that really underpins Guattari's whole project:

> The opening to the real, along with creative and perceptive affectivity, is subservient to the employment of matters of expression in their disparity, in their heterogeneity. What singularizes such and such a trait of a matter of expression is its nature of excess in relation to the significant redundancies; it is the fact that a difference can never be completely recuperated by any correspondence grid. What passes from one component to another is not simply a 'message', a measurable quantity of form or information, but an effect which escapes from the dominant significations
>
> (Guattari, 2011: 270)

Thus, the unconscious is precisely about the production of another reality. It exists in those singularising events of thinking that escape meaning – the significant redundancy – and leaves us as disparate and heterogeneous and consistently unmade. If we can allow ourselves to remain with this disparation and heterogeneity we can experiment with becoming different through the simultaneous associations that our current matters of expression determine, or indeed, through the alternative realities of being simultaneously sexual, musical, and social in the now fragmenting correspondence grids that Guattari's reading of Proust provides.

Consequentially so: abstract machines for subjective reversal

Thinking something new, as in both the modest sense of the singularising events of thinking that become us, and in the grand sense of thinking social science differently for another world of possibilities, is perhaps only truly possible if you remain open to the force of its deterritorialisation. It is a matter of nurturing what deterritorialisation does to the dominant representational schemas and institutional bureaucracies that define our age, our culture, and our politics. In each instance of thinking something new there are matters of expression that can be put into play within different assemblages creating refrains of themselves within new contexts. These refrains bring new disparate and heterogeneous associations that drive and describe the abstract machinism at work in Guattari. The concept of machinism is an act of abstraction, not towards a logic or system of ideas, but as that which abstracts from certain stratifications: that phrase of music that evokes an invisible reality that is supposedly redundant to the significance of our world; that grin that heralds so many past-future desires and yet is nothing more than what it was. There are many legitimate categories of meaning to give such events, but what Guattari's diagramming through Proust shows, as does the art of Proust itself, is that these categories (in the world of *In Search of Lost Time* the sexual, musical, and social) are productive on condition that they are not envisaged "as eternal entities irreducibly separated from one another" (2011: 308). Music, sex, and the socius are always present on the same scene, but they are awaiting an assemblage to make them work together. It is in this regard, that Guattari would argue that *In Search of Lost Time* is not directed towards the past, as a kind of psychoanalytical anamnesis, "but towards a construction of things to come, towards a proliferation of the future in the act of creation" (2011: 313).

If the social sciences were to become another laboratory akin to the arts and equipped with the means to express the machinic unconscious it could do worse than start emulating how the science of the individual in the space of literature operates as a creation machine changing the perception of the cosmos. This would involve not seeing change in the linear sense but via approximation, advance and retreat; by highlighting alternative assemblage pathways that loop, and just might, one day, trigger that crossing of the threshold of their "affective enactment", that miraculous split second that then compounds and produces the matter of expression, which acts as "the diagrammatic stimulus", that reframes the sense of the subject and the understanding of the cosmos, and then the cosmos as such (Guattari, 2011: 319). It would not orientate its understandings first through the individual either, as Guattari, following Proust, regards the individual as a trope of individuation that he takes and leaves to the chance development of the rhizome of subjectification. Thus, Swann, the Narrator, Proust, you the reader, all are the subject "individuated, collective, or machinic" that "captures refrains within fields of all sorts, assembles heterogeneous matters of expression beginning from which it produces times, spaces, significant redundancies, and above all else the components of subjectification" (2011: 323). There is no individual subject to orientate this meaning making, that is why for the building blocks of such subject orientations – categories, forms, durability – Guattari substitutes rhythms, intensities, and their proliferation. In the end, then, to think something new in this way is to fundamentally ask us as social scientists to check both the categorical coordinates and formulas we use to determine our problem, project, or understanding and the coordinates of our subjectivity as produced in the resolution or achievement that results. If there is no sense of "a subjective 'reversal'" involving "the entry of a new type of refrain that works to empty the dominant redundancies of their content and deterritorializes and disindividuates the enunciation" (320), it is not a question of failure but of needing to ask yourself as to why you have allowed the status quo to succeed. That status quo is all the things you fear it is, whether that be in your day-to-day rhythms, the people you associate with, the politics you rail against, or, indeed, the philosophers you read.

Notes

1 This is not meant to be a hopeful sentiment with tendencies to defer action increased, rather it is demanding action based on different expectations of potential.

2 To be clear, this is about affirming surprise itself; this is not about indifference nor about being sad with yourself for not knowing what was about to happen. It is a critique of no longer considering "any event except in terms of its probability of occurrence", of proclaiming "that everything must tend towards a state of equilibrium" necessarily evolving "towards a reduction of tensions and disorder" (Guattari, 2011: 13).

References

Deleuze, G. and Guattari, F. (1987) *A Thousand Plateaus: Capitalism and Schizophrenia II* (trans. B. Massumi). Minneapolis, MN: University of Minnesota Press.

Genosko, G. (2002) *Félix Guattari: An Aberrant Introduction*. London: Continuum.

Genosko, G. (2015) Micropolitics of hope. *Spheres: Journal of Digital Culture*, 2: n.p (online only).

Gerlach, J. and Jellis, T. (2015) Guattari: Impractical philosophy. *Dialogues in Human Geography*, 5(2): 131–148.

Glowczewski, B. (2015) Resisting the disaster: Between exhaustion and creation. *Spheres: Journal of Digital Culture*, 2: n.p. (online only).

Guattari, F. (1995) *Chaosmosis: An Ethico-Aesthetic Paradigm* (trans. Paul Bains and Julian Pefanis). Indianapolis, IN: Indiana University Press.

Guattari, F. (2011) *The Machinic Unconscious: Essays in Schizoanalysis* (trans. T. Adkins). Los Angeles, CA: Semiotext(e).

Guattari, F. (2012) *Schizoanalytic Cartographies* (trans. A. Goffey). London: Bloomsbury.

Guattari, F. (2016) *Lines of Flight: For Another World of Possibilities* (trans. A. Goffey). London: Bloomsbury.

Hynes, M. (2013) The ethico-aesthetics of life: Guattari and the problem of bioethics. *Environment & Planning A: Economy and Space*, 45(8): 1929–1943.

Jellis, T. (2014) Schizoanalytic cartographies, by Félix Guattari. *Society and Space*. Available online: http://societyandspace.org/2014/07/19/schizoanalytic-cartographies-by-felix-guattari-reviewed-by-thomas-jellis/

Lapoujade, D. (2017) *Aberrant Movements: The Philosophy of Gilles Deleuze* (trans. Joshua David Jordan). Los Angeles, CA: Semiotext(e).

Massumi, B. (2002) *Parables of the Virtual*. Durham, NC: Duke University Press.

Negri, A. (2017) *Marx and Foucault* (trans. E. Emery). Cambridge: Polity Press.

Polack, J-C. (2011) Analysis, Between Psycho and Schizo. In: Alliez, E. and Goffey, A. (eds.) *The Guattari Effect*. London: Continuum.

Proust, M. (2003) *In Search of Lost Time: Volume 1 the Way by Swann's* (trans. L. Davis). London: Penguin Classics.

Sharpe, S. (2014) Potentiality and impotentiality in J. K. Gibson-Graham. *Rethinking Marxism*, 26(1): 27–43.

Part II

Ecologies

Ecology's canopy, as an umbrella term addressing the multi-faceted ways organisms are related to their environment, generally occludes the nuance and definition of the very terms of relation that make it a meaningful concept. Guattari's thought of ecology ran counter to this 'normal' ordering of organic beings shaping and being shaped by a material world. Nonetheless, in a rare moment of conceptual simplicity, Guattari presented three ecological registers – "the environment, social relations and human subjectivity" (2000: 28) – to address "ways of living on this planet" through the creation of "new existential configurations" (2000: 45). As three interrelated ecologies of natural, social, and mental forces, this triumvirate foregrounds the equal presence of social and mental deficiencies alongside that of the physical evidence of so many natural catastrophes presented in equally numerous newscasts. Indeed, isn't it the failure of our social ecologies to provide infrastructures for organising lives in better synergy with the natural world? And isn't it the failure of our mental ecologies to prevent the stupid thinking that aligns our desires with faulty aspirations and global warming, that scripts the catastrophes of climate change, inequality, and resentment?

The power of Guattari's thinking on ecology comes in his argument that ecological crisis should not be exclusively read and resolved through concerns with "visible relations of force on a grand scale" but must equally "take into account molecular domains of sensibility, intelligence and desire" (2000: 28). Therefore, whilst the three ecologies are related in Guattari's associative thinking,[1] the force of thinking the three ecologies together is in the emphasis of tackling global problems through micro-transformations. Guattari is quick to rebuke any tendency towards using ecology as a conservative term, disdainfully commenting that ecological movements are often reductionist when they need to be heterogenetic (see Guattari, 2011: 38). Many of the chapters here are cautions against putting the concept of ecology to the service of maintaining already existing structures.

Although Guattari had hardly used the term 'ecology' until the late 1980s, his earliest work effectively focused upon mental and social ecologies (Watson, 2009: 182). Perhaps the more explicit mode of ecological thinking in Guattari had its roots in Gregory Bateson's idea of an 'ecology of the mind' (Glowczewski, 2011: 99) following Guattari's attendance of a conference on Schizoanalysis in 1985 entitled "Topia: Centre for the Ecology of the Mind"; indeed, Beradi calls schizoanalysis "an ecology of the mind" (2008: 146, 114). Such rooting is rhizomatically remodelled by Guattari toward the construction of ecologies of machinic assemblage full of material and semiotic fluxes. The principal aim, and energy behind Guattari's use of the concept of ecology, is, like schizoanalysis, for the liberation of the unconscious. The fluxing boundaries of such ecologies billow in and out repositioning the points where interior and exterior folds meet in constructing subjectivity; thus, we can see group formations, transversal associations, as equally the core to what we would call the subject (O'Sullivan, 2006: 91). This play on expanding our thinking of the production of subjectivity is underwritten by the sense that thinking ecologies in Guattari's terms means learning to think 'transversally' (2000: 43). The concept of transversality (Guattari, 2015) captures "the communication and relationship of the unconscious" (Deleuze, 2000: 188n.5) allowing us to explicate the transformation from one ecology to the next without undermining the specificity of each.

Taken together the three ecologies comprise "an ethico-political articulation" termed 'ecosophy' (Guattari, 2000: 28); "a pluralist ontology" (Guattari, 2011: 38); and "an environmental consciousness adequate to the technological complexity of late modernity" (Berardi, 2008: 34). In continuing the lines of ecosophy in *Chaosmosis*, Guattari figures in an ecology of the virtual. Here we have the ecologist as a defender of incorporeal species – poetry, music, the plastic arts, the cinema – whose qualities propose a new paradigm of reference to aid the composition of new social and analytic practices (1995: 91), that enact the ecosophical responsibility for being as creativity (2011: 38).

> It is quite wrong to make a distinction between action on the psyche, the socius and the environment. Refusal to face up to the erosion of these three areas, as the media would have us do, verges on a strategic infantilization of opinion and a destructive neutralization of democracy.
>
> (Guattari, 2000: 41)

Ecologies are also a diagnostic tool for assessing our contemporary condition. We can better understand why an impoverished social ecology "leads to the promotion of charismatic leaders" when linking it to the impoverishing mental ecologies that depend "on stereotypical order-words" (Guattari, 2000: 34). It is easy to find social relations based on "impotent catchphrases of resentment" (Guattari and Negri, 1996: 252), and just as easy to find the production of mental ecologies manufactured by mass media "synonymous with distress and despair" (Guattari, 2000: 34). As much as we see plastic in the ocean, further floods, fires, and drought, we see dead migrants on beaches, increasing homelessness, gender inequality, and suicidal youth, and we see infantilisation on all our screens, proliferating non-communication, miscommunication, and simply stupid communication. The rangers of these ecologies need skills to act against the grain of this zeitgeist, namely skills like being versed in the recomposition of collective assemblages of subjectivity (2011: 29) and in trapping the danger of those deathly ecologies sprouting from philosophies like Heidegger's (Guattari, 2011: 37).

In the chapters in this section we have many of these ecosophic skills in evidence, where we see the question of subjectivity installed "simultaneously in the realms of the environment, in the major social and institutional assemblages, and symmetrically in the landscapes and fantasies of the most intimate spheres of the individual" (Guattari, 2000: 69). The ethos of Guattari's three ecologies is very much in evidence too, with critical arguments made against the capitalist ecology of innovation, the microfascism of paranoid politics, the production of marginal subjects, the clichéd markers of community engagement, and the groundedness of traditional research convictions – all of course, and to give Guattari the last word, "to counter the pervasive atmosphere of dullness and passivity" (2000: 69).

Maria Hynes skillfully syntheses Guattari's schizoanalytical method to offer a different account of the burgeoning institutional ecology of art-science as arranged in our sloganised times where everything is a part of our age of innovation. Rather than simply bemoaning this state of affairs, Hynes pushes aside "the stultifying cliches of the innovative paradigm" to give us space to think for the "moments of genuine transversality" that are also being brought into being (this volume: 106). Central to her arguments is the challenge to live by a different ecology of valorisation, one where we see multiple expressions of "a much more dynamic libidinal economy" (this volume: 114). Focusing in particular upon Guattari's short essay on "Mary Barnes' Trip", Hynes uses the practical analysis of desiring production to be found here to set out the transversal relations that are behind the creation of mutating subjectivities and their assemblages of enunciation that institutions have the potential to give rise to. Resisting the scientific paradigm's desire for interpretation throughout, Hynes's chapter gives us much needed energy to see the "flows of creativity and value" in the machinic enterprise of our art-science ecologies.

Mahoro Murasawa and Stéphane Nadaud find in Guattari's concept of 'ecosophy' a way of thinking the necessary subjective revolution for a non-fascistic living in the contemporary world. Arguing for a revolution that is molecular rather than contra, they use ecosophy as a conceptual tool to fragment the subjectivity of the Japanese people of Tohoku in the political and social situation after the Fukushima nuclear disaster in 2011. Central to Murasawa and Nadaud's argument is an avocation for an affective ethos composed from these fragments that they term 'subjective micro-cosmoses'. Operating their analysis through an orientation to schizophrenic processes, they stage the struggle for the mental transformation of Japan in the aftermath of Fukushima between the philosophical rather than the medical understanding of the neurotic as macropolitical and the psychotic as micropolitical. In rendering these mental pathologies apart, they emphasise the microphysics of quanta, modulation, and dissonance to show how a "localised change – in a domino effect (but with this difference, the imitations are not identical to the original invention) – will modify the entire society" (this volume: 129). In thus taking seriously the physics at play in Guattari's molecular revolution, they show how to effect a subjective escape from the paranoiac machine of scientific-political-economic discourse.

Michele Lancione investigates the performance of the subject through thinking through Guattari's machinic dimensions of subjectivation in marginalised urban communities. Implicit to his arguments, Lancione links the critical force of Guattari's ecological project to spotlight 'anti-capitalist politics at its roots' and its role in the production of specific assemblages of 'marginal' subjects. Writing out of recent methodological debates in Geography Lancione operates performative research pathways that implicitly trace the contours and caustic aspects of our social ecologies, and the different ways they produce us as subjects, through the production of his own research subjectivity. Vital to his enterprise, in rendering visible the abstract machines of capitalism, is the art of "recollecting performances in order to challenge – through the fine and gritty nuances of the social field – the established notion of the 'truth'" (this volume: 142). Whilst Lancione's performative angle opens up multiple mappings of association, capturing the "art, activism, and engagement" (this volume: 137) of ecosophy, his ethnography perspective immersively captures some of the overlooked forces at play in the contemporary social ecologies that define people's place in the world.

Rebecca Catarelli beautifully stages her essay through a chaos-rhythm of poetic refrains that at once exemplifies both the ethos of Guattari's ecosophic project and its conceptual critique of our general incapacity to "mobilize our minds, sensibilities, and wills" (Guattari, 1995: 119) to the world's suffering. In Caterelli's chapter, this suffering is a hurricane: its event, its memorialisation, its reduction to natural disaster, its betrayal in being named, to name a few of the 12 refrains of the hurricane staged. Underneath her careful prose, Caterelli echoes Guattari's caustic attack on the cynicism and infantilisation of capitalism; she questions the trend towards barer relationality, general relations poisoned marked as they are by the values of consumption, personal relations ossified by cliché relationship evaluation, and miserly time shared in community engagements. The refrains themselves are the experiment of which she talks throughout, and as experimental entanglements, threads, and tall tales, they find, desire, and produce new territories or existential refrains, becoming new modes of intervention in themselves. The ecologies that get stitched together, often of accident, bifurcation, and drift, suggest things no longer as certain as they had once appeared. Together these refrains present some of the logics of ecosophy in expressing productive points of negotiation between complexity and chaos, and, in underlining the intensities of the moment of sensation, affinity, and impression, present a way of thinking through and with Guattari's three ecologies.

Sasha Engelmann presents flight as an atmospheric journey both literal and poetic, and one which acts as a moving ecology of practices "that detach and deterritorialise a segment of the real" (Guattari, 1995: 131), ungrounding the certainties of researchers as they relearn how to float. In tracing out the lines of flight that come together in the aesthetic-research into *Aerocene Gemini*, Engelmann's chapter uses a personal account of aerostatic snapshots to show the many "aesthetic relations between science, technology, art, and the social" (this volume: 160). The project that she outlines shows that it is a sense of mutual urgency that underwrites "the forces of obligation, attachment and – crucially – imagination" (this volume: 164) that seems so essential to the successful enunciative territories that Guattari hoped ecosophy would help us achieve. Holding onto the tethers of friendship, thrill, risk, and love, Engelmann finds her orientation with Guattari through his devising of "a vocabulary for such alter-adventures in practice, sensing, and politics" (this volume: 164). And in presenting this orientation as a new aspect in the ecology of practice that is research into art, science, and politics, which Engelmann terms 'cosmological aesthetics', her chapter gives us a means for grounding ourselves on our next disorienting research journey.

Note

1 For some even operating in a scalar manner: "the environment corresponding to the macroscopic, mental ecology to the molecular, and socius to an intermediary scale" (Watson, 2009: 183). This perhaps dilutes, however, the point of the fluidity and different emphasis of processual thinking that Guattari offers us in his late work.

References

Berardi (Bifo), F. (2008) *Félix Guattari: Thought, Friendship and Visionary Cartography* (trans. G. Mecchia and C.J. Stivale). Basingstoke: Palgrave Macmillan.

Deleuze, G. (2000) *Proust & Signs* (trans. R. Howard). London: The Athlone Press.

Glowczewski, B. (2011) Guattari and anthropology: Existential territories among Indigenous Australians. In: Alliez, E. and Goffey, A. (eds.) *The Guattari Effect*. London: Continuum. pp. 99–111.

Guattari, F. (1995) *Chaomosis: An Ethico-Aesthetic Paradigm* (trans. P. Bains and J. Pefanis). Sydney: Power Publications.

Guattari, F. (2000) *The Three Ecologies*. London: The Athlone Press.

Guattari, F. (2011) The vertigo of immanence: Interview with John Johnston, June 1992 (trans. A. Goffey). In: Alliez, E. and Goffey, A. (eds.) *The Guattari Effect*. London: Continuum. pp. 25–39.

Guattari, F. (2015) *Psychoanalysis and Transversality: Texts and Interviews 1955–1972* (trans. A. Hodges). Los Angeles, CA: Semiotext(e).

Guattari, F. and Negri, T. (1996) Communist propositions. In: Genosko, G. (ed.) *The Guattari Reader*. Oxford: Blackwell Publishers. pp. 248–258.

O'Sullivan, S. (2006) *Art Encounters Deleuze and Guattari: Thought Beyond Representation*. Basingstoke: Palgrave Macmillan.

Watson, J. (2009) *Guattari's Diagrammatic Thought: Writing Between Lacan and Deleuze*. London: Continuum.

7 The (schizo)analysis of value in the 'Age of Innovation'

Maria Hynes

At a time when a well-branded concept is celebrated as an innovation, it is tempting to consider this the 'Age of Innovation'. We daily hear talk of innovation nations and innovation economies, and everywhere the value of innovative ideas, technologies, and practices are extolled. Amidst the hyperbolic claims made in the name of innovation, what is evident is that contemporary economies are mining ideas with the same fervour that they had previously mined the earth. New institutions are born – innovation hubs, networks, and virtual ecosystems – with an eye to cultivating the kind of problem-based research that universities are still learning to produce. The conventional bounds set by disciplinary thinking are deemed a hindrance to the innovation paradigm. Such a paradigm recognises experimentation in the production of both ideas and things as value-adding at the level of process, as well as outcome. The interdisciplinary hybrid now familiarly known as 'Art-Science' is a case in point, with the meeting of art and science increasingly acclaimed as precisely the kind of cross-perspectival thinking necessary for solving complex problems (Wilson et al., 2014). It is the disparate nature of art and science that is celebrated in popular formulations of such interdisciplinarity, since heterogenous knowledges are identified as the source of the 'creative leaps' so crucial to innovation (Hacklin and Wallin, 2013).

Of course, we know how readily such interdisciplinarity can be reterritorialised in line with those uncompromisable social axioms that make the opportunistic mutations of capitalism work. Art finds itself the handmaiden of science; discovery is reduced to the completion of a picture; partnerships in research become synthetic affairs oriented to the transcendent universals of science.[1] And a mere novice in critical thinking can quickly deconstruct the very notion of innovation as the wish fulfilment of laissez-faire entrepreneurialism; that which is innovative is that which is profitable, useful, exchangeable. Yet to the extent that the pseudo-scientific paradigm that has dominated in the social sciences prevails, the evaluation of these new social forms will invariably succumb to either "progressivist illusions or visions which are systematically pessimistic" (Guattari, 1995: 5), missing much about the peculiar economy of desire and value expressed in the notion of innovation and materialised in its emerging social forms.

In seeking to understand the kinds of emerging institutional and social arrangements exemplified in Art-Science, we could learn much from the schizoanalytic method that Guattari developed in his sole and joint authored work with Deleuze. In the many contexts in which the notion of schizoanalysis is deployed by Guattari and Deleuze and Guattari, the relationship between the psychoanalytic method and capitalism, with its contrary schizophrenic and paranoid tendencies, is at the fore (Guattari, 1995, 2000; Deleuze and Guattari, 1983; Guattari and Rolnik, 2008). In the light of these inharmonious tendencies, schizoanalysis functions as a way of avoiding those "petrified systems of modelisation", which characteristically refuse such dynamism and err on the side of the paranoid reassertion of obsolete forms (Guattari, 1995: 68). While Guattari's own context was clearly a clinical one, the method of schizoanalysis represents, more generally, an orientation toward the "chaotic vertigo" which, while it finds its "privileged expression in madness", is the very engine of ontologically heterogenous reality (Guattari, 1995: 77). Schizoanalysis seeks to make "direct contact with the unconscious in the social field", which is to say, with the interrelational and intensive fabric of life (Guattari, 2009: 251). In its orientation toward a differential ontology, the schizoanalytic method shares transcendental empiricism's concern with orienting method, not to the model, but to the encounter with the groundlessness of the ground (see Todd and Hynes, 2017). Or, rather, schizoanalysis is, as Ian Buchanan (2013: 163) notes, a form of 'meta-modelling', in the sense that it "tries to grapple with the realm of 'what might happen' that constantly dogs the realm of 'what is happening'". As such, the schizoanalytic method provides a means of analysing institutional and extra-institutional mutations, mapping their emergent assemblages of enunciation, the processes of subjectification to which they give rise and the fields of future possibility they engender, within and beyond the dominant imperatives of valorisation.

Insofar as innovation involves, firstly, the production of something new and, secondly, the production of something new that has value (Gibson, 2014), what is at stake is how we might generate novel modes and forms of life that can find sufficient context within existing and emerging structures, habits, and refrains to actualise a surplus, which is productive of new territorialisations and subjective possibilities. While the task of innovative capitalism is to extract a surplus from the deterritorialised flows of capital and labour, there is an immanent processual potential that remains strictly irreducible to dominant reterritorialisations. Put differently, while innovative capitalism is oriented to the exploitation of the creativity of the commons, it runs the risk of abetting a potentiality which escapes privitisation and aids liberation (Lazzarato, 2014). And this immanent and irreducible potential might, to the extent that it operates within an ethico-aesthetic paradigm, push beyond the "dominant redundancies" that normalise the production of subjective and objective worlds (Guattari, 1995). For all the stultifying clichés of the innovative paradigm, there will be moments of genuine transversality, in which new universes of reference are created and new existential territories brought into being. These moments of machinic heterogenesis involve flashes of novelty which demonstrate that "the work of capitalism and scientific paradigms – their ongoing attempts to tame affective intensity – will never be completed" (Bertelsen and Murphie, 2010: 154).

Between the scientific and ethico-aesthetic paradigms

Acclaiming the inventive potentials of individuals and the rewards of interdisciplinary collaboration, the innovation paradigm pursues a new image of the institution, promising with it, a veritable unleashing of creativity. While emerging institutions of art-science, for example, have plural genealogies, from public engagement to more ontologically experimental ventures (Born and Barry, 2010), each is fashioned within this undertaking of institutional remodelling and the release of creative potential. But what this discourse of innovative collaboration obscures is the sense in which these emerging institutions are factories in the production of subjectivity. Discourse on innovation describes an objective process of liberating human creativity towards the solution of ideal and material problems. Yet, such discourse is, of course, the effect of a collective assemblage of enunciation – "a collective saying (*dire collectif*) that brings together machinic elements of every kind: human, semiotic, technical, scientific etc." (Guattari, 1996: 152). And the signs and things associated with the name of innovation ultimately "combine with one another independently of the subjective 'hold' that the agents of individuated enunciation claim to have over them" (Guattari, 1996: 151).

While Maurizio Lazzarato (2014) is right that an entrepreneurial mode of subjectivity is the privileged form within this subjective production, Guattari's schizoanalytic method better grapples with its less determinant dimensions. While it may be true that the processes of capitalist innovation and the associated production of techno-objects serve as routes to the attainment of subjectivity in the social, an understanding of the actuality of intensities and relations of force cannot take refuge in any existing model of the money economy of capitalism. There can be, as Guattari (1996: 51–52) suggests, no analytical unravelling "of the micropolitical conflicts in which the subject is imprisoned" under capitalism, nor does one discover the secret of a primal libidinality, since the unconscious has to be built, albeit within the conflictual tendencies of capitalist society. The method of schizoanalysis, then, is certainly not a romanticisation of the psychology of the schizophrenic, but is an effort to elaborate upon the impact of the quantitative calculations of the market on psychic and social functioning. What is at issue here are the schizophrenic tendencies of capitalism as a semiotic operator. Eugene Holland (1999: 2) writes:

> In a first approximation, then, we could define schizophrenia as a form of "unlimited semiosis" – in the psyche as well as throughout society – that emerges when fixed meanings and beliefs are subverted by the cash nexus under capitalism. . . . What is at issue is not the reduction of praxis to cash, but the ways in which the schizophrenic tendencies of capitalism are overcoded by the more paranoic ones. . . . Put differently, the ways in which unlimited semiosis clashes with the resuscitation of obsolete, or traditional, belief-based modes of social organisation.

Of course, many of Guattari's insights on the transversal relations between diverse social domains stem from his own clinical practice. Indeed, it is here that he explores pragmatically the tension between the ethico-aesthetic and scientific paradigms that is evident today in the emerging institutions of art-science, while being strictly irreducible to the disciplinary encounter between art and science. With this in mind, I want to pursue the analysis of contemporary innovation economies via a brief discussion of a peculiarly 1960s experiment in the unleashing of creative energies, which Guattari outlines in his short paper "Mary Barnes' Trip", originally published as *Le Voyage de Mary Barnes* in 1973.

"In 1965, a community of about 20 people formed around Ronald Laing . . ." (Guattari, 1996: 46). Thus begins Guattari's fascinating analysis of the events of Kingsley Hall, East London, where R.D. Laing's anti-psychiatry movement embarked on a radical counter-cultural experiment. Guattari (1996: 46) explains that "over the course of five years the leaders of the anti-psychiatry movement and the patients who, as they say, 'made a career of schizophrenia', collectively explored the world of madness". This was not, he stresses "the madness of the mental hospital, but the madness that dwells in each of us", and the point of the Kingsley Hall experiments was not to cure such tendencies toward delirium but to liberate them "in order to release inhibitions and symptoms of every kind" (Guattari, 1996: 46). The Kingsley Hall community, as it conceived itself, was not insensible of the significance of its abode. As one of the founding members, psychiatrist Joseph Berke, recorded in his diary, "Kingsley Hall had a long and honourable history as a centre for social experiment and radical political activity", among which the six-month period in which Mahatma Gandhi shared a room with a goat was considered exemplary (Barnes and Berke, 1973). Happily burdened by this auspicious history, the 1965 residents undertook an audacious ethico-aesthetic experiment, in which the physical state of the three-storey brown-brick building served as the best barometer of the good or bad health of the community. When the kitchen was "clean, or freshly painted, the milk bottles taken down, the garbage removed and food stacked high in the larder, then things were going well"; when "the walls looked like a Jackson Pollock abstraction, but with egg yolks substituted for paint . . . then matters weren't too good" (Barnes and Berke, 1973: 227–228). Either way, in this adventure into and out of a kind of collective psychosis, led, it seems, by the self-appointed "guinea pig", Mary Barnes, the variable health of the community was grist to the mill (Barnes and Berke, 1973).

Without evident malice, Guattari registers the shortcomings of this genuinely practical research into the collective dimension of desiring production and the liberation of desire. Despite the efforts of Laing and his colleagues to break with the structures of institutionalised psychiatry, they ultimately fall prey to the paranoia peculiar to the scientific paradigm, within which their own habits of thought and practice had been nurtured. Efforts to disregard the division of labour "among patients, psychiatrists, nurses, etc." are undermined by the persistence of the interpretative screen that is cast over every thought and action of the residents, which reins the chaos of delirium back into the comfort of the interpretative framework and its orientation to familial structures. Yet, paradoxically, Guattari notes, this very paranoia threatens to unleash the delirium that it aims to manage. Attentive to the necessity of being there with desire in its very unfolding, one of the psychiatrists regresses with his patient "in a way that is still rare among typical psychoanalysts: he grunts, acts like a crocodile, bites and pinches her, rolls her in her bed" and in doing so almost opens up to the unlimited semiosis that interpretation had aimed to keep at bay (Guattari, 1996: 50).

Ultimately, however, it is not the analysts who held sway at Kingsley Hall, and Guattari makes much of the way that his paper's namesake, Mary Barnes, has both the other patients and the practitioners running in circles in response to her infantile regressions and the "vampirism of 'human warmth'" that she cultivates in all her relations. Mary is an artist and whether she covers the wall with paint or with shit, what is important to her is that she be allowed to be expressive, that she not "push her feelings down" (Barnes and Berke, 1973).

Yet, it is the brief mention that Guattari makes of Mary's brother, Peter, that I find especially catalytic. Guattari tells of how, in response to his sister's various harassments, Peter settles in Kingsley Hall, but he loathes its "bohemian style": there is, for him, "too much noise, too much chaos, and besides, he wants to hold onto his job" (Guattari, 1996: 46). Where the rest of the community "seems to live on air", being more preoccupied with the mad "flows of shit, piss, milk and paint" than with the flow of money, Peter's madness is considered by Guattari to be "far more disturbing" and ultimately far more radical. Guattari (1996: 48–49) writes:

> Here perhaps one can understand the difference between a real schizo trip and a petit-bourgeouis style of familialist regression. The schizo is not all that interested in "human warmth". His dealings are elsewhere, on the side of the most deterritorialized fluxes: the flux of the "miraculating" cosmic signs, but also the flux of monetary signs. The schizo is not unaware of the reality of money – even if it is put to extraordinary uses – any more than he is unaware of any other reality. The schizo does not behave like a child. For him money is a reference point like any other and he needs to have at his disposal a maximum number of systems of reference, precisely to enable him to keep his distance.

Amidst this, at least apparently chaotic, anti-institutional experiment, Guattari (1996) espies in the figure of Peter the possibility of a genuinely schizo-becoming. While the other residents and staff of the hall immerse themselves in the web of essentially familial relations characterising their avowedly anti-institutional enterprise, the "real schizo trip" proves to be a far more singular undertaking. Because of course the anti-institutional stance cannot prevent the emergence of the institution that Kingsley Hall turns out to be, though the manner in which the experimentation with desire is conditioned by this emerging institution must remain opaque to those who refuse its existence.

In his introduction to the *Guattari Reader* in which "Mary Barnes' Trip" is published, Gary Genosko (1996: 16) discusses the role played in Guattari's thought by conceptual "lightning conductors": events, thought experiments, titbits, and tales that spark off connections, strike to the heart of a problem, and energise its transformation into a new form. Fittingly perhaps, Guattari's account of the experiments at Kingsley Hall is my lightning conductor. In attempting to consider what kind of analysis the art-science institutions of today's innovation economies demand, the scenario that Guattari describes opens up conceptual possibilities, not through the synchrony of a structural analysis but through the transversality of a machinic one. More on that later. For the moment, let us leave Peter and his fellow residents of Kingsley Hall in history, so as to paint the contours of a significantly different and very 'now' social form, with its distinct enunciative assemblages and panoply of subjectivities.

From the entreuprenurialism of the familiar to experimentation with desire

At an institution already much vaunted for its innovative thought and practice, fantastic forms and plastic possibilities are produced through the meeting of art and science: fully recyclable and multifunctional structures are made out of chitin, the most abundant polymer on the planet, printed with a robotically controlled extrusion system, and used for architectural purposes; new lifeforms, "computationally grown, additively manufactured and biologically augmented", are formed into sustainable clothing suitable for interplanatary voyages (Oxman et al., 2015); sculptural helmets are 3D printed from algorithms modelled on the natural suturing processes in skull tissue; a silk pavillion is spun by 6,500 computer guided silkworms in a "unique hybrid of structural and biological engineering" (Flaherty, 2013); a photosynthetic, wearable digestive system is designed not to mimic nature, but to augment it. Working out of the MIT Media Lab, Sony Corporation Career Development Professor and Associate Professor of Media Arts and Sciences Neri Oxman and her team in the Mediated Matter research group combine the principles of computational design, additive manufacturing, synthetic biology, and materials engineering to produce objects and architectural forms that are biologically informed and digitally engineered "for, with, and by Nature" (Oxman, 2015). In other experimental mergings of art and science, artists use scientific data to create aesthetic works, as a way of converting the objectivity of scientific knowledge into experiential form (Ryiochi Kurokawa; teamLab; Stelarc), communities form to "promote the synergy between art and science for innovation and creativity" (Art-Science Node, Berlin), museums educate on the future potentials of interdisciplinary encounters (ArtScience, Singapore) and work toward "a community of lifelong explorers inspired by art and science" (Louisiana Art and Science Museum), while laboratories give institutional form to the essentially experimental character of art-science innovation (Spektrum, Berlin; Symbiotica, Perth; The Laboratory, Harvard).

Among the diverse claims made for these institutions by those keen to stake a claim for art-science within today's innovation economies, some common themes emerge. In what might well become a watchword of the innovation age, the concept of 'STEAM' represents the recognition of the significance of art (and specifically design) to the crucial work done in the STEM disciplines "to ensure a prosperous future" (stemtosteam.org, 2017). Among art's celebrated potentials is its capacity to provide a new angle on the solving of complex problems in a dynamic world in which pattern oriented or rote learning are no longer adequate (Born and Barry, 2010). Champions observe in art-science, then, the emergence of a third culture, a "transition between the two, rather than a sharp disciplinary boundary", which gives rise to a productive discomfort, as artists and scientists push each other beyond their conventional purviews and territories (Milroy et al., 2015: 2–3). The conventional opposition "between the functionality of science and the impracticality of art" is deemed increasingly redundant, since the experimental conditions in which each encounters their problems effectively puts them "on the same footing" in the dynamic, speculative process that is art-science practice (Milroy et al., 2015: 17).

Beyond the obvious and no doubt predictable recruitment of art-science towards techno-design and the production of new realms of consumable objects, there is much that could be said about these acclaimed potentials of the 'third culture'. Arguments for the complementarity of art and science in the production of innovation characteristically admit little real difference between the two modes of thinking. Artistic "intuition" is rendered a supplement to science's more "analytic focus", while the "conceptually multiple" approach of artists is recruited as a corrective to science's more "singular focus" (Leach, 2011: 153). In any event, the role attributed to art is essentially to add perspective and imaginative problem-solving to problems that are already determined in scientific terms. It is hard not to feel fatigued in the wake of the proliferation of clichéd formulations of the new harmony between art and science, which unashamedly reduce art to a scientific paradigm oriented to the transcendent universals of human subjects and their objects of knowledge and desire (Hynes, 2013). The attribution of a communicative function to art does little to disrupt the dominant "ordinary, bland apprehension of the world" (Guattari, 1995: 77) and recalls Deleuze and Guattari's (1994: 108) frustrated declaration that "we do not lack communication . . . on the contrary, we have too much of it". Finally, the idea that art-science can serve as a catalyst for public reason in a changing world rarely shakes the hierarchical ordering of the disciplines, art serving largely as a means of giving "immediate experiential impact" to the ideas of scientific experts (Fraser et al., 2015). In the face of these vertical hierarchies and a generalised scientism, claims to a "special relationship" between the arts and sciences can be as "opportunistic" as they are "reductive" and efforts to highlight the structural homology between the arts and sciences are arguably "well-meaning" forms of "intellectual neocolonialism" (Biggs, 2014: 47).

There is little point, then, in offering prescriptions or even ideal models for an alternative relation between the disciplines, which would have nothing to do with actual relations in any case, the multiperspectivalism of dominant versions of art-science being premised in most cases on the scientific paradigm's absorption of both science and art into the instrumentalism of the object. More important than the production of ideal models is, Guattari (1995: 61) reminds us, an analysis of the adequacy of existing processes of modelisation and the refusal of "reductionist modelisations which simplify the complex". More positively, a properly schizoanalysis works toward "processual enrichment", with an eye to the "ontological heterogeneity" of complex reality (Guattari, 1995: 61). Schizoanalysis is a *meta*modelisation because it is always more than "pure objective description" (Guattari, 1995: 61). Rather, in seeking out "the degree zero of all possible modelisation" (1995: 63), it explores the chaosmotic conditions of the "mutations of ontological textures releasing new lines of the possible" (1995: 62).

For Guattari, of course, such mutations are principally subjective mutations. And certainly, what is obscured by a focus on the structural homology between the disciplines of art and science in the attempt to drive innovation is the machinic production of subjectivity taking place here. For while the production of innovative ideas and objects is increasingly celebrated as a victory for the human relation to objective nature, what is clearly also being birthed in the institutions of art-science are lines of possibility for the "creative" actualisation of subjectivity. In this respect, diagnoses of a structural inequality between art and science and promises of structural homology are no different; both miss the machinic operations that, as Guattari (1995; 2011; 2015a; 2015b) insists, are always primary with respect to structure. If, from the point of view of structure, we are condemned to an affective polarisation of the engagement with art-science innovation – hope for its potentials or despair at its poverty – a machinic perspective may give rise to a potentiated and less crude grasp of its actuality.

The machine, Dosse (2011: 224) reminds us, is the operator "destined to think the repressed of structuralism". Where structure is "defined by its ability to exchange specific elements", the machine is driven by the more primary operations of repetition and difference (Dosse, 2011: 224). In this respect, the discourse of innovation is not without ambiguity. If economic discourse has assumed something of a monopoly over the concept of innovation in recent years, its hold on this notion, as Gibson (2014) notes, is a curious one. On the one hand, innovation is "a kind of entrepreneurialism of the familiar, as distinct from creating the new", since the value of any innovation derives not from "the creativity of the product" but from "the imagination of the market" (Gibson, 2014: 4). In this sense, innovation is conditioned by its capacity to imprint itself upon the familiar, to embed itself within existing practices in order to meet a need and find a use. On the other hand, "[i]n so far as innovation is a desirable value, it is value not in a product, but in change itself; it is the difference and production in repetition". In this sense, innovation is "incomparable, indefinable and unrepresentable, recognised only by its effects" (Gibson, 2014: 6). This paradoxical character may not be unique to innovation; what marks the heyday of the discourse of innovation is the dramatic extension of the sense of the term to make explicit its operator as value, sited not merely in "the objectified artefact of creativity", but in "the hyper-proliferative experience of use" (Gibson, 2014: 7).

As the accounts of Kingsley Hall make evident, even the most radical and seemingly least encumbered experiments in the production of subjectivity can become exercises in the entrepreneurialism of the familiar. But their value as experiments in desiring production is not thereby made redundant. Between and beside their efforts to produce the properly creative individual or to realise the self-realising community, a machinic unconscious is doing its thing. And the value it produces is too oriented toward the reality of actual situations to be captured by either individual or collective structures. What is needed, then, is an analysis that can be attentive to the production of machinic values, which is to say to those "values of aesthetic and scientific creation, which do not have an immediate effect on exchange values" (Guattari, 2015b: 31). Perhaps through this attentiveness, a matrix of sorts might be laid out, within which new habits and refrains might be possible. The point is not to seek to identify a model for such analysis, at least insofar as a model is understood as an essentially normative pattern or structure that can be transferred from circumstance to circumstance (cf. the Freudian transference). The notion of metamodelisation clearly mobilises a much more transversal and cartographic sense of the model, which is essentially constructive. In this, a silk pavilion constructed through the recognisably 'innovative' partnership between the Mediated Matter research group, a computer-numerically controlled machine and 6,500 silkworms belongs among other models, perhaps equally capable of moving from things to events (Roberts, 2014), developing "possible openings on the virtual and onto creative processuality" (Guattari, 1995: 31). Like any mode of experimentation, nothing is given here, and already capitalist refrains have stepped in as the most convincing, most familiar, model by which to measure the machinic values at play. But it may also be within this very same machinery that schizo becomings are made possible. A mutation in one practice or domain can have effects upon another, to the extent that the "virtual line of bifurcation and differentiation" can be given consistency (Guattari, 1995: 60) and an event extracted from it.

Schizoanalysing art-science, or how to maximise points of reference

Faced with these rapidly emerging and patently diverse couplings of art and science, the social scientist might best consider themselves a schizoanalyst, whose cartographies construct rather than interpret their object, for "analytical cartographies reach beyond the existential territories to which they are assigned" (Guattari, 2000: 33). As Guattari insists in *Chaosmosis* and elsewhere, it is the shift from a scientific paradigm to an ethico-aesthetic paradigm that is crucial in the effort to push a budding bifurcation towards the production of new Universes of reference. This is, as Guattari (1995) puts it, the first ethical choice. With the ethico-aesthetic paradigm, it is sensation and creation that are primary; the structures that found the scientific paradigm are denounced as particular modes of expression, which are falsely posited as transcendent universals. If the cries of the innovation economy have become the "neuroleptic ditties" (Guattari, 2015a) of our time, the question here, following Scott Sharpe (2013), concerns what it means to activate the anti-anaesthetising capacities of the aesthetic – its ability to produce, amidst the dominant regimes of sensation, new sensations that operate directly upon the nervous system of the always collective subject in its chaosmotic mutations. Guattari's (2000: 131) reference in *The Three Ecologies* to the "scientistic super-ego", which has condemned the human and social sciences "to overlooking the intrinsically developmental, creative, and self-positioning dimensions of processes of subjectification" is a timely rebuke to the familiarly pseudo-scientific habits of these modes of thinking. Schizoanalysis, in contrast, surrenders the comforts of the scientific paradigm and its tendency to take the objectal world of functions to which it orients itself as transcendent structures. And "what does schizoanalysis ask? Nothing more than a bit of a *relation to the outside*, a little real reality" (Guattari, 1996: 89).

In an interview conducted in 1980, Guattari (2015b: 31) makes the "utopian" suggestion that machinic values of creation and invention that "do not have an immediate effect on exchange values . . . actually deserve to be financed", being so poorly recognised in "the societies in which we find ourselves", which tend to "only valorize a certain form of production". Financing such values, he suggests, might give us an articulation point from which to "understand and critique the mode of capitalist valorization". Perhaps this 'vision' looks to us less utopian from within the heart of the innovation economy and from the purview of the financing of teams of artists and scientists to experiment in the production, not merely of innovative ideas and objects, but of innovation itself. But the complexity of the answer should not lead us to believe ourselves to be somehow above the money question: "what is at issue is not the reduction of praxis to cash, but the ways in which the schizophrenic tendencies of capitalism are overcoded by the more paranoic ones" (Holland, 1999: 2). For the analyst committed to a structural understanding of the experiments of Kingsley Hall, Peter's apparent inability to leave behind the economic concerns that characterise ordinary life is readily explicable: guided by the scientific paradigm orienting classical psychoanalysis, we can confidently diagnose his concern with money as an obvious inability to transcend the anal stage of development! But, viewed machinically, economic processes are strictly irreducible to such (psycho)analytic postulates. The exchange of money is (just) one point of reference, which might still play a role in connecting us with the intensive relations of the actual, rather than withdrawing us into the dominant social codings of capitalism.

To suggest that there is no purely extra-capitalist refrain is not a gesture of resignation, but a recognition that the dominant money economy is only one expression of a much more dynamic libidinal economy. Metamodelisation, Guattari insists, does not aim to gain an external grasp on the pure objectivity of the situation, but rather gets into the middle of emergent forces and *tries to maximise the points of reference*. Because metamodelisation is never merely a question of grasping the actual but also points toward the virtual, and here again the 'economic question' is crucial. The dealings of "the schizo are . . . elsewhere, on the side of the most deterritorialized fluxes: the flux of the 'miraculating' cosmic signs, but also the flux of monetary signs" (Guattari, 1996: 48–49). So "the schizo is not unaware of the reality of money – even if it is put to extraordinary uses – any more than he is unaware of any other reality" and indeed who is to say to what extraordinary uses investment in art-science innovation might give rise? Perhaps, then, the social sciences can play a role in giving the machinic values operating through these mergings of science and art a point of reference, which can in turn play a role in relativising exchange-value as *the* point of reference. It is a question of seeing money as a "reference point like any other", a way of keeping "a maximum number of systems of reference" in play, precisely to enable us to keep our distance.

As Guattari insists of the real schizo trip, "*the schizo is not all that interested in 'human warmth'*" and certainly cartographies of art-science will need to bring about a 'de-rationalization' of the group model if they are to map genuine lines of horizontality and transversality. At the very least it is necessary to explode the myth of togetherness underpinning the falsely harmonious group, "through reference to its instinctual, unconscious desires", which are certainly not the same as the wishes of the human beings within it (Genosko, 1996: 113). Lazzarato (2014) is right that the entrepreneurial model of subjectivity privileged in the discourses and institutions of the innovation economy tends toward a disintegration of the social, though it does not entirely preclude the possibilities of "new complexes of subjectivation: multiple exchanges between individual-group-machine" (Guattari, 1995: 5). But such new possibilities need to be constructed, through cartographies capable of otherwise unfolding the event.

No analysis of these emerging institutions can operate from a point of absolute exteriority; it too is defined by the exclamation mark of innovation in ideas. Still, alongside the diversity of emerging institutions designated by the moniker, "art-science", a schizoanalysis will construct equally diverse ways of relating the "finite modes" of the materials and problems that are engaged and the "infinite attributes of the Universes of the possible" that are implied (Guattari, 1995: 100). Each will map out its own immanent field of events, its own mutatory effects on other practices and areas of life. And while such mutations will no doubt be subject to the flattening effects of the universalisation of time under capitalism, they will also produce their own peculiar movement "from infinity to the passage of time" (1995: 101) – this is the 'fallout' that accompanies any mutation. What makes the logic of innovation so alluringly confusing is that the schizo and paranoic tendencies, the fallout and the desired end, are at once so close, and yet worlds apart. The whole point of innovation in practice is to produce creative effects, yet such creation is at least partly reterritorialised within the very delimited set of Universes of reference peculiar to capitalism, with its efficacy in erecting "a transcendent autonomised pole of reference" (Guattari, 1995: 103) over any multiplicity, and its peculiarly delimited forms of social subjection and machinic enslavement. The point is not that there is no escaping the paranoic tendencies of capitalist and scientific institutional arrangements but that these might also provide the conditions for novel becomings. If an artist's use of astronomical data to create a synaesthetic experience becomes an occasion for the communication of science, it might also provide "a space of ontological encounter", opening "a creative interval in disciplinary habits of thought" and a transformation in the fabric of experience (Lapworth, 2015: 1).[2]

The schizo does not behave like a child (I want an unappropriated, unqualified creativity!). There is no pure desire to be liberated and, in any case, as Mary Barnes' brother senses acutely, the 'bohemian style' of anti-institutional experimentation is tired. Better to work within the division of labour which will in any event give form to the flows of creativity and value to their machinic operations:

> Schizoanalysis neither legitimates the significations of the dominant codes nor accepts the impositions resulting from their overcoding. The schizoanalyst won't, in other words, open a technical consultancy in pragmatics. Still, even schizoanalysts have to make a living at a social pragmatics aimed at detecting micopolitical orientations, freeing them up, and making connections between them.
>
> (Genosko, 1996: 17)

Given the susceptibility of innovation to the paranoic tendencies of the scientific paradigm, the anti-anaesthetising capacities of the schizoanalytic method are critical here. To the extent that the social sciences can orient themselves, not to the dominant codes, but, like the 'real schizo trip', to 'the most deterritorialised fluxes', social scientific enquiry has a role to play in resingularising the present. In their efforts to produce referents for the chaotic forces of the world, the social sciences might be able to provide some sort of context by which the machinic surplus value of the innovation age can find a place within the pragmatics of existence. As Dewsbury (2015) reminds us, after Guattari, the 'place' created by any genuinely transversal movement is at best also a rupture in time, a 'resingularization of existence' in which the vertiginous affirmation of new possibilities is not merely a cliché, but an existential reality.

Acknowledgements

I am very grateful to Scott Sharpe and JD Dewsbury for their enthusiastic and insightful contributions, for Philippa Barter and Rohan Todd for the gift of *Lines of Flight* and Andrew Lapworth for his equally timely gift of *Machinic Eros in Japan*. Thanks also to the members of the Difference Lab for their generousity with ideas and for their feedback on an earlier draft.

Notes

1 Barry et al.'s (2008) delineation of the logics and modes of practice characterising contemporary interdisciplinary encounters usefully registers that, while there are forms of practice that productively engage disciplinary differences toward radical ontological enquiry, interdisciplinarity just as likely operates through practices of integration and subordination.

2 In this respect, as Lapworth (2016) points out, the art-science practices enacted at an institution such as SymbioticA (University of Western Australia) represent both an enquiry into our ontological constitution and a material reconfiguration of our ontogenetic capacities.

References

Barnes, M. and Berke, J. (1973) *Two Accounts of a Journey Through Madness*. Melbourne: Penguin Books.

Barry, A., Born, G. and Weszkalnys, G. (2008) Logics of interdisciplinarity. *Economy and Society*, 27(1): 20–49.

Bertelsen, L. and Murphie, A. (2010) An ethics of everyday infinities and powers: Félix Guattari on affect and the refrain. In: Gregg, M. and Seigworth, G. (eds.) *The Affect Reader*. London: Routledge. pp. 138–157.

Biggs, I. (2014) Beyond aestheticism and scientism: Notes toward an 'Ecosophical' praxis. In: Wilson, B., Hawkins, B. and Sim, S. (eds.) *Art Science and Cultural Understanding*. Champaign, IL: The Arts in Society. pp. 44–65.

Born, G. and Barry, A. (2010) Art-science: From public understanding to public experiment. *Journal of Cultural Economy*, 3(1): 103–119.

Buchanan, I. (2013) Schizoanalysis: An incomplete project. In: Dillet, B., McKenzie, I. and Porter, R. (eds.) *The Edinburgh Companion to Poststructuralism*. Edinburgh: Edinburgh University Press. pp. 163–185.

Deleuze, G. and Guattari, F. (1983) *Anti-Oedipus: Capitalism and Schizophrenia*. Minneapolis, MN: University of Minnesota Press.

Deleuze, G. and Guattari, F. (1994) *What Is Philosophy*? Brooklyn, NY: Verso.

Dewsbury, J.D. (2015) Guattari's resingularization of existence: Pooling uncertainties. *Dialogues in Human Geography*, 5(2): 155–161.

Dosse, F. (2011) *Gilles Deleuze and Félix Guattari: Intersecting Lives* (trans. D. Glassman). New York, NY: Columbia University Press.

Flaherty, J. (2013) Mind-blowing domes made by 6,500 computer guided silkworms. *WIRED*, July. Available online at: www.wired.com/2013/07/your-next-3-d-printer-might-be-filled-with-worms/

Fraser, J., McDonald, F.P. and Ardalan, N. (2015) Reflections on public art + science reasoning. *Transformations*, 26.

Genosko, G. (1996) Introduction. In: Genosko, G. (ed.) *The Guattari Reader*. Oxford: Blackwell Publishers. pp. 1–37.

Gibson, J. (2014) *The Logic of Innovation: Intellectual Property, and What the User Found There*. London: Routledge.

Guattari, F. (1995) *Chaosmosis: An Ethico-Aesthetic Paradigm* (trans. P. Bains and J. Pefanis). Bloomington, IN: Indiana University Press.

Guattari, F. (1996) Mary Barnes' trip. In: Genosko, G. (ed.) *The Guattari Reader*. Oxford: Blackwell Publishers. pp. 46–55.

Guattari, F. (2000) *The Three Ecologies* (trans. I. Pindar and P. Sutton). London: Athlone Press.

Guattari, F. (2009) *Chaosophy*. Los Angeles, CA: Semiotext(e).

Guattari, F. (2011) *The Machinic Unconscious: Essays in Schizoanalysis*. Los Angeles, CA: Semiotext(e).

Guattari, F. (2015a) *Lines of Flight: For Another World of Possibilities* (trans. A. Goffey). London: Bloomsbury.

Guattari, F. (2015b) Translocal: Tetsuo Kogawa interviews Félix Guattari. Part I: October 18, 1980. In: Guattari, F. *Machinic Eros: Writings on Japan* (eds. G. Genosko and J. Hetrick). Minneapolis, MN: Univocal Press.

Guattari, F. and Rolnik, S. (2008) *Molecular Revolution in Brazil*. Los Angeles, CA: Semiotext(e).

Hacklin, F. and Wallin, M.W. (2013) Convergence and interdisciplinarity in innovation management: A review, critique, and future directions. *The Service Industries Journal*, 33(7–8): 774–788.

Hynes, M. (2013) The ethico-aesthetics of life: Guattari and the problem of bioethics. *Environment and Planning A*, 45(12): 1929–1943.

Holland, E. (1999) *Deleuze and Guattari's Anti-Oedipus: Introduction to Schizoanalysis*. London: Routledge.

Lapworth, A. (2015) Beyond bifurcation: Thinking the abstractions of art-science after A.N. Whitehead. *Transformations*, 26.

Lapworth, A. (2016) Theorising bioart encounters after Gilbert Simondon. *Theory, Culture & Society*, 33(3): 123–150.

Lazzarato, M. (2014) *Signs and Machines: Capitalism and the Production of Subjectivity*. Los Angeles, CA: Semiotext(e).

Leach, J. (2011) The self of the scientist, material for the artist: Emerging distinctions in an interdisciplinary collaboration. *Social Analysis*, 55(3): 143–163.

Milroy, A., Wegener, M. and Holmes, A. (2015) Labpunk – curiosity, intra-action and creativeness in a physics-art collaboration. *Transformations*, 26.

Oxman, N. (2015) *Design at the Intersection of Technology and Biology*. TED Talk. Available online at: www.ted.com/talks/neri_oxman_design_at_the_intersection_of_technology_and_biology

Oxman, N., Ortiz, C., Gramazio, F. and Kohler, M. (2015) Editorial: Material ecology. *Computer-Aided Design*, 60: 1–2.

Roberts, T. (2014) From things to events: Whitehead and the materiality of process. *Environment and Planning D*, 32(6): 968–983.

Sharpe, S. (2013) The aesthetics of urban movement: Habits, mobility and resistance. *Geographical Research*, 51(2): 166–172.

Stemtosteam.org. (2017) *What Is STEAM?* Available online: http://stemtosteam.org

Todd, R. and Hynes, M. (2017) Encountering the animal: Temple Grandin, slaughterhouses and the possibilities of a differential ontology. *Sociological Review*, 65(4): 729–744.

Wilson, B., Hawkins, B. and Sim, S. (2014) *Art, Science and Cultural Understanding*. Champaign, IL: Common Ground.

8 Ecosophy as an ethical mode of existence[1]

Mahoro Murasawa and Stéphane Nadaud

A philosophical concept is interesting because it is a tool with which to shape the world. Félix Guattari, philosopher and psychoanalyst, invented a concept which he named 'ecosophy'. This philosophical concept is only interesting because it is a tool which we, the authors of this chapter, think we need in the contemporary world in which we live. In the lines which follow, we want to show why this tool is particularly well suited to this world, rather like a saw for cutting a board, or a hammer for driving a nail (even if a serial killer may use these in quite different ways). Our particular focus is Japan following 11 March 2011.

Neither 'hantai' nor 'hantai-hantai' but molecular

In light of the Fukushima catastrophe, a sort of anti-nuclear revolution appeared in Japan: 2012 was indeed a year the like of which Japan had not known since the student movements in the early 1970s. On 19 September 2012, 60,000 people were on the streets of Tokyo. Unimaginable! However, this movement very quickly ran out of steam. In fact, a neoliberal 'counter-revolution' (as much political as economic) emerged which set itself in opposition to the anti-nuclear revolution, like fire hoses in the face of a growing blaze.

In short, after Fukushima, there was a revolutionary opposition to nuclear power. In Japanese, to *oppose* [*contre*] is written as *hantai* (反対),[2] and we can say that the anti-nuclear revolution *opposed* itself – as the name suggests – to the nuclear; to be against nuclear power was a *hantai* revolution. It was later that this *hantai* revolution was crushed by an inverse movement, *opposing* the revolution against nuclear power: a sort of capitalist revolution *hantai-hantai* which was that of the official propaganda, supported by doctors, scientists, the agricultural industry, with speeches finding their way into each home, thanks to the media (and particularly television). In Japan, they are called the community of privileged economic-political-expert-media [*sankangakumedia no riken kyodotai*], an invented term to show the entanglement of these different social actors in the organisation of power in Japan. It exists in France but it is rather difficult to circumscribe, and it must obviously not be reduced to the gutter press that flourishes in the windows of newsagents (*Le Point, L'Express, Marianne*, etc.).

But another revolution took place unknown to the politicians and the revolutionary capitalist media (*hantai-hantai*), and also, it must be said, unknown to the militant revolutionaries against nuclear power (*hantai*). It is this that we will, drawing on Félix Guattari (2012), call a 'molecular revolution'. Indeed, following the Fukushima accident, Japanese families changed from within. Couples who live in the region of the nuclear power plant, the Tohoku, are torn on the decision they have to make. Mothers are worried for their children and want to leave to go as far away as possible. This obliges fathers to find work elsewhere, which is far from easy. Some are, therefore, forced to change social status (with all the difficulties that this brings) because they are rehoused in a different part of Japan, thousands of kilometres from their home which has become toxic. In a way, the 'deterriorialisation' – a Deleuzo-Guattarian concept which indicates that an experiment/experience does not necessarily need a geographically determined place to unfold – is very real here. Even if they stay within the family's territory, the changes brought about by the Fukushima tragedy are truly unfolding for people. We might say that this is a *subjective* revolution.

In other words, instead of a classic battlefront 'against' *versus* 'against-against', there is a new battlefront which does not have a location as easily determinable as two designated territories – one for the *hantai* group, and one for the other group (*hantai-hantai*). The fight is beyond . . . Of course, the official discourse (doctors, government, . . .) continues to function in a classic way, that is to say they are 'territorialised' – they have an identifiable territory (the university, the government, the HQ of Tepco – who own the plant, etc.). And in opposition to this state territory, another territory confronts it (the anti-nuclear power associations, the antiglobalisation groups, etc.). But beyond the territorialised battlefront, there is also a non-official discourse (internet, friends, families, etc.) which does not have time to find itself a territory, and nor does it seek one. It expresses itself only by its necessity. These are movements which we call, after Deleuze and Guattari, 'deterritorialised'.

A subjective revolution

We can describe, almost clinically, this deterritorialised subjective revolution which is taking place at the heart of the subjectivity of the Japanese people of Tohoku. Let's take one of them[3]: first of all, this person believes the official discourse which tells him that everything is ok, that everything is under control, that the radioactivity is limited, that the power station is no longer leaking. . . . He believes, then, in what he does not yet see as a lie, which allows him to live in the illusion, and therefore to think that, really, all is well. But the reality is louder than the lie: the deterritorialised information – at the very heart of his life, of his own subjectivity of father (or mother) of the family – accumulates like clouds raining their news in contradiction to the unified discourse found on in the television or in newspapers. These clouds disappear and reform elsewhere, raining yet more contradictory drops. This Japanese person from Tohoku hears talk of polluted food and water, thyroid problems in children. . . . If he doubts enough, if he has let himself (unconsciously, obviously) become deterritorialised, he will try to question himself more deeply, without being territorialised as 'anti-government' (by adhering to a group who fight against nuclear power, for example). He too, then, becomes a deterritorialised information, which results in others taking him for what he becomes, a cloud which can only rain, and to which they assign a code (Deleuze and Guattari talk of 'recoding'): 'madman', 'liar'. These others still believe in the territorialised official lie; they remain firmly hooked to their own territory. This Japanese person will then live a life of exclusion which, even if the others try to *re*territorialise him (assigning him a code which stands for territory) like an individual outside of the norm, will in reality deterritorialise him even more. If this deterritorialisation is a banishment (according to the territorialised official discourse of those we call 'the others'), it is also an emancipation. Compared to the world of media, of business, of territorialised government, and even compared to the anti-nuclear officials, this Japanese person transformed into a cloud escapes the 'stabilised' world, which dictates (according to meteorology) the weather that we will have.

This deterritorialisation forces our Japanese person to make a subjective choice – a choice which is, let's remember, unconscious:

1 Either he represses the anguish caused by the official discourse because he continues to believe in them.
2 Or he does not believe in the official discourse yet trusts in the honesty of those who speak it and he tries to convince himself that the discourse he knows to be false, is perhaps true.
3 Or he does not believe the official discourse but he believes in the unlimited power of those who speak it. The feeling of powerlessness and injustice is therefore the strongest. This confrontation can reach social death (*hikikomori*, staying outside of all social life), even up to physical death (suicide).
4 The last possibility, and the most extreme: he tries in one way or another to flee or to fight *completely*.

To make sense of this subjective movement, we can use the psychoanalytic categories of neurosis and psychosis. For psychoanalysis, the neurotic remains an individual, his self is complete (in the Deleuzo-Guattari jargon, he is deterritorialised), and even if his relation to reality appears difficult and painful, he never loses consciousness of his own reality, which is different (contradictory, even) to that of others. The psychotic does not have his own self and he is confused with all of reality, including that which is not what he is taught to perceive as himself: what he constructs in his mind and the outside world are not ontologically different. This is what we call, clinically, delirium. This is where a thought has as much reality as an external object to the individual. According to well-known psychoanalyst, Jacques Lacan, it is not only the dog which can bite the psychotic, but also the word 'dog' which qualifies him and which he has in his head: 'the word dog bites'. Pierre Deproges (1986: 7) thus explains the difference between a neurotic and a psychotic: "A psychotic is someone who believes so strongly that 2 plus 2 makes 5, and is completely satisfied with this. A neurotic is someone who knows that 2 plus 2 makes 4, and that he makes him sick!"

If we return to the four positions of the Japanese person of Tohoku, in light of these psychoanalytic categories:

(1), (2) and (3): one is in neurosis. The individual still has a link to reality. We can say that, from a subjective point of view, he is still individualised, territorialised.

(4): one is in psychosis. There is no subjective territorialisation and the elements of common reality do not take on value for the subject, who is interchangeable with the society (Deleuze and Guattari use the term '*socius*') in which he bathes.

Put differently:

(1), (2) and (3) (neurosis): the subject believes that he can still change society by positioning himself against the movements of the state carried by the capitalist system. This way he might sign up to a counter movement, *hantai* (for example in a movement *against* nuclear power). Capitalism, and the society which promotes it, is considered here by the individuated subject as an abstract substance which transcends him. For him, there exists but one paradoxical solution: he will combat this abstract substance while fleeing it. Suicide or *hikikomori*, which are forms of fighting at the same time as they are flights of escapes, can without difficult be placed in parallel with the investment in an association which fights against nuclear power. This paradoxical solution is due to the fact that the abstract substance is in some ways the ideal of the neurotic, but transposed onto capitalist society. It is an ideal which the neurotic *desires* at the same time as he tries to *both* fight it *and* to flee. The neurotic *desires*, therefore, this abstract substance, a desire made up of love and hatred. Montaigne's friend, La Boétie, would say that the neurotic calls for his chains as much as he abhors them. Capitalist society, which promotes nuclear power, is nothing but the necessary object to desire and to detest at the same time for the neurotic.

(4) (psychosis): this kind of change is not programmable because there is no individualised subject. The Japanese person who finds himself in this subjective situation leaves behind all territory, dematerialises, and becomes himself the abstract substance, Capitalism as a whole, outside of all idealisation, in the most conscious way. The subject becomes schizophrenic, melts into Capitalism, and dissolves by following a process of disindividualising subjectivity. We have here, then, 'Capitalism and Schizophrenia' which is the subtitle of the two most famous books by Deleuze and Guattari (2004a, 2004b), *Anti-Oedipus* and *A Thousand Plateaus*.

In other words, the struggle of the neurotic is macropolitical and the struggle of the psychotic is micropolitical.

If we use the terms 'neurotic' and 'psychotic' (or 'schizophrenic'), they have nothing to do with *psychiatry*. We are not saying that the Japanese person who is in position (4) is a psychotic in the clinical sense of the term, no more than positions (1), (2), and (3) qualify neurotics. We are saying, though, that whatever the psychic structure of an individual – if we believe in the concept of psychopathological structure – he/she can behave either on a neurotic side, or a psychotic side, concepts which for us are philosophical – rather than medical – tools.

Micro rather than macro

Currently many of those from Tohoku – couples, refugees – are experiencing a psychotic deterritorialisation (even if, surely, most of them are neurotic). The subjective struggle pursued by the Japanese people is therefore not only macropolitical – it no longer is a combat of neurotics against (*hantai*) an idealised Capitalism, but a micropolitical combat (beyond *hantai* versus *hantai-hantai*). This experience of the transformation of subjectivity is what Deleuze-Guattari call the 'schizophrenic process' and what Guattari calls the 'molecular revolution'. This schizophrenic process advances particularly rapidly in Japan. At the forefront of the model of capitalist society, a schizophrenic process experienced by people who are excluded from this is gaining ground.

But this schizophrenic process involves a reaction of the whole social body which does not how to face these schizophrenics – who are, we insist, not all schizophrenic in the clinical sense of the word – without territory, who rain everywhere. The capitalist society understood as a totality, as an abstract substance, hopes to control once more this fleeing subjectivity, and to *reterritorialise* it by force. Deleuze and Guattari (2004a) call this mechanism the 'paranoiac machine'. The subject who experiences this schizophrenic process, who is ever more deterritorialised, will find himself pushed, by the socius, to become paranoid. If neurosis is a mechanism that is transcendent from the start, since the subject considers himself as different the 'Whole' (1–3), psychosis (schizophrenia) is, by contrast, immanent from the start; it resists all territorialised defence mechanisms that are like antibodies desperately trying to encircle what cannot be grasped, because it has no fixed territory. The paranoiac process is the ultimate defence movement of capitalist society which seeks to return the transcendent where there is none. In a way, the paranoiac process is the *neurosisation* (transcendentalisation) of schizophrenia operated by the capitalist socius in order to give a territory (reterritorialisation) to those who have none in order to better integrate them to their place. Guattari often said, at the psychiatric clinic at La Borde, that the only ethical act of caring consisted of psychotising neurosis, rather than neurosising psychosis – which is the role of the police (antibodies). We could translate it in this way: immanentising (to render immanent) the transcendent is more revolutionary than transcendentalising the immanent.

Guattari and Deleuze named the 'oedipal folding' the paranoiac machine which tries to return the transcendent where deterritorialisation tends to render everything immanent. Indeed, even if the Japanese schizophrenic (position (4)) seems to escape this totality as abstract and ideal substance, he never ceases becoming caught up by this totality which dawns on him like a transcendent element looking to reassign him a territory, rather like a hunter tirelessly pursuing its fleeing prey. This hunt is paranoiac and Oedipal: *paranoiac* because the fugitive will react by displaying paranoia ('they are all against me, and I interpret every event in my life in this way') and *Oedipal* because the forced reterritorialisation always operates on the 'germinal unit of society', to use Freud's (2010) famous turn of phrase (which he used to describe the family). Marx explains that the true nature of Capital is not made up of concrete existences and that all that exists is an ideal category without objective substance, just as mammals and dicotyledons do not exist in and of themselves but are sub-species of an ideal category dictated by the natural sciences.[4] Following Marx, we argue that the 'abstract substance totality' is a ghost which controls and which constantly possesses people and – and as we know – ghosts do not exist! This is why we cannot pulverise this totality, this ghost, at a macro level. No revolutionary operation can take place except on a deterritorialised level, at a molecular level. This is why it is necessary to abandon concepts of abstract totality like 'Nation', 'People', or 'Society'. This is why, in any case, we must not lose sight that the appearance of an ultimate category which we call 'Capitalism' – thereby detaching it from the totality 'Humanity' – is to assume the necessity of multiple becomings: minoritarian, mineral, vegetal, fox or bear, child or woman.

Fragment(s) everywhere

Note: it is not about splitting the whole into parts, but to assume a process of fragment(s) that do not make up a totality. If there is a link between the fragment(s) and this totality, it is not that of the parts to a 'Whole'.[5] If the parts are the mirror reflections of this totality, each fragment(s) is/are one (or many) mirror(s) in which other fragment(s) reflect themselves – that's why we write 'fragment(s)' and not 'fragment' or 'fragments'. For example, each citizen is a part constructed from the whole nation, but as the whole mirror of the infinite of the terrestrial world is reflected in the moon of the night sky, each citizen is the mirror which reflects the entire nation. If the subjective fragment(s) altogether do not compose a totality, each fragment nevertheless reflects all subjectivity.

This is why these subjective fragment(s) cannot be limited to humanity alone. These fragment(s) are also animal, vegetal, mineral – subjectivities of all that exists, what Guattari calls the 'cosmos'. The cosmos is not a totality that transcends everything. Fragment(s) are the possibility of existence of all things; there is an infinite number of fragment(s), as it were, within all fragment(s). In other words, we can find the cosmos in a park 50 metres away from home, just as we can find it in the gaze of a child or of a friend or even – why not – in the infinity of space.

But these small cosmoses are ephemeral. In the same way that the light of a firefly, visible in the middle of the night, can no longer be seen in the light of the midday sun, the universe of interlaced fragment(s) disappears in the face of the action of the transcendent and ideal 'Whole'. There are fireflies which hide from the rays of the sun, in a shaded spot, in order to continue shining, which flee to continue fighting against (*hantai*) the 'Whole', or who decide to group together to shine more brightly, knowing that they will never have the luminescence of the sun. This type of firefly, neurotic enough, knows that the powerful sun which prevents it from expressing its own light – they are in the positions (1–3). Completely different is the action of firefly – schizophrenic (4) – which stays out under the midday sun and shines its luminosity back towards it, and which succeeds – miraculously? – to put in pieces the 'Whole' because of its deterritorialised molecular becoming. This firefly, while it is dying by dint of shining with all its might under the relentless sun rays which mortally irradiate it, will continue to shine in this way, without knowing that others see it despite the hyperluminosity of the all powerful sun, and warm to its power. In these difficult times, following the political gesture of the firefly/fragment(s) that we are reminds us not to yield to our desire – its phenomenal will to power, and to always express such a strong emotion that all the fragment(s) remain intertwined with an infinite force, without disconnecting. There are two terms for qualifying emotions in Japanese: *Kanjo* (感情), the feeling that we consciously control in any given social situation, and *Jodo* (情動), the uncontrollable, unconscious, emotion which expresses itself even if the individual rejects or refutes it. Here, luminosity – the emotion which most readily associates subjective fragment(s) with each other – is *Jodo*. *Jodo* is not the result of a conscious will of individuals (which can then no longer be described as fragment(s)) wishing to remain together. This sentiment is, instead, *Kanjo*. *Jodo* is the force which assures the continuity of subjective fragment(s) with each others, and which constitute these small comoses.

We can easily understand the force and the fragility of this emotion, *Jodo*, which, if it can impress on the sun itself can just as easily be broken by the smallest of knocks. Faced with the dazzling brilliance of the 'Whole', individuals tend to only allow their *Kanjo* feelings to be expressed, leading to a fragmentation of fragment(s), which no longer express their *Jodo* emotion at all. In each individual, the paranoiac – oedipal – reterritorialisation condenses the feeling of self – the *Kanjo* – which always gives ever more force to what is known as a 'Whole'. This is why Guattari calls the feeling (*Kanjo*) microfascist; this is what we called the transcendentalisation of the paranoiac machine. Within the family, at the workplace, in school, or in society – that's to say throughout the world – this microfascist *Kanjo* feeling augments the rage that violently transforms fragment(s) of these micro-cosmoses in as many 'parts' working for the 'Whole', which makes them the pieces of the paranoiac machine which is the capitalist society.

We said above that we are witnessing an awakening, throughout Japan, of those who refuse to become 'parts' of this 'Whole', who flee or fight to protect the fragment(s) of this cosmos. For example, women who choose to protect their children who are micro-cosmoses by divorcing their husbands – cowardly office workers – who continue to serve the cause of this shining 'Whole'. Similarly, young 'freeters'[6] cannot bear having to be in sync with their school or place of work. It is among them that starts a movement demanding to continue to associate new fragment(s) – through *Jodo* – and which stops pushing them back – with *Kanjo* feelings, in as many as 'parts' of the 'Whole'. We could also write of the famers who return to live in the micro-cosmoses of their own fields which have been cleaned of deadly ash. Or we could write of the residents who do all they can to return home, having been displaced to other towns due to the earthquake which hit Fukushima in 2011, thousands of miles away – as far as Okinawa. In short, it is as if Japan itself was changing its appearance and was becoming a sort of Guattarian *heterotopia*. A concept developed by Foucault (1986), heterotopia can be defined as a different society within society itself, a kind of anti-place (deterritorialised) which overturns the reality of society.

Ecosophy as life ethics

The 'Whole' called Capitalism is a concept which becomes ingrained without us wanting it to, like a ghost, but it is not our concrete existence. The problem is that if we continue to consider the 'Whole' as real and concrete, we will each remain a part of that Whole. The very substance of Capitalism tries to violate the small cosmoses of fragment(s), and this is precisely the microfacism that Félix Guattari talks of. This microfascism is the dizziness which makes shine the world of commodities, the *Kanjo* emotion which weaves the illusion that we have of ourselves (reterritorialisation) as part of a 'Whole' and, let's say it how it is, as a commodity which is forever exchangeable in light of the new God of Money.

Each individual becomes himself a 'Whole' which controls all the other goods to which he is linked, and frees himself, across this feeling (*Kanjo*), from the various fragment(s) which are truly the little cosmoses of fragment(s). These are, however, the cosmoses of fragment(s) which truly constitute the cosmos – that's to say: ourselves. These goods which individuals have become take the place of family or neighbours, of flowers or animals close at hand, of mountains where you played when you were a child. In other words, these goods take the place of an existence which is truly our own.

What Guattari calls subjectivity are the fragment(s) of these small cosmoses. And the political problem for Guattari is the reduction, the impoverishment, the destruction of these subjectivities. For him, what drives these concrete phenomena of the destruction of subjectivities are the wide variety of mental pathologies, the fascistisation of organisations, the collapse of community movements, and the pollution and destruction of the natural environment. The remaining subjectivities, which no longer have anything to do with the multitude of fragment(s) of the small cosmos, are reduced to being pieces ('parts') of the paranoiac machine, subsumed by an illusory 'Whole' known as 'Capitalism'.

In his book, *The Three Ecologies*, Guattari (2008) conceptualises three aspects of the cosmoses where subjectivities, which are at the same time grains (fragment(s)) of these cosmoses, are woven: these fragment(s) are sensitive – they shine like fireflies – *mentally*, *socially*, and *environmentally*. This is what he calls *ecosophy*. But we must be careful not to understand ecosophy as a method for producing a straightforward theory which would synthesise the three ecologies.

There is no ecology in the sense that the 'Whole' would be the synthesis of the three ecologies (which would be the 'parts'). On the contrary, ecosophy is a cosmos in which all ecology – of whatever register – is fragmentary. In contrast to a policy meant to be efficient and built on the feeling of *Kanjo* (of which humanism is the most pretentious harbinger), ecosophy is an ethic based on the protective emotion (*Jodo*), to the end, of small cosmoses emerging from the combination of fragment(s), rather than the control of the *Kanjo* emotion of the 'Whole' (Capitalism). In other words, ecosophy is a practical way of thinking, a way of living – put simply, an ethic.

The Marxist principle of opposing capitalism based on the analysis of economic and social systems (which, in effect, explains capitalism as the emergence of money generated by the world of commodities) has taken place historically in the communist party and the class struggle. If we understand capitalism in this way, the dialectic understanding of the 'parts' of this 'Whole' ultimately leads to a collection of microfascisms (think, for instance, how Stalinism is the dogmatism of the communist party) which are the very substance of capitalism. And it is this collection which strengthens capitalism. This, then, explains the cooperation of the so-called communist states with the basic harmony of capitalism. Take, for example, the cooperation of the West with Stalin's USSR or, today, the cooperation between China and world capitalism. Even if Guattari is one of Marx's successors, he still shows where the Marxist and Freudian (Lacanian) doctrines ran into difficulties. The trap into which they unwittingly fell into is that they considered the 'Whole' necessary – and forgot that Capitalism is but a ghost which does not exist.

On the other hand, taking another position – that's to say by turning our back on the illusion known as Capitalism – we can offer a new method, one which is both theoretical and practical, in order to fight not against this Capitalism but by focusing resolutely on the micropolitics of desire.

Molecular undulations: social change from subjective revolutions

If Capitalism, as a 'Whole', has no real existence, and if we argue that only the fragment(s) matter, what is the monster that ruins our lives and which – truly – irradiates the plants and children of Tohoku? Can the micropolitical practice that we advocate, and which some will no doubt think is too idealistic (religious, even), function at the macro level of the capitalist society against this monster? Put differently, is a real opposition to globalised neoliberalism possible? Is the reign of the nuclear mafia, of the community of privileges that we spoke of earlier, endless? Does this leave us as delusionary philosophers writing about 'small cosmoses'?

When Deleuze and Guattari published *Anti-Oedipus* in 1972, explaining that capitalism is "always expanding its own borders, and always finds itself in a situation where it must close off new escape routes at its borders, pushing them back once more. It has resolved none of its fundamental problems. It can't even foresee the monetary increase in a country over a year. It is endlessly crossing its own limits which keep reappearing farther out" (Deleuze and Guattari, 2004a: 270), they were violently criticised by a great part of the Left of the time, which reproached them for having given up – for having argued that it was impossible to KO capitalism. These critics are right, in one sense, but for the wrong reason. Indeed, when one considers capitalism as a macro system or an actually existing 'Whole', one continues to give life to a ghost. Therefore, the critics are right to say that – following Deleuze and Guattari – capitalism cannot be KO'd; this is because from the moment we give an actual existence to this virtual monster, all combat against it becomes impossible. Such a confrontation would be tantamount to trying to fill the barrel of the Danaids, knowing that it is pierced at the bottom, and being surprised not to see the level rise. But they are wrong to draw the conclusion that no struggle is possible. For Deleuze and Guattari explain that managing to see Capitalism as it is – to know it is an illusion – is already a way of knocking it out. And this can be seen in the multitude of ways in which resistance across the world has managed to impose something other than a globalised uniformisation advocated by this monster.

In other words, the molecular confrontation/escape at the level of all fragment(s) (confrontation and escape happen at the same time as the mechanism, as we said, is psychotic) is the only possibility of dismantling microfascisms. At that moment, the 'Whole' that is capitalism (which therefore does not really exists as a 'Whole' but to which it is given this consistency) is not so much KO'd as it is transformed. We can say that these molecular revolutions never allow the crystallisation of the cosmos to extent to the point of becoming a 'Whole' which we have become accustomed – after Marx – to calling 'Capitalism'. A little like the surface of the water that we might prevent from freezing completely (and becoming a solid surface) by breaking it, by heating it up (thanks to the *Jodo* emotion which is warming). Where the water has returned to its liquid form, microfascism would recommence to solidify the water once more. But the *Jodo* emotion will be heating up another spot. And this is without end. Is that not what one would truly call a real revolution . . . one which is permanent?

Here, we must take seriously the way in which Guattari understands this *molecular* revolution. What Guattari terms 'molecule' is not the same as what physics would normally call a molecule (which is composed of atoms). It is, instead, at an unending micro level, something which in physics is known as the quantum level. At this infinitesimally small level, quanta are both matter and waves, which explains how the same quantum of energy/matter can find itself at various locations at the same time. These phenomena can obviously not be experienced at the level of our own macroscopic physical bodies but this does not, however, invalidate the quantum reality – indeed, certain computers are now made based on these principles. Like these quanta of energy/matter, molecular revolutions – as subjective revolutions – are at a different level to what we call social revolution, which only possesses – if we follow this quantic analogy – the material aspect which all philosophers since the 18th century have taught us to know and which we can easily apprehend. But at the molecular (micro) level, there are revolutions which have a material *and* undulatary (energetic) quality at the same time. It is this incredible skill, difficult to apprehend at our macro level, which allows both escape and combat.

Quanta are composed of differences and repetitions which, following the sociologist Gabriel Tarde (1903), we might call 'imitation'. According to him, an invention which is a new for a given society, after having been original for a while, it soon finds itself imitated and extends rapidly across said society. In other words, a localised change – in a domino effect (but with this difference, the imitations are not identical to the original invention) – will modify the entire society. In the same way, a molecular revolution, understood as original difference at a micro level, is potentially a social revolution understood as repetition at a macro level. Tarde calls this first invention (which in Deleuzo-Guattarian terms we might call 'desiring singularity'), which extends across society through its imitations, the *fashion* – in the sense that it is a phenomenon that changes the lifestyle of society. Similarly, macro social practices such as those of the 'nation', the 'foreign exchange market', etc., are nothing other than fashions which spread out over a long period of time. In reality, these are molecular micro-revolutions which, extending by difference and repetition, have colonised the whole socius. For example, in relation to the tsunami which struck Tohoku in March 2011, we must understand that all the water molecules (in the physics sense of the term) did not, individually, travel the Pacific Ocean. It was by transmitting at their modest level – thanks to the intermolecular forces which connect them to one another, an imperceptible movement from top to bottom – that a gigantic amount of energy was transmitted across the Ocean. In the same way, when each of the numerous subjective fragment(s) are emotionally connected (*Jodo*), the revolution awoken by a single atom becomes an undulation which transmits a gigantic energy to the whole Socius. Rather than 'overthrowing' Capitalism, it would be more accurate to say that such revolutions 'change the tone'; they are modulations. As Capitalism repeats its actions in an ever automatic fashion (like a Deleuzo-Guattarian machine which at the quantum level, always repeats the same operations without ever producing the same result), these molecular revolutions bring about not only an extended fashion in the popular sense of the term (the expansion of an original model) but also a fashion as a style which is always reinventing the model (the double movement that Tarde calls 'imitation').

These modulations repeat themselves as long as there are molecular revolutions, and in the end, the forces increase: the face of Capitalism, even if it seems the same, has become different. This is why Guattari says that every molecular revolution possesses "both a social character and a revolution of desire" (Guattari, 2012: 168–169) and that "collective competence in this matter happens along the way without it being necessary to support 'social projects'" (2012: 149).

Bartlebys everywhere

In music, dissonance always goes hand-in-hand with modulation. Modulation is not a movement which is in opposition to the system (it is not *against*, *hantai*) which would try to put another system of rules in place of the one that governs the economic and political world. Each molecular revolution is a movement which refuses the system, which consists of not obeying the 'Whole', the system of rules. Or, more precisely, it consists of becoming aware that the 'Whole' is an illusion. It is not so much a counter-power than a movement which seeks to achieve a non-Power, an anti-Power. We say anti-Power but not *anti-power*: following Foucault we differentiate between 'power to' (with a small p), at a micro level, and 'Power over' (with a capital P), at a macro level.[7] The first is diffuse and without a fixed territory; the second is a manifestation of the sovereignty of the one who possesses it.[8] Foucault borrows these terms from Nietszche but we differentiate power (which is Foucault's '*power*' and Nietszche's '*will to power*'), which we name *chikara* (力). We differentiate, therefore, the *power chikara* from the *Power Kenryoku* (権力). The molecular revolution that we are talking about is a *Kenryoku* anti-Power at the macro level (therefore, in terms of sovereignty) but is not the search for a *Kenryoku* Power which would oppose itself to an already-established *Kenryoku* Power (not, therefore, a counter-Power which would always remain a *Kenryoku* Power). Every molecular revolution is a *chikara* power which seeks to grow ever more. Indeed, if every molecular revolution is an anti-Power *Kenryoku* (and not, let us repeat, a counter-Power *Kenryoku*), it does not at all correspond, however, to an anti-power *chikara*; in fact, it is the opposite. If we use Deleuze and Guattari's terms, a molecular revolution is not an oppositional plan founded on the choice between A or B (like in logic, in the law of contradictions) but a movement hovering on the plane of consistency where heterogeneous elements are not opposed to one another. This is where we rethink the quantum mechanics: on this plane of consistency, the fragment(s) of the *chikara power* are both matter *and* energy. As we find in *Anti-Oedipus* (Deleuze and Guattari, 2004a), we can be both man and woman, parent and child. If we put this another way, this movement of a molecular revolution is the interiorisation of the will of those who have *Kenryoku Power*, and the transformation of the *Kenryoku Power* into *chikara power*. It is not an oppositional movement to this power (an anti (*hantai*) *Kenryoku Power*). A molecular revolution is not the taking of *Kenryoku Power* from those who have it – it consists of transforming the transcendent *Kenryoku Power* into an immanent *chikara power*. Like the hero of Melville's (2016: 25) story, *Bartleby, the Scrivener*, we might say, "I would prefer not to".

A multitude of Bartlebys are already in the process of emerging around the whole world. For instance, and specifically in Japan, they are emerging even in supermarkets where vegetables are sold. We allude to not wanting children to have to eat vegetables polluted by radioactivity, as far as possible. In thinking this way, the mothers who choose the least dangerous food are like students who – every spring, when they finish their studies – refuse to look for work: "I'd prefer not to do that job". More specifically, they refuse to endorse those companies that have no ethics; see, for example, Matsumoto Hajime and *his revolt of amateurs* (*shirotonoran*).[9] In the springtime, like the fireflies which all begin to shine, the students do not waste their *Jodo* emotion on unproductive *Kanjo* feelings but express their *chikara* power beyond any anti-establishment sentiment (beyond all *hantai Kenryoku*). These parents in supermarkets are suspicious of the media and government propaganda, and yet they do not think that they are disobeying it. They no longer believe in the *Kenryoku power*, nor in the scientific discourse, and they ignore the information they are bombarded with all day long. They behave like fragment(s) of *chikara power* and are no longer the parts of the 'Whole' which is Capitalism or Japan. They are, however, not disengaged; they remain related to the *Jodo* emotion of their children and the youth who shine like fireflies. And in making this small ethical decision – 'small' at the macro level, but already a tsunami – they make an ecosophic choice, an ethical choice.

Notes

1 Translated by Thomas Jellis.
2 In the French, *contre* can mean 'against', 'versus', 'anti'; it is oppositional. [Translator's Note]
3 In the original text, Murasawa and Naduad talk of *'le Japonais'* (the Japanese person) or *'ce Japonais de Tohoku'* (the Japanese person from Tohoku). They go on to refer to this person as *'il'* (he) and so for the sake of clarity and fidelity, I have followed suit. [TN]
4 "It is as if alongside and external to lions, tigers, rabbits, and all other actual animals, which form when grouped together the various kinds, species, subspecies, families etc. of the animal kingdom, there existed also in addition the animal, the individual incarnation of the entire animal kingdom" (Marx, 1976: 27).
5 On this question, see Nadaud (2011).
6 The term *'freeters'*, an abbreviation of the English for *free* and the German *arbeiter* (worker) originally referred to the young Japanese who had chosen to consider their relationship to work differently, in particular by choosing to have real free time. Therefore, they were opposed to the model that having a place in Japanese society necessarily involves a life exclusively devoted to work. This term, now much less positive, refers to the almost 2 million young people living in a very precarious state (who, many of them without a home, sleep in the internet cafes found in the cities of Japan).
7 As Massumi (2004: xviii) notes in his translation of Deleuze and Guattari's (2004b) *A Thousand Plateaus*, "Two words for 'power' exist in French, *puissance* and *pouvoir*". *Puissance* is often understood as a capacity, a power to do something; *pouvoir* is normally understood as a power over (see Bogue, 2007). I have followed Murasawa and Nadaud in their usage of a lower- or upper-case *P* to distinguish between the two. [TN]
8 "We must cease once and for all to describe the effects of power in negative terms: it 'excludes', it 'represses', it 'censors', it 'abstracts', it 'masks', it 'conceals'. In fact, power produces; it produces reality; it produces domains of objects and rituals of truth. The individual and the knowledge that may be gained of him belong to this production" (Foucault, 1991: 194).
9 See: http://keita.trio4.nobody.jp/index_com.html.

References

Bogue, R. (2007) *Deleuze's Way: Essays in Transverse Ethics and Aesthetics*. Abingdon: Routledge.

Deleuze, G. and Guattari, F. (2004a) *Anti-Oedipus* (trans. R. Hurley, M. Seem and H.R. Lane). London: Continuum.

Deleuze, G. and Guattari, F. (2004b) *A Thousand Plateaus* (trans. B. Massumi). London: Continuum.

Deproges, P. (1986) *Pierre Desproges se donne en spectacle*. Paris: Papiers.
Foucault, M. (1986) Of other spaces (trans. J. Miskowiec). *Diacritics*, 16(1): 22–27.
Foucault, M. (1991) *Discipline and Punish: The Birth of the Prison* (trans. Alan Sheridan). London: Penguin Books.
Freud, S. (2010) *Civilization and Its Discontents* (trans. James Strachey). New York, NY: W.W. Norton & Company.
Guattari, F. (2008) *The Three Ecologies* (trans. I. Pindar and P. Sutton). London: Continnum.
Guattari, F. (2012) *La révolution moléculaire* (ed. S. Nadaud). Paris: Les prairies ordinaires.
Marx, K. (1976) *Value: Studies* (trans. A. Dragstedt). London: New Park Publications.
Massumi, B. (2004) Notes on the translation and acknowledgements. In, Deleuze, G. and Guattari, F. *A Thousand Plateaus* (trans. B. Massumi). London: Continuum.
Melville, H. (2016) Bartleby, the scriverner: A story of wall street. In: *Billy Budd, Bartleby, and Other Stories*. New York, NY: Penguin Books.
Nadaud, S. (2011) *Fragment(s) subjectif(s): un voyage dans les îles enchantées nietzschée-nnes*. Paris: L'Unebévue.
Tarde, G. (1903) *The Laws of Imitation* (trans. E.C. Parsons). New York, NY: Henry Holt and Company.

9 Pathways to the machinic subject

Michele Lancione

Beyond oneself

What is the subject? We have been taught that the subject is that which accomplishes the action, the one triggering the event, and the one who bears responsibilities for it. The subject is the self-reliant human being; the Vitruvian *man* who is carving himself out of the magma of history; the entrepreneur who can control how things unfold and manage them for his own sake. We have been told that the subject is the one who comes into the world to define it by his own perception and cognitive abilities and to inhabit it as an extension of his will. This is a human-centred (and a gendered and racialised) view of the subject: the Cartesian rational man. Despite its woes, and notwithstanding post-colonialism, feminism, and decades of critical social thinking, this is still largely the canon upon which the 'autonomous self' is constituted and social responsibilities are attributed in the West and beyond. Rhetoric of the self-made man, neoliberal takes on one's own responsibility for wealth and success, blame placed on the poor and the diverse for their misfortunes are just few of the contemporary stances descending from that very specific understanding of what being human is and what it can supposedly portend (Braidotti, 2011; Amin and Thrift, 2013).

Yet, perhaps confronted with the incumbent self-destruction that this logic of the subject brings for the subject themselves, in recent years human subjectivity has undergone a process of profound scrutiny and rethinking in the humanities, arts, and social sciences alike. The canonic ideas of the subject have been contested and rethought, alongside the associated politics of being in time, in space and with the different other. Such rethinking owes much to different philosophical traditions and approaches. The list is long but here are just a few such approaches: for Derrida there is nothing like 'being' – "no complete, coherent and self-present subject who speaks, acts and senses" (Wylie, 2010: 107); for Foucault the subject is determined by historically contingent practices and the related 'truth' that have little to do with consciousness or rationality (Foucault, 1990; Kelly, 2013); for Badiou the external event is conceived as the foundation of the subject and the associate politics (Badiou, 2009); while Lefebvre (1991) signals a distress with bounded understanding of the self and opens the subject up to "the underlying material, spatial and political forces that have the possibility to transcend the visual domain" (Simonsen, 2005: 5).

What these and many other thinkers do, each one in their own way, is detach understanding of the subject from Cartesian boundaries, in order to open it to a wider set of interests and circumstances (Blum and Nast, 2000). The subject, again to follow Lefebvre, becomes 'spatialised':

> space – my space – is not the context of which I constitute 'textuality': instead, it is first of all my body, and then it is my body's counterpart or 'other', its mirror-image or shadow; it is the shifting intersection between that which touches, penetrates, threatens or benefits my body on the one hand, and all other bodies on the other.
>
> (Lefebvre, 1991: 184)

In this chapter I engage with one of the most productive of these rethinkings, namely the vitalist take on subjectivity of Guattari and Deleuze. My aim is to offer a brief introduction to it and then to propose some methodological reflections on the actualisation of the vitalist take on subjectivity. The solo writings of Félix Guattari are particularly inspirational in this sense. They are distinguished from Deleuze's writing since they are, in essence, even more openly political: they are about drawing alternative cartographies (Guattari, 2013); about engagement with micropolitical social forces (think of Guattari's involvement in Italian affairs, Berardi, 2008); and about alternative practices of subject articulation and care (at La Borde, but also in their critical engagement with the wider anti-psychiatric movement; see for instance 'Part II' of Guattari, 2009a). Guattari's solo works offer a politically engaged way to trace the assemblage of subjectivity within the broader molar and molecular articulation of the social field, which to me, as it will become clearer in a moment, offers poignant insights on the specific assemblage of the 'marginal' subject (Lancione, 2016; see also Patrick, 2016). In the second part of this chapter I present a number of pathways to the machinic subject that may be seen (and read as) an attempt to translate the sophistications of a vitalist and Guattarian approach to subjectivity into fieldwork practice. These reflections are carved out of my own ethnography work with marginalised urbanites, where I attempt the deconstruction of dominant (normative and stigmatising) narratives of the self through a grounded engagement with the performances and potentials of life at the margins (Lancione, 2013, 2015, 2016).

A vitalist subject

Thinking subjectivity matters. It is the story and the methodology of how we think who we are and, consequentially, of how we conceive both our being in the world, and associated political imaginaries. As Simon O'Sullivan clearly states in his lucid *On the Production of Subjectivity*, thinking about the subject matters because that is "the site and locus of a kind of battle against the homogenizing powers of capitalism, and especially its reduction and standardization of heterogeneity" (O'Sullivan, 2012: 2). The collaborative work of Deleuze and Guattari – which move from a critique of the Lacanian subject to the radical politics of *Anti-Oedipus*, *A Thousand Plateaus* and *What Is Philosophy?* – is essentially about the problematisation of subject-production under contemporary capitalism. According to them the subject is not only the 'subject of' capitalism but also 'subject to' and 'subject through' capitalism: it is master, slave, and liberator; player, played, and referee; character, spectator, and director at the same time and more – because both the subject and capitalism are constantly (re)constructed in new forms and fashions (see, for instance, the plateau '7000 B.C.: Apparatus of Capture'; see also Lazzarato, 2006).

Capitalism here is not merely understood as the engine of peculiar forms of exploitation and expulsion. It is, rather, a system of decoding that creates new axioms of experience: namely new commodified relationships that encompass economy to reach every aspect of life (Massumi, 1992; Deleuze and Guattari, 1987). This is a system of relations that are constantly (re)produced and created anew: the transversal production of new decoding and axioms is the true generative power of capitalism. Such ubiquity and power constitute the plane where (and through which) normalised forms of subjectivities are constituted *but also* the plane through which alternatives exist and more can be foreseen. Investigating subjectivity is key, in other words, because the subject experiences capitalism in the most extended – yet private – possible form. The rhizomatic 'social-field' that Deleuze and Guattari investigate (un)fold in and through the subject: how we conceive the latter is pivotal in imagining ways of decoding life without axioms.

A key aspect of the 'social-field' is that *life*, within and through the rhizome, is not something held by *some*body (the subject) counterposed to somebody else (the object), but a plane shared by *any*body. Vitalism, in this sense, is about the immanent life that anybody shares (Bennett, 2010), the common ground which occurs "across human and non-human materialities and in-between distinctions between body and soul, materiality and incorporeality" (Anderson and Harrison, 2010: 13). A human-centred narrative of the subject does not make any sense on the vitalist plane: there is no 'self' and no singular story to be told. Rather, Deleuze, and particularly Guattari, understand the subject as a desiring-machine assembled with and through a plethora of other (human and non-human) desiring-machines (Deleuze and Guattari, 1977). The boundaries of the self are provisional, contested and always (although mainly unconsciously) (re)assembled. There are three fundamental aspects of this understanding of subjectivity that are worth highlighting here. First, understanding the social as a force field of desiring-machines does not mean to abstract a general motion – or pattern, or theory – about how things go. Desiring-machines are transcendent (in their possibilities – see below) but are first and foremost immanent and material (in their articulations): they are matter of stuff and speech and more, which in their relation *produce* further relations and new desiring-machines (Bonta and Protevi, 2004; Buchanan, 2008).

Within this production the subject is not 'something' but a 'recollection', a 'collectivity', a 'refrain', a 'moment' (Braidotti, 2011; Guattari, 1996): "*humans constitute a machine* as soon as this nature is communicated by recurrence to the ensemble of which they form a part under specific conditions" (Deleuze and Guattari, 2009: 91). The subject in this sense could be at best thought of as a provisional and ongoing cartography of some of these machines, but a cartography that exceeds what the brain can think of it. As Thrift puts it: "Human subjects are [. . .] momentary creations [. . .], conjured into existence according to the summonings of particular contexts, and working to positional and dispositional ethics that they are often only vaguely conscious of" (Thrift, 2003: 2021). This does not deny the cognitive abilities of any human beings, but decentres those capacities, and understands humans as 'dividuals': "who for the most of the time are simply part of a combination of bodies or parts of bodies, resonating around a particular matter of concern" (Amin and Thrift, 2013: 50).

Second, what follows from the productive aspect of desiring-machinism is that the subject is never achieved but always in process. It is a matter of continuous entanglements and rearrangements that lead to the production of *constant* difference (becoming) (Deleuze, 1994; Colebrook, 2004). Deleuze and Guattari point to the fact that capitalism enhances the production of this difference (liberating coded – territorialised – assemblages), but at the same time it renders difference axiomatic in a continuous and repetitive form. In other words it makes life – any form of life – exchangeable, no matter its coded form. What follows is that through this reterritorialisation-deterritorialisation process *life is in becoming* and the subject, being part and content of that life, is also constantly moulded.

Third, if this processual aspect of (de)codification-axiomatisation is part of the 'actual-real' side of life (of what we see, perceive and can imagine), Deleuze and Guattari show how this is just half of what is possible. Attached to that 'actual-real' there is the presence of the 'virtual-real' side of life, the realm of what it is but does not emerge; of what is fluids but is not actualised; of the eventful and potentiality (Anderson et al., 2012; De Landa, 2005; Deleuze and Guattari, 1987). Importantly, the relation between the two (between the finite and the infinite) is continuous, a 'reciprocal determination': "both the actual and the virtual *together* constitute reality: the virtual needs its actualization, just as the actual is not complete without an account of its virtual side" (O'Sullivan, 2012: 139, emphasis in original). This portends something uniquely relevant: the subject, as any other recollection of a desiring-machine, can always obtain (and to a certain extent, it always already has) other possible articulations (specifications of content and expression). In other words it always has the potential to be (and to a certain extent it always already is) something else.

The productive aspect of desiring-machines, their processual nature, and their immanent actuality and virtuality can be considered as the basic tenets of a Deleuzian-Guattarian take on subjectivity. This is a post-human, vitalist take on who we are and what we portend. It decentralises our cognitive abilities to re-centre ourselves in the midst of things. Importantly, this understanding is not detached from reality, but it is intrinsically political (Patton, 2000): first, because such an understanding of the subject brings to the fore a specifically post-human ethics of being human, namely a contextualised ethics of partial objects that is spurred by generalisations and abstractions (Grosz, 1993; Bennett, 2010; Bignall,

2010); second, because such an understanding of the subject already contains a revolutionary anti-capitalist politics at its roots. This is a politics of what Guattari calls 'liberated desire'. The aim of investigating the post-human subject is to understand how to liberate it from its own annihilation (microfascisms) and to rework it in anti-axiomatic ways. Desire (the power of life) must be liberated: "it is not a question of adapting it, socializing it, disciplining it, but of plugging it in such a way that its process not be interrupted in the social body, and that its expression be collective" (Guattari, 2009a: 43).

Investigating subjectivity allows for an understanding of those molar and molecular forces *within which* alternatives can be crafted. It is Guattari, in this regard, that pushes the agenda of how to imagine different forms of subjective actualisation through art, activism, and engagement (Guattari, 1995). In the remaining part of this chapter I aim to contribute to this political ethos in relation to my job and what I do – namely, ethnographic research. If subjectivity is not about self-centred individuals, but about a micropolitical cartography, and given the relevance of subjectivity in constructing alternative forms of life to normative axiomatising, how can ethnography contribute to tracing that cartography and, therefore, contribute also to establishing an alternative political project? How, in other words, can we give justice to the suggestions derived from a vitalist take on subjectivity when it comes to a concrete piece of research and to actualised, grounded encounters in the field?

On methodological pathways

Subjectivity is formed by (and a producer of) machines of various forms and fashions, which exceed 'techné'[1] to encompass discourses, emotions, designs, infrastructures, flows, and more (Raunig, 2010; Lazzarato, 2006). Translating this understanding of subjectivity into broader research agendas demands a methodological attunement to the complexity of life (understood in a vitalist sense) and its irreducibility to clear-cut definitions, theories, or policies. Arguably, Guattari's schizo-cartographies of the subject are about this methodological overture, this attempt to let multiplicity flow without being reduced to calculus and interests (Guattari, 2013). As he puts it: "Recognition of these machinic dimensions of subjectivation leads us to insist, in our attempt at redefinition, on the heterogeneity of the components leading to the production of subjectivity" (Guattari, 1995: 4).

In my work I contend that academic research (and in particular ethnographic enquiries at the urban margins) bear a responsibility in not reducing the heterogeneity Guattari discusses to a set of manageable (i.e. axiomatising) particles. Field enquiries need to be attuned to a particular sensibility of the subject in order to produce knowledge relevant for the kind of antifascist self and society the two French philosophers argue for. So for all that the literature has begun to address methodological questions related to rhizome, becoming, and assemblage (Coleman and Ringrose, 2013), more can be said specifically about how to investigate the machinic subject. In what follows I offer a number of suggestions in this regard.

Performing and performed subjects

It is worth starting with the notion of performativity that, as the non-representational school in geography has clearly shown (Thrift, 2004; Gregson and Rose, 2000; Nash, 2000), helps in grasping the dynamics of the vitalist subject. Dewsbury (2000: 475) offers a clear analytic in this sense, with his definition of performativity as "the gap, the rupture, the spacing that unfolds the next moment allowing change to happen". Here performativity is not understood as a representation on stage or enacted only through discourse (Nelson, 1999), but rather as a spatio-temporal moment characterised by five aspects (Dewsbury, 2000). First, performativity is a matter of connections, between human and non-human alike. It is not about tactics and strategies, but about coping unconsciously, with continuous lines of interventions that intersect with oneself by nature, even when it does not seem so. Second, performativity is a heterogeneity of possible worlds coexisting within it and, third, a multiplicity of matters intersecting in such a way that boundaries between things are unavoidably melted. Fourth, nothing is pre-given in performativity: the world is constructed through body and language, where the latter is itself a bodily performance embedded in the world and relationally constituted through it. Fifth, performativity is a matter of breaks: it "is announced out of the diffuseness of the ineffably unexpected, out of subtle suggestions emerging out of an encounter itself" (Dewsbury, 2000: 477). It requires, as such, an openness to the unexpected in order to be registered.

Performativity understood in this way does not only act as a bridging concept (condensing more elaborate ideas and notions into a convenient theoretical tool), but it insinuates also a peculiar methodological sensibility. Performativity invites one to let things (both humans and non-humans) unfold, rather than to capture them still; it orientates us toward a processual attentiveness to the construction of the world, rather than an assumption of what it is or what it is not; and lastly it alerts us to the unexpected and the potential of life (Anderson and Harrison, 2010). If one takes these stances seriously, a genuinely different account of the subject can emerge. The subject is a rhizomatic Janus, on the one side performer and on the other performed by, in a process where the two side co-constitute themselves and exceed – always and at any point – their actualised representation. Approaching the subject through performativity is therefore about asking "the right questions in the right way" (Thrift, 2004: 132): it is about tracing the two sides of the Janus not in their distinctive (illusionary) forms, but in their constitutive (immanent and virtual) interaction.

In order to look at the subject through the performative lens, one needs a methodological approach able to listen, more than to speak; to record, more than to explain; and to being attentive to sudden, and mostly small, changes, more than to fixities. In other words, performativity needs to be rendered visible and traceable by a specific methodological attunement to the complexity and nuances of the post-human world. This is not about reinventing methods – but about rethinking *how* one might approach what one does in the field. In my research with homeless people, evicted people, and drug users in European cities I have attempted to map the performative subject by relying on three methodological stances, which I have freely worked out from Pile and Thrift's 'six pathways', contained in their inspiring edited book *Mapping the Subject* (Pile and Thrift, 1995). The six ways proposed by Pile and Thrift are directly related to the facets that compose performativity: positioning, movement, practices, encounters, visuality, and aesthetics/ethics.

In my research I have reworked these pathways combining them, hopefully without reducing their scope. In what follows I offer an overview of this methodological approach – based on the three stances of *committed positioning, reporting,* and *poetic mapping* – using my research as a brief point of illustration.

Committed positioning

The first methodological stance is positioning, which for Pile and Thrift is a way to "allow people to speak for themselves" (1995: 17). Positioning is related to one of the fundamental points in performativity theory: performativity is embodied, and thus only the one performing can – eventually – say or perform something about her or himself (Bell, 1999). Since the subject is never individualised but collective, positioning take place always in relation to something else, and it is revelatory only in the 'emergence' of these relations. Here I am referring both to the positioning of the researcher in the field and to the positioning of the 'researched' in respect to the former (Rose, 1997), but also to the positioning of the researcher in relation to the broader mechanosphere of human and non-human populating the field at that moment in space and time. At play here is an encounter that is in itself a machine: a meaningful positioning is one that allows this machine to be free to generate flows of information and desire, not one that tames them for the sake of 'data to be collected' or 'research outputs' to be achieved.

In order to be coherent with the anti-fascist politics evoked by the Guattarian-Deleuzian take on the subject, on the part of the researcher positionality should be attended to by the aim of producing practice-based, contextually relevant, and committed knowledge (Flyvbjerg et al., 2012). By 'committed' I mean a form of knowledge that is oriented toward – and meaningfully assembled with – the people, issues, and wider assemblages making up the research context. I intend this in at least two senses.

Firstly, in a postcolonial vein, committed positioning aims to produce knowledge relevant for the people and spaces it pretends to represent. This form of committed positioning owes a lot to feminist radical scholarship and a situated form of knowledge (Haraway, 1988; Gibson-Graham, 1994; Katz, 1994). It is a positionality that follows Haraway when she claims that the only way to be objective about something "turns out to be about particular and specific embodiment and definitely not about the false vision promising transcendence of all limits and responsibility" (Haraway, 1988: 582). Secondly, this time in a vitalist sense, to produce that kind of 'relevant knowledge' positioning must allow for collective subjects to emerge, without trying to harness them within the gridlock of mainstream methodologies. I argue that the solo work of Félix Guattari, combined with its radical activism in Italy and France, was attuned to this form of positioning: one made of travels, relations, chats, engagements, shared ideas, notes, bodies, cigarettes, flows, patched rhizomatic entanglements, and more. The commitment here is not only – as it is in situated knowledge approaches – to embrace (human) subjectivity as a mean to produce meaningful scientific knowledge; but it is also to produce (or allow for the production of) new collective subjects, assemblages, and affectivities *precisely* because only allowing for their production, a non-dualistic, non-axiomatic, and possibly empowering knowledge can emerge (see Lancione, 2017 for the practicalities involved in the production of such form of knowledge).

Committed positioning is not done in the realm of abstraction but in that of immanence, for the sake of fostering a "collective, unformed search, from above and below, for another kind of politics", a "micropolitics" or "molecular revolution" (Guattari, 2009b: 138). In the field this requires a political embodied form of performativity. It is about announcing one's own aim, scope, and political orientation – but to do so mainly via bodily gestures, presence, careful explications, patience, attention; rather than consent forms and prepared speech. I am not arguing against the latter, but evoking the necessity for a broader, collective, and committed kind of positioning. One that requires time, effort, patience, and a peculiar 'poetic' sensibility. In one of my recent research endeavours, with injectable drug users in Bucharest (Romania), it took me more than four months of constant presence and observations in order to feel 'able' to ask for a semi-structured interview. Positioning in that case has been matter of standing for hours in the cold, of facing provocations and suspicion, and of wondering around a needle-exchange service with the constant feeling of being out of place, unnecessary, obtruding. It was more about exposing my body and allowing it to become of public interest than about talking about my research. If in the end I was able to have meaningful encounters with users and service providers, it is mainly because they saw – as many of them told me – that I did not give up. That, in a sense, I allowed enough time and enough space for our machinic bodies to encounter: for my figure to become familiar to the other, for the audio recorder to become a constitutive piece of the relation, as much as the used syringes on the floor too. This does not mean that positioning came without mistakes on my side, or that I achieved that kind of 'constitutive negotiation' argued by Rose (1997). It means, however, that I tried to get closer to it via an embodied effort to position myself as a means to achieve a clear political end that was explicitly or implicitly *relevant* for the other (in that case, telling a story of discrimination, lack of service, pitiful politics, and avoidable death).

Such careful, embodied, and time-consuming form of positioning is hard to attain, and it is always provisional. It is, however, the only way to allow the 'researched' and the broader field to position in a meaningful way – namely in a way that allows them to 'speak for themselves'. This means that most of the time people will speak about (and do) things that seem utterly irrelevant for one's own research 'purposes'. However, this is not only unavoidable, but welcome; because it is in those things that most of the time one can find ways to deconstruct dominant imaginaries of the subject. It is thanks to the countless and apparently meaningless hours that I spent with homeless people in Turin (Italy) – hours spent talking about the weather, sport, and politics – that I gain insights into their subjective experience of homelessness. It has been in following them into public parks without any apparent reason, and in waiting for and with them outside institutional places just because they asked me to do so, that I got a (still superficial) grasp of how their subjectivities were assembled with the machine of the city. It is because I allowed for their schizophrenic cartographies of scavenging, rough sleeping, smoking, rambling around, and collecting stuff on the go to unfold in front of me, without trying to containing them, that at the certain point my body was allowed to become contaminated and to contaminate. This was done through a commitment: the body needed to be exhausted and re-assembled because that was the only way to let the subject of 'my' research to emerge. Such a form of open, committed positioning is key to investigate the post-human subject because it is about maintaining an open assemblage of research, to allow for its artefacts, its authors, and its public to position within it in hopefully empowering ways.

Re-porting

The subject does not, however, assume a position and remain fixed. The second methodological insight is the most fundamental one, and it is about movement/practice. If positioning allows the researcher to get a first hint of the power dynamics of subjectification in a particular loci, it is only following the course of action that subjectification can be fully unfolded. If Actor-Network approaches are undoubtedly helpful in this sense – because of their sensibility toward non-human agencies – it is the traditional ethnographic method implied by Robert Park and the Chicago School that I found most inspiring. Take, for instance, Anderson's *The Hobo* (Anderson, 1923). One of the most interesting features of this work, as well as of others that explicitly refer to Anderson's approach (Harper, 2006), consists in the fact that Anderson seems to act more like a reporter than a sociological researcher (Rauty, 1999). In following hobos and tramps on (and through) the streets of Chicago, Anderson was following direct advice received from Park – *go out there and report everything*:

> As an explorer, the reporter develops research techniques which correspond both to the image of the adventurer and to the altered conditions in the world of the big city: observation and interview, on-the-spot investigation and undercover research. The big city gives free rein to the art of observation [. . .] Just like the ethnologist, the reporter has his sources, 'key persons', like the concierge, the hotel porter, the bartender and his 'native' informants in the ethnic quarters.
>
> (Lindner, 2006: 29)

In reporting, the 'vitalist ethnographer' shows an uncanny attention to detail, a curiosity toward the smallest devices, a propensity to unfold the rich texture of reality and virtuality through description (Harrison et al., 2004; see also the works of AbdouMaliq Simone for a concrete exemplification of this method, 2010, 2018). It is worth nothing that 'reporting' should not be understood in a positivist sense.

There is no aim to 'report' the 'truth', but 'to re-port' in its original etymological sense: to bring back something, what can be seen and understood, which will always be mediated, and always partial, but nonetheless insightful if grounded and conscious of its limitations. Reporting is essentially about Latour's call to 'follow the action' (2005), which requires and attentiveness and a sensibility toward the 'messy', not easily definable, materialities of the social (Law, 2004). Such attention to details and openness to improbable associations, divergent patterns, and lines of flights resonates with Guattari experimental mapping of enunciations and unfolding 'chaos' (Guattari, 2013). Re-porting is a never finished mapping of the subject: it is about mapping it while, at least to a certain extend, one becomes *other* with and through it (at least if one is committed to such becoming). Ethnography is, after all, about the quest to write about the other while becoming *other*.

Once again following a politically charged Guattarian ethos, if reporting is not about representing the 'truth', it must be about recollecting performances in order to challenge – through the fine and gritty nuances of the social field – the established notion of the 'truth' (*Anti-Oedipus* is in the end, as Foucault reminds us in the introduction, precisely about this kind of 'political action', Foucault, 1977). To offer a short example in this sense I will borrow again from my ethnography with homeless people in Italy. The example is related to the assemblage of *clothing*. Homeless people were getting free clothes from several faith-based organisations in Turin. At first sight, one could register this as a charitable act allowing urbanites without a home to survive in the street. But in following the practices and movements of both the clothes and the people one can reveal a more complex machine. Few homeless people were using the above clothes to dress in, since they were mainly bartered in exchange for favours or sold on the black market in the city. In order to get the best clothes from the Faith-Based Organisation's (FBO) distribution – namely in order to get the only items worth bartering or selling – people were queuing as early as five-in-the-morning outside the service's door (in a city, Turin, where the average temperature in winter falls below zero degrees Celsius). All sorts of performance were put in place in order to get the best clothes – flattering, cheating, smuggling – and all sorts of urban assemblage were activated to sell them – studying the topography of the black market, choosing the right display for the clothes, connecting with people who could grant access to the market, etc.

What is important to notice in the production of this assemblage of clothing is that homeless people were able to reframe the assemblage of 'help', which was normatively direct to them, into something different: an assemblage of work, which allowed them to get-by autonomously; to become more than the simple recipient of help. Their subjectivity – in this plugging and unplugging of machines – was (re)constituted: at certain moments they were no longer only 'homeless' as morally subjectified by the charitable machine ('individuals in need of help') but hustlers as molecularly performed by themselves in their relations to the black market (Thieme, 2015). Re-porting these performances, I do not mean to romanticise homelessness. The reconfiguration of the subject just described was only a 'relative' deterritorialisation (Deleuze and Guattari, 1987: 135): the machine of help was still subjectifying them as 'homeless' in the bare moment they returned to the FBO to eat; to queue for more clothes; to ask for alms. However, re-porting practices, movements, and their nuanced entanglements with multiple bodies and potentialities allows one to reveal the tension between the constructions of these contrasting subjectivities. That tensioned politics is all that the margins are about. It is within that play, between bodies, strategies, charitable actions, second-hand clothes, normative discourses, informal economies, and more, that life unfolds and the cartography of the subject comes to the fore. Exploring and re-porting that complexity is of fundamental importance because it gives unprecedented insights into how (in this case) marginalised subjectivities are constituted and into the role, and responsibilities, of policies in the process (Lancione, 2013).

Poetic mapping

The research around the vitalist subject is a political act only if it is able to connect – as Deleuze and Guattari reminds us – the molecular to the molar.[2] That can be done only if in reporting the positioning and performances of the subject we are able to avoid the reproduction of stigmatising, disempowering, normalising, and fascist knowledge. In other words, one needs to be committed to map and then to foster the lines of change already present in the collectivity of the subject, though still largely unseen (because of the axioms at play). This is a committed form of geography, a mapping, as Pile and Thrift put it, which leads in three interrelated directions:

> First, towards redrawing the old maps in ways [. . .] that delegitimate the claims of truth of those maps which rely on an unspoken universal and universalised subject; second, toward the resymbolisation, resignification and parodic repetition of the maps that we already have; and, third, towards new maps of the subject [. . .] in order to re-establish tolerance towards different practices of body and subjectivity.
>
> (Pile and Thrift, 1995: 50)

In order to draw such a map one needs an 'ethico-aesthetic' paradigm (Guattari, 1995), a *poetic* inclination to life and its nomadic subjects (Braidotti, 2011). Poiesis is about the potentiality inherent in the performative subject and – in methodological terms – could be understood as an encounter with the world that exceeds the world itself (Dematteis, 1985). Becoming poetic – i.e. to perform (as researcher) in poiesis – does not mean to be particularly artistic. Rather, it is a tuning with the infiniteness of life, an opening to its unpredictability. It is a way to grasp the poetics of human movement, the "tacit choreographies of everyday events [that] create situational, personal and cultural expressions of life" (Engel, 2007: 20; Lanne, 2016).

Poiesis is therefore more an affective inclination toward the field rather than a methodological schema. It consists in being open to the unexpected, in recognising it, in acknowledging its importance, as well as to be emphatically tuned with the otherness of the world, being ready to follow it and being followed by it, re-considering the theoretical grid that usually binds field research. This kind of opening allowed me to see homelessness and drug users in a poetic way: not in the sense of romanticising them, forgetting the hardness and their trauma, but rather in the sense of being ready to grasp the nuances of that same trauma, to give justice to its disturbing movements and openings. I therefore learned that a discarded cigarette on the ground is never rubbish; that a traffic light is better red than green because the former allows one to beg at the stopping cars; that sometimes it is best not to sleep in a shelter but to face the cold of the night, because the shelter is too charged with negative affective atmospheres; that mixing heroin and 'legal' highs may be the only way to not feel one's own bones cracking; that there is a lot at play in an invisible difference between the size of two needles, pressure, vein, blood, and the gesture of a social worker explaining to you how to inject; and so on. *The commitment to map a reality that exceeds what is visible and what is understandable at first sight* – this is a politically charged approach to the subject fuelled by poiesis. To paraphrase Lindner, mentioning poiesis as one of the foundational elements of a vitalist approach to the subject "is no accident. In a culture distorted by norms, it is the empathy, intuition and sensitivity of the poetic person that can successfully break through the ossified shells of conventional thought. After all, the Greek poietes was the creative man per se" (Lindner, 2006: 204).

Concluding remarks

What is the subject? Guattari and Deleuze do not offer a reply to this question. They do however sketch insightful diagrams of thought and action. Altogether these diagrams disperse subjectivity in the collective stream of life but in doing so, they re-centre politics as a commitment to render visible and then dismantle the abstract machines of capitalism. In this chapter I enumerated a number of elements that may allow ethnographers to contribute to such an oeuvre. Starting from a *perfomative* understanding of the assemblage of life, I proposed *committed positioning*, *re-porting,* and *poetic mapping* as concrete tools to carry on critical explorations of the subject.

If the solo work of Guattari offers a number of fruitful insights to re-explore subjectivity theoretically, its real contribution relies in its offering of cartographic escapes for concrete action. To appreciate those, one should connect what Guattari wrote on the blank page to his committed positioning in the rooms of La Borde clinique, on the streets of Bologna, in the overcrowded seminar rooms of universities across Europe. It is the extension and openness of that performative body – made of speech, actions, and matter – that allowed for and contributed to the production of numerous collective subjects made of further thoughts, events, and action. This is what I find intellectually and affectively inspiring in the Guattarian subject, despite its limitations: it is the invitation to explore, to juxtapose, to transverse, and thus to seek pathways for emancipated – in the sense of collectively liberated – subjects.

Such explorations, once again, will always be partial and constantly (re) configured – but to undertake them is the best chance we have to understand and combat "the fascism in us all, in our heads and in our everyday behaviour, the fascism that causes us to love power, to desire the very thing that dominates and exploits us" (Foucault, 1977: xiii). Guattari's extensive cartography of the subject is all about this – increasingly needed – fight.

Notes

1 One of the three arts, or knowledges, discussed by Aristotle refers to technical knowledge ('techne'), oriented toward concrete applications and instrumentalisation. It is opposed to 'episteme' – as a broader way of knowing – and 'phronesis' – which can be understood as a form of wisdom deriving from praxis.

2 "Molecular escapes and movements would be nothing if they did not return to the molar organizations to reshuffle their segments, their binary distributions of sexes, classes, and parties" (Deleuze and Guattari, 1987: 219).

References

Amin, A. and Thrift, N. (2013) *Arts of the Political: New Openings for the Left*. London: Duke University Press.

Anderson, B. and Harrison, P. (eds.) (2010) *Taking-Place: Non-Representational Theories and Geography*. Farnham: Ashgate.

Anderson, B., Kearnes, M., McFarlane, C. and Swanton, D. (2012) On assemblages and geography. *Dialogues in Human Geography*, 2(2): 171–189.

Anderson, N. (1923) *The Hobo: The Sociology of the Homeless Man*. Chicago: University of Chicago Press.

Badiou, A. (2009) *Infinite Thought Truth and the Return to Philosophy*. London: Continuum.

Bell, V. (ed.) (1999) *Performativity & Belonging*. London: Sage.

Bennett, J. (2010) *Vibrant Matter: A Political Ecology of Things*. Durham, NC: Duke University Press.

Berardi, F.B. (2008) *Félix Guattari: Thought, Friendship, and Visionary Cartography*. London: Palgrave Macmillan.

Bignall, S. (2010) Affective assemblages: Ethics beyond enjoyment. In: Bignall, S. and Patton, P. (eds.) *Deleuze and the Postcolonial*. Edinburgh: Edinburgh University Press.

Blum, V. and Nast, H. (2000) Jacques Lacan's two-dimensional subjectivity. In: Crang, M. and Thrift, N. (eds.), *Thinking Space*. London: Routledge.

Bonta, M. and Protevi, J. (2004) *Deleuze and Geophilosophy: A Guide and Glossary*. Edinburgh: Edinburgh University Press.

Braidotti, R. (2011) *Nomadic Subjects*, 2nd ed. New York, NY: Columbia University Press.

Buchanan, I. (2008). *Deleuze and Guattari's Anti-Oedipus*. London and New York: Continuum.

Colebrook, C. (2004) The sense of space: On the specificity of affect in Deleuze and Guattari. *Postmodern Culture*, 15(1): 1053–1920.

Coleman, R. and Ringrose, J. (eds.) (2013) *Deleuze and Research Methodologies*. Edinburgh: Edinburgh University Press.

De Landa, M. (2005) Space: Extensive and intensive, actual and virtual. In: Buchanan, I. and Lambert, G. (eds.) *Deleuze and Space*. Edinburgh: Edinburgh University Press. pp. 80–88.

Deleuze, G. (1994) *Difference and Repetition*. London: Continuum.

Deleuze, G. and Guattari, F. (1977) *Anti Oedipus: Capitalism and Schizophrenia*. New York, NY: Penguin Books.

Deleuze, G. and Guattari, F. (1987) *A Thousand Plateaus*. New York, NY: Continuum.

Deleuze, G. and Guattari, F. (2009) Balance-sheet for 'desiring-machines'. In: Guattari, F. (ed.) *Chaosophy. Texts and Interviews 1972–1977*. Los Angeles: Semiotext(e). pp. 90–115.

Dematteis, G. (1985) *Le metafore della Terra*. Milano: Feltrinelli.
Dewsbury, J-D. (2000) Performativity and the event: Enacting a philosophy of difference. *Environment and Planning D: Society and Space*, 18(4): 469–473.
Engel, L. (2007) The dance of the now – poetics of everyday human movement. *Forum: Qualitative Social Research*, 9(2).
Flyvbjerg, B., Landman, T. and Schram, S. (eds.) (2012) *Real Social Science: Applied Phronesis*. Cambridge: Cambridge University Press.
Foucault, M. (1977) Preface. In: Deleuze, G. and Guattari, F. (eds.) *Anti-Oedipus: Capitalism and Schizophrenia*. New York, NY: Penguin Books. pp. xi–xiv.
Foucault, M. (1990) *The History of Sexuality, Vol. 1: An Introduction*. New York, NY: Vintage Books.
Gibson-Graham, J.K. (1994) 'Stuffed if I know!' Reflections on post-modern feminist social research. *Gender, Place & Culture*, 1(2): 205–224.
Gregson, N. and Rose, G. (2000) Taking Butler elsewhere: Performativities, spatialities and subjectivities. *Environment and Planning D: Society and Space*, 18(4): 433–452.
Grosz, E. (1993) A thousand tiny sexes : Feminism and rhizomatics. *Topoi*, 12(2): 167–179.
Guattari, F. (1995) *Chaosmosis: An Ethico-Aesthetic Paradigm*. Indianapolis, IN: Indiana University Press.
Guattari, F. (1996) Subjectivities: For better and for worse. In: Genosko, G. (ed.) *The Guattari Reader*. Oxford: Blackwell Publishers.
Guattari, F. (2009a) *Chaosophy: Texts and Interviews 1972–1977*. Los Angeles, CA: Semiotext(e).
Guattari, F. (2009b) *Soft Subversions*. Los Angeles, CA: Semiotext(e).
Guattari, F. (2013) *Schizoanalytic Cartographies*. London: Bloomsbury.
Haraway, D.J. (1988) Situated knowledges: The science question in feminism and the privilege of partial perspective. *Feminist Studies*, 14(3): 575–599.
Harper, D. (2006) *Good Company: A Tramp Life*. Boulder: Paradigm.
Harrison, S., Pile, S. and Thrift, N. (eds.) (2004) *Patterned Ground*. London: Reaktion Books.
Katz, C. (1994) Playing the field: Questions of fieldwork in geography. *The Professional Geographer*, 46(1): 67–72.
Kelly, M.G.E. (2013) Foucault, subjectivity, and technologies of the self. In: Falzon, C., O'Leary, T. and Sawicki, J. (eds.) *A Companion to Foucault*. London: Blackwell Publishers. pp. 510–525.
Lancione, M. (2013) Homeless people and the city of abstract machines: Assemblage thinking and the performative approach to homelessness. *Area*, 45(3): 358–364.
Lancione, M. (2015) The city and 'the homeless': Machinic subjects. In: Frichot, H., Gabrielsson, C. and J. Metzger (eds.) *Deleuze and the City*. Edinburgh: Edinburgh University Press.
Lancione, M. (ed.) (2016) *Rethinking Life at the Margins: The Assemblage of Contexts, Subjects and Politics*. London: Routledge.
Lancione, M. (2017) The ethnographic novel as activist mode of existence: Translating the field with homeless people and beyond. *Social & Cultural Geography*, 18(7): 994–1015.
Lanne, J-B. (2016) The machine and the poet: A tale about how the subject goes into the field (and how it comes back). In: Lancione, M. (ed.) *Rethinking Life at the Margins: The Assemblage of Contexts, Subjects and Politics*. London: Routledge. pp. 136–150.
Latour, B. (2005) *Reassembling the Social*. Oxford: Oxford University Press.
Law, J. (2004) *After Method: Mess in Social Science Research*. Abingdon: Routledge.

Lazzarato, M. (2006) The machine. *Transversal*. Available online at: http://eipcp.net/transversal/1106/lazzarato/en/print

Lefebvre, H. (1991) *The Production of Space*. Oxford: Blackwell Publishers.

Lindner, R. (2006) *The Reportage of the Urban Culture: Robert Park and the Chicago School*. Cambridge: Cambridge University Press.

Massumi, B. (1992) *A User's Guide to Capitalism and Schizophrenia: Deviations from Deleuze and Guattari*. Cambridge, MA: The MIT Press.

Nash, C. (2000) Performativity in practice: Some recent work in cultural geography. *Progress in Human Geography*, 24(4): 653–664.

Nelson, L. (1999) Bodies (and spaces) do matter: The limits of performativity. *Gender, Place and Culture: A Journal of Feminist Geography*, 6(4): 331–353.

O'Sullivan, S. (2012) *On the Production of Subjectivity: Five Diagrams of the Finite-Infinite Relation*. New York, NY: Palgrave Macmillan.

Patrick, D.J. (2016) Between the fool and the world: Towards a (re)contextualisation of assemblage thinking in the contemporary university. In: Lancione, M. (ed.) *Rethinking Life at the Margins: The Assemblage of Contexts, Subjects and Politics*. London: Routledge. pp. 215–228.

Patton, P. (2000) *Deleuze and the Political*. London: Routledge.

Pile, S. and Thrift, N. (eds.) (1995) *Mapping the Subject: Geographies of Cultural Transformation*. London: Routledge.

Raunig, G. (2010) *A Thousand Machines*. Los Angeles, CA: Semiotext(e).

Rauty, R. (1999) Introduction. In: Rauty, R. (ed.) *On Hobo and Homelessness*. Chicago, IL: University of Chicago Press.

Rose, G. (1997) Situating knowledges: Positionality, reflexivities and other tactics. *Progress in Human Geography*, 21(3): 305–320.

Simone, A. (2010). *City Life from Jakarta to Dakar: Movements at the Crossroads*. London: Routledge.

Simone, A. (2018). *Improvised Lives: Rhythms of Endurance in an Urban South*. Cambridge: Polity press.

Simonsen, K. (2005) Bodies, sensations, space and time: The contribution from Henri Lefebvre. *Geografiska Annaler, Series B: Human Geography*, 87(1): 1–14.

Thieme, T. (2015) Turning hustlers into entrepreneurs, and social needs into market demands: Corporate-community encounters in Nairobi, Kenya. *Geoforum*, 59: 228–239.

Thrift, N. (2003) Performance and. . . *Environment and Planning A*, 35(11): 2019–2024.

Thrift, N. (2004) Performance and performativity: A geography of unknown lands. In: Duncan, J.S., Johnson, N.C. and Shein, R.H. (eds.) *A Companion to Cultural Geography*. Malden: Blackwell Publishers.

Wylie, J. (2010) Non-representational subjects? In: Anderson, B. and Harrison, P. (eds.) *Taking-Place: Non-Representational Theories and Geography*. Farnham: Ashgate. pp. 99–114.

10 Memorial persistence

A hurricane in twelve refrains

Rebecca Catarelli

I

The hurricane arrived in late August, 2011, moving quickly up the East Coast of the USA, a Category 1 storm expected to make landfall in New York. At our home on a remote Vermont mountain, we spent some time the day before securing items in the yard, expecting moderate wind damage and utility disruption. Power outages and road closures are routine in New England during the harsh winter; a foot of snow can easily leave Vermonters housebound without phone or electricity for several days. But so far inland, and so accustomed to weathering these events in relative isolation, there seemed no reason to panic *when the hurricane arrived.*

The hurricane arrived on an unassuming late-summer morning, warm with a light rain. My husband left the house early and I took our son outside to play in the puddles, unconcerned by the minor downpour. Yet soon the rain grew heavier and we were forced inside, drenched through our raincoats and boots. We watched anxiously from the windows as the weather became more alarming. The lawn disappeared under a muddy river and deep channels were cut into the gravel driveway as the torrent grew relentlessly. Before long, the power went out and we sat in the dim house for hours, watching the world around us flood *as the hurricane arrived.*

After the hurricane arrived, we ventured cautiously outside to assess the damage. The road up the mountain had been washed away, making it impossible for a vehicle to pass. Visits to neighbours revealed universal panic and confusion. In the early evening, my husband finally returned home, having made his way home through the woods, avoiding chasms newly cut in the roads. He never arrived at his destination that morning, instead waiting out the storm in a bar that was kind enough to open its doors on a Sunday morning. He finally made it back to the bottom of the mountain after convincing emergency crews to let him pass over paved roads left barely intact after *the hurricane arrived.*

II

Guattari's concepts "*crucially depend upon actually being put to work*" (Stengers, 2011: 141) having been created not for theoretical application alone but in the interest of unrestrained proliferation. Stengers calls his ideas 'operative constructs' because they require users to involve themselves in "experimenting with new tales and modes of intervention" (2011: 141). Here, I put Guattari's concept of ecosophy to work to engage the storm as an event that is both inexpressible and inexhaustible, to deliberately participate in the singularity of Hurricane Irene through the cultivation of new connections and affinities. Ecosophy, an "ethico-political articulation" (Guattari, 2000: 28) between three mutually influencing ecological registers (the natural environment, social relations, and the individual human subject), understands subjectivity as always collective, with affective intensities forging transversal links between often-diverse elements so that ecology not only "questions the whole of subjectivity" (Guattari, 2000: 52), but is accompanied by an imperative to prolong the expression of an event because operative constructs *crucially depend upon actually being put to work.*

Ecosophy *crucially depends upon actually being put to work* as it aims to generate a responsiveness that expands upon a singular event without claiming any representative powers over it. As "dissident vectors" (Guattari, 2000: 45), hurricanes form outside our field of awareness, their arrival is difficult to predict and even after landfall they remain idiosyncratic. With their surprising force and destructive potential, hurricanes are 'ecosophic objects', through which subjectivity as a machinic process can be mapped in any conceivable manner or direction. Efforts to respond to the storm must therefore avail themselves of the storm's virtuality rather than simply attempting to explain its circumstances, responding to a world that always exceeds our powers of representation and sustaining rather than halting the progression of the machine that *depends upon actually being put to work.*

As ecosophy *depends upon actually being put to work*, I have taken up Guattari's call to "re-learn ritornello games" (1996: 165). Prompted by the idea that "an environment responding to chaos gives rise to a 'chaos-rhythm'" (Dosse, 2011: 253), this chapter is the outcome of a semi-poetic experiment using refrains, a game that aims to produce experience as much as analysis and to call attention to the process through which subjectivity continues to reorganise in the wake of the storm. Ideas must be accompanied somewhere regardless of outcome, a process Guattari describes as setting a machine in motion that is attempted here through a game responding to a world that defies our efforts of capture. The machine, instead, draws us along on a perceptually inventive encounter through which theory *is actually put to work.*

III

A poem is a "*momentary stay against confusion*", according to Robert Frost, himself a notable Vermont resident, who wrote one of the most celebrated refrains in American poetry in "Stopping by Woods on a Snowy Evening". Frost's words on confusion were quoted by the executive director of the Vermont Humanities Council to explain the proliferation of art in the state following Hurricane Irene. The creative process, he explained, helps the individual cope with the experience "in a tentative way" (Lindholm, 2012), so that art is an effort to "enframe chaos" (Grosz, 2008: 9). By narrating a chaos-rhythm into the lingering aftermath of the storm, Hurricane Irene is enframed briefly here against its ongoing chaotic trajectory, generating a *momentary stay against confusion.*

This momentary stay against confusion is a deliberate effort towards "sensible finitude" (Guattari, 1995: 112) in response to the storm's chaos; it is a representation drawing on and participating in the storm's continued creative potential. As Frost tells us in *The Figure a Poem Makes*, written a year after the 1938 Hurricane also decimated the state of Vermont, "a poem can have wildness and at the same time a subject that shall be fulfilled" (2007: 131). Although it ends in a concrete composition, a poem functions as a sensory anchor that must be allowed "the greatest freedom of the material to move about in it and to establish relations in it regardless of time and space, previous relation, and everything but affinity" (Frost, 2007: 131). The hurricane's affective immediacy is diminished over time but sensations, impressions, and affinities remain to be recast as *momentary stays against confusion.*

This momentary stay against confusion is assembled through the prolonged machinic force of the storm, along the "point of negotiation between complexity and chaos" (Guattari, 1995: 111). It is a poetic effort that extracts from and simultaneously participates in the passage of the storm, deploying refrains that carry the hurricane in a new direction. According to Frost, "Like a piece of ice on a hot stove the poem must ride on its own melting" (2007: 131): poems are expressions that move along with the world while also attempting to move the world in a small way. In the end, regardless of the effect of these refrains, "Its most precious quality . . . will remain its having run itself and carried away the poet with it" (Frost, 2007: 131), leaving in its wake *a momentary stay against confusion.*

IV

A disturbance in the atmosphere was first recorded by a ship south of the Cape Verde Islands on 9 September 1938. Gaining speed in the following days, the loosely organised system developed the central rotation of a tropical storm as it travelled across the Atlantic. It strengthened into a powerful hurricane as it neared the Bahamas and, feeding off the winds of two other strong weather systems, its rotation intensified as it travelled up the Eastern seaboard of the US. The hurricane was forced westward by high pressure near Bermuda and it "veered inland and smote the North Atlantic seaboard" (The Brattleboro Daily Reformer, 1938: 1), making landfall in New York with a devastating storm surge. A number of seaside communities in its path were utterly destroyed by this unparalleled *disturbance in the atmosphere.*

This disturbance in the atmosphere continued northward over land so rapidly that the temporarily calm conditions caused by the passage of the storm's vertical eye were recorded as far inland as southern Vermont. The hurricane finally lost momentum as it moved northward and crossed the Canadian border, but not before killing 680 people throughout the region and destroying countless roads, bridges, and homes. With the passage of the storm, as many as a quarter billion valuable trees "fell like jacksaws and defied enumeration" (The Brattleboro Daily Reformer, 1938: 1). According to a local newspaper, "The value of trees destroyed was incalculable, aside from the havoc they wrought by falling" (Terrific toll, 1938: 1), with the apple and maple sugar crops ruined for years by this unparalleled *disturbance in the atmosphere.*

This disturbance in the atmosphere, brought South Newfane, Vermont, "the most severe flood damage of any part of the county" (The Brattleboro Daily Reformer, 1938: 10), much as it would again during Hurricane Irene in 2011 when I would be stranded above the town for several days. "House after house . . . either fell or was badly damaged by the onslaught of water which tore out all bridges and the highway. . . . Former residents returned to find the homes they once occupied not even in sight, wide expanses of stone, lumber and debris being scattered in their places" (The Brattleboro Daily Reformer, 1938: 10). A photo from a 1938 publication commemorating the storm shows damage identical to that wrought by Hurricane Irene, with the river re-routing itself through and then behind a road full of houses, taking a number of them with it in the powerful surge. In 2011 as well, locals were stunned as they returned to survey the damage: their town altered enduringly by *this disturbance in the atmosphere.*

V

The content of a natural disaster is beyond capture because catastrophic events are *surreal,* meaning that they are super-, above-, or beyond-real. This word was used frequently to describe Hurricane Irene, with one witness observing, "It was so surreal. I'm here and I'm watching this, but I have nothing to compare it with. I am watching nature at a scale and level I have not seen before. I was looking at it and listening to it, but I wasn't accepting it" (Brandon et al., 2012: 66). Another resident offered a similar description: "Those first couple of days everything seemed so surreal that I don't think I was processing" (Brandon et al., 2012: 70–71); another explained that "at first it was surreal, like a movie – everyone was living on an adrenaline rush" (Brandon et al., 2012: 98). The hurricane is not a reality, but a *surreality*: it occurs in a place above, over, and beyond the representable world, as an event that is experienced by the body but that also exceeds it. Yet within this excess, affinities can be developed, following Frost, connections made "regardless of time and space, [and] previous relation" (2007: 131) that partake of the disaster but preserve *its content in a place beyond capture.*

The content of a natural disaster can never be captured, as artist Kara Walker notes in her work on Hurricane Katrina. She describes the months following the storm during which "the narrative of Hurricane Katrina had shifted . . . to a more assimilable legend. . . . And always at the end of these tales, reported on the news, in newspapers, and by word of mouth, always there is a puddle – a murky, unnavigable space that is overcrowded with intangibles" (Walker, 2007: 7). One way this unnavigable space can be addressed is through the production of discursive expressions that retain a sense of intangibility. As an alternative to containment, affinities produce sufficiently and deliberately precarious representations that endeavour to make contact with what Whitehead has described as nature's "ragged edge" (1978: 50) where we find a chaos to which words cannot ascribe meaning, a chaos that *can never be captured.*

The content of a natural disaster can never be captured, and rather than attempting to portray the singularity with any fidelity we can instead be carried away by the force of the storm as a "nucleus of chaosmosis" (Guattari, 1995: 80) through which new configurations continue to emerge. The storm in its immediacy has no analytic agency; existing at a nexus of the virtual and the actual, it is an ecosophic object, at the centre of which the unexpected, unfamiliar, and unthinkable can only be experienced. According to Guattari, "The primary purpose of ecosophic cartography is thus not to signify and communicate but to produce assemblages of enunciation capable of capturing the points of singularity of a situation" (1995: 128). Cartography, from this perspective, does not communicate the storm but enunciates the existence and resulting influence of the rupture, which is an event of pure sensation minus the subsequent understanding, evaluation, or imagination. We can relate our experiences of the singularity, but any account does little to recreate the original intensity, the *content of which can never be captured.*

VI

The expression of a natural disaster can never be completely exhausted, and ecosophy offers a way to keep the 'ragged edge' in sight without relying on the problematic notion of accuracy. It allows us to cautiously revisit the singularity itself, not returning to the moment of the event, but apprehending the event as a new event in the current present, engaging nature's "unexhaustiveness" (Whitehead, 1978: 14) to *create* the world rather than simply account for it. This approach emphasises the ongoing process of change and extension, rather than the consistency of a limited event. As the storm continues by way of new sensations, impulses, and affinities, ecosophy invites us to participate in the newly occurring existential refrains that constitute the shifting storm territory and through which *the expression of a natural disaster is never exhausted.*

The expression of a natural disaster can never be completely exhausted, for a singularity of this magnitude continues to resonate in unforeseen ways long after the physical event itself has run its course. Following the 1938 Hurricane in Wilmington, VT, a flood marker was painted on the side of the police station at a height of five feet above the sidewalk to commemorate the town's complete inundation when the Deerfield River overflowed its banks during the storm. The marker is located next to a stoplight at the heavily trafficked corner of the town's main intersection. Stopped travellers, including thousands of tourists each year, have the opportunity to notice the sign and encounter the extended presence of this historic storm as a newly iterated event that fails to *exhaust the expression of the natural disaster.*

The expression of a natural disaster can never be completely exhausted because we can continue to experiment with different ways to work around and through the point of rupture to prolong the force of the singularity even as this force is altered through varied repetition. By addressing the 'unnavigable space' of an event, we may be able to avoid what Guattari calls "semiotic enslavement" (1996: 143) by generating a deliberate suspension along the storm's continuity that calls attention to the refrains through which the event continues to unfold. This involves re-enlivening the original event, less in an attempt to represent and more an effort to *re*-present, to make present again as a minor affective experience, the passage of expression that the storm machine set in motion, which *can never be completely exhausted.*

VII

The production of subjectivity was discernible in the sparse mountaintop community the morning after Hurricane Irene. In an area where the neighbouring houses are barely, if at all, visible through the dense forest, everyone was out on what was left of the roads, sharing news of damage sustained, speculating about the condition of the town below and determining the extent of food and fresh water supplies. Through this routine of frenzied information sharing we collectively established the territory of a singularity newly occurred. Between the 15 or so households scattered across the mountain, we pieced together an initial hurricane narrative, a story enacted through refrains and *productive of subjectivity.*

Our subjectivity was produced on a long hike down the mountain as we scrambled through the woods in places where the road had been damaged. Every so often we encountered others out on the broken road, like us astounded by the damage and unsettled by the fact that our knowledge of the world extended in only a two-mile radius. One industrious neighbour was already out on his small excavator filling in some of the smaller rifts in the road. It was becoming clear that we would be able to leave the mountain in a day or so, although where we might be able to go beyond that remained unknown. Still, our ongoing movement helped us cope with the confusion brought on by the *production of our subjectivity.*

The production of subjectivity altered the contours of life throughout Vermont as residents lost their homes and livelihoods in Irene's flooding. For months after the storm, there were still possessions, washed out of ruined homes, stuck in the branches of trees along the rivers. One interviewee on a Vermont Public Radio program described these as "things and variations of everybody's life that went down the river" (Cohen, 2011). The storm undercut very stable elements in the 'life' of Vermonters: it is a life in the singular, as he described it, a collective existence that has been altered irreversibly through the force of the storm. Those affected did not simply lose something valuable, but their common understanding of the environment's agency has been transformed by the hurricane machine and caught up in *its production of subjectivity.*

VIII

"*Betrayal becomes something to be affirmed*" (Alliez and Goffey, 2011: 10) when a hurricane suddenly and drastically alters the boundaries of the familiar world. Vermonters were shocked and disturbed by the surprise of Hurricane

Irene, alarmed by the changes it produced, both physical and subjective. Yet what was at first a sudden anomaly has become an enduring part of our selves and communities, a naturalised element of our lives. We were forcefully overwhelmed by the environment and as the unfamiliar event was experienced and processed we became a subject group that both participates in the storm's refraining as much as we are shaped by it as the *betrayal became something to be affirmed.*

Betrayal becomes something to be affirmed after the storm's occurrence, as dissident vectors reorient the world according to their trajectories. Events and experiences, and by extension communities and individuals, are defined by their threshold characteristics, their contrary and incongruous features rather than their consistencies, so that "what holds an assemblage together is not the play of framing forms or linear causalities but, actually or potentially, its most deterritorialized component" (Deleuze and Guattari, 1987: 336). The world is normalised around this deterritorialisation so that even after we were stunned by the arrival of Irene, unnerved by its confusion and disruption, we were unexpectedly affirmed as a community through *the storm's betrayal.*

The storm's betrayal was definitively affirmed when the flood marker was painted in Wilmington following the 1938 Hurricane, fixing in place a rupture by which the world was thereafter organised. The hurricane was captured as an indisputable event on the outer wall of the police station, a very concrete reminder that the town had once been completely underwater during this aberrant event. Hurricane Irene similarly inundated the town in 2011, altering the flood marker in meaning and magnifying its significance. Suddenly, the original marker called to mind a repeated event rather than a staggering anomaly. As the town recovered, a second flood marker commemorating Irene was painted on the outer wall of the police station, several inches above the original. The 1938 storm was multiplied and amplified so that the original storm became an event refrained, a *betrayal that was definitively affirmed.*

IX

Refrains achieve consistency within the world, changing subjective milieus through "intensive repetition, as piercing existential affirmation" (Guattari, 1995: 28). It is through refrains that singularities endure as they configure the territory of an event: news stories, works of art, nightmares, the simple act of driving past a desolate patch of land where a house once stood. In December 2013, Vermont Public Radio broadcast a refrain simultaneously celebrating and lamenting the departure of the Federal Emergency Management Agency (FEMA) as it finally closed its Vermont office, opened in the wake of Irene. While FEMA helped secure $23 million in recovery aid for the state, it also left Vermonters outraged at the bureaucratic struggle of rebuilding (Cohen, 2013). This story both generates a fixed record of the hurricane's aftermath while also deploying the affects of shock, frustration, and helplessness, the murky intangibles that continue to linger after the storm as *refrains achieve consistency.*

Refrains achieve consistency within the world even as they alter the world, reinforcing the hurricane's domain of influence while simultaneously reterritorialising the margins of this domain. An event intensifies through this repetition, developing the unrepeatable quality required for it to reach a "threshold crossing . . . that legitimates a relation of full existential belonging to a group-subject" (Guattari, 1996: 165), whereby the storm bypasses the point at which it can still be considered just another storm and becomes the incontrovertible event of Hurricane Irene. An event must cross this threshold of impact before it has the power to forge subjectivities, even while the details of the impact itself continue to play out endlessly as the storm's *refrains achieve consistency within the world.*

Refrains achieve consistency within the world, manifesting in the feeling of being constantly confronted with storm damage as much as in an article in the local newspaper marking the anniversary of Irene, dealing in the actual and generating the virtual. These sustaining rhythms make the storm at once a consistent *and* an evolving event around which local subjectivity aligns and continuously realigns, establishing the storm's territory through the occurrence of "existential ritornellos" (Guattari, 1996: 165). Irene is at once an accrual of dialogue, action, and analysis, while also a felt excess beyond our capacity for communication. It is an atmosphere of signs and affects to which we all belong, which also belongs to us, and through the rhythm of which *our world achieves consistency.*

X

A "*residual horizon of discursive time*" (Guattari, 1995: 102) emerges in the wake of a natural disaster as recovery efforts progress, present circumstances align with past events, and subject groups cohere around the singularity. According to Guattari, "the ecosophic problematic is that of the production of human existence itself in new historical contexts" (2000: 34) so that affected groups are produced anew as a major hurricane continues to reterritorialise the narrative of history. Over three years after the storm devastated Vermont, local media continues to regularly report on Irene's prolonged aftermath, while nearly every day we are confronted with the sight of empty lots where homes were washed away and others are still marked for destruction. The hurricane persists as a machine in-transit, riding on its own melting as its narrative continues to unfold along a *residual horizon of discursive time.*

As a residual horizon of discursive time, history can be understood as a mapping of singularities through which time doesn't exclusively move forward, it moves equally backwards and in fact all around us, as our subjectivity incorporates new events and realigns with old events so that space and time are "never neutral receptacles; they must be accomplished, engendered by productions of subjectivity" (Guattari, 1995: 103). As Guattari explains, it is at the intersection of "radical innovations and once forgotten, then reactivated, ancestral machinic lines, that the movement of history singularises itself" (1995: 41). In this way, the 1938 Hurricane and Hurricane Irene, both considerable singularities in their own right, converge across time, combining strength to form a larger and much more formidable storm presence as they unite along the *residual horizon of discursive time.*

Along this *residual horizon of discursive time*, temporality realigns according to transversal movements across the three registers through forces of chaos that don't necessarily advance in foreseeable ways. It is an activity of constant cognitive and affective reorientation, leaving in its path a horizon along which we can now expect another hurricane. Vermont is reconstructing its towns based on a new orientation in which catastrophic storms are no longer atypical but rather periodic devastating events. Town planners are prohibiting development in flood zones, reinforcing bridges and riverbanks and enacting mitigation directives to deal with the next 'big one'. We anticipate a changed and changeable future based on a newly signified past, aligned along a *residual horizon of discursive time.*

XI

Guattari's call to experiment the world through a consideration of the incorporeal results in a project predisposed to failure because affect, its driving force, is "manifested by unlocatable transferences" (1996: 158). Attending to the production of subjectivity is therefore one way to locate affect, to address the unnavigable spaces emerging throughout the event of a hurricane. An ecosophic cartography considers how the three registers are transversally linked through a singularity of this magnitude by engaging with the nucleus of this linkage: the affective intensity. This cartography therefore requires the creation of a new event that prolongs the expression of the hurricane itself, not locating the original singularity but continuing it through new affective pathways in response to *Guattari's call to experiment.*

Guattari's call to experiment is accompanied by an imperative to work with his philosophies in terms closer to his own. As Stengers reminds us, we are not to 'use' the philosophy of Guattari, but to 'do' the philosophy of Guattari. This chapter, in other words, must be "animated by a mutant creationism" (Guattari, 1995: 116), however provisional, that provides a way to think through natural disasters not as static events but as abstract machines both altering the parameters of subjectivity and continuing on as limitless agents of expression. This modestly experimental narration, though its words may be insufficient against the surreality of the event-in-action, serves to prolong, 'in a tentative way', the expression of the storm in response to Guattari's *call to experiment.*

Guattari's call to experiment would be ignored if this writing were not complicated by a determined gesture towards the non-discursive. Mapping the transversality of a hurricane can only be a post-emptive action, but one in which creative impulses, themselves transversal, can continue the map in any direction. Here, it is movement that makes a map more than representation, and though the event of a storm may seem temporally and spatially bounded, it possesses the itinerancy of a poem in that "The possibilities for tune from the dramatic tones of meaning struck across the rigidity of a limited meter are endless" (Frost, 2007: 131). Poetic enunciation, therefore, cultivates a restless tension between the virtual and the actual, a chaosmic stasis that is developed through refrains but must ultimately ride on its own melting as it responds to this *call to experiment.*

XII

As the narrative evolved in the days and weeks following the storm, stories emerged from all over Vermont of towns isolated for days and lives ripped apart by the flooding. It continued to rain heavily off and on for several weeks after Irene passed and many locals described a visceral feeling experienced with each downpour. As one woman whose home was destroyed by the flood explained, "we no longer trust the river. She took something away from us . . . when it rains you get really nervous. And you watch the water closely" (Cohen and Daniels, 2011). Another who also lost her home described the panicked 'water dreams' that "haunted her" during the first several weeks after the storm (Brandon et al., 2012: 72). Water became a source of anxiety and an object of mistrust *as the narrative evolved.*

As the narrative evolved, the residents of South Newfane organised a parade on the first anniversary of the storm to commemorate Hurricane Irene. Beginning at a huge pile of rubble where a house once stood, participants marched along newly reconstructed roads and through an historic covered bridge that miraculously survived the flood. Tractors pulling farm animals, pick-up trucks full of preschoolers and locals on foot carried signs declaring "Together We Survived Irene", "We still love our river", and "No bridges, no roads, no problem!" My own family made it through the long, transient years of graduate school and we finally found a place to settle in an old red farmhouse just outside of town. The house needs work, with its peeling paint and sagging beams, but there is a swimming pool in the yard and a field big enough to play baseball. We'll fix it up over the years, add an apartment for my mother-in-law, watch the kids grow into teenagers *as the narrative evolves.*

The narrative continues to evolve because ecosophy doesn't ever conclude: it is successful only if it has "found being" (Guattari, 2011: 35), producing sensation as much as sense, remaining open at its ends, affirmative but inconclusive. To achieve such a momentary stay against confusion we must assemble "a narration whose primary function is not to engender a rational explanation but to promote complex refrains, supports of an intensive, memorial persistence" (Guattari, 1995: 61). Writing does not contain affect but it is a tool for generating it, for transferring singularities anew, involving the reader in the persisting hurricane machine as it continues to refrain through our lives. After the sellers accept our offer, we learn that our red farmhouse is in a 'Fluvial Erosion Hazard Area', though the tiny brook across the road never seems to contain more than a trickle of water. The area experienced only minor erosion during Hurricane Irene, yet the water dreams remain, even five years later. Unsettled again by the hurricane, we give up on the farmhouse and continue our search for a home, riding on the storm's melting *as the narrative evolves.*

References

Alliez, E. and Goffey, A. (2011) Introduction. In: Alliez, E. and Goffey, A. (eds). *The Guattari Effect*. London: Continuum. pp. 8–16.

Brandon, C., Garman, N. and Ryan, M. (2012) *Goodnight Irene*. Keene, NH: Surrey Cottage Books.

The Brattleboro Daily Reformer. (1938) *The Flood and Hurricane of 1938*. Brattleboro, VT: Brattleboro Daily Reformer.

Cohen, N. (2011) Picking up debris months after tropical storm Irene. *Vermont Public Radio News*, radio transcript, Vermont Public Radio, 17 November. Available online: www.vpr.net/news_detail/92521/picking-up-debris-months-after-tropical- storm-irene/

Cohen, N. (2013) FEMA closes vt. office, ending chapter in post-irene recovery. *Vermont Public Radio News*, radio transcript, Vermont Public Radio, 19 December. Available online at: http://digital.vpr.net/post/fema-closes-vt-office-ending-chapter-post-irene-recovery#stream/0

Cohen, N. and Daniels, P. (2011) Coping emotionally after tropical storm Irene. *Vermont Public Radio News*, radio transcript, Vermont Public Radio, 6 December. Available online at: www.vpr.net/news_detail/92672/coping-emotionally-after-tropical-storm-irene/

Deleuze, G. and Guattari, F. (1987) *A Thousand Plateaus* (trans. B. Massumi). Minneapolis, MN: University of Minnesota Press.

Dosse, F. (2011) *Gilles Deleuze and Félix Guattari: Intersecting Lives*. New York, NY: Columbia University Press.

Frost, R. (2007) *The Collected Prose of Robert Frost*. Cambridge, MA: Harvard University Press.

Grosz, E. (2008) *Chaos, Territory, Art: Deleuze and the Framing of the Earth*. New York, NY: Columbia University Press.

Guattari, F. (1995) *Chaosmosis*. Bloomington, IN: University of Indiana Press.

Guattari, F. (1996) Ritornellos and existential affects. In: Guattari, *The Guattari Reader* (ed. G. Genosko). Oxford: Blackwell Publishers. pp. 158–171.

Guattari, F. (2000) *The Three Ecologies*. London: The Athlone Press.

Guattari, F. (2011) The vertigo of immanence: Interview with John Johnston, 1992. In: Alliez, E. and Goffey, A. (eds). *The Guattari Effect*. London: Continuum. pp. 25–39.

Lindholm, J. (2012) How Irene inspired the arts. *Vermont Edition*, radio transcript, Vermont Public Radio, 27 August. Available online at: www.vpr.net/episode/54269/how-irene-inspired-arts/

Stengers, I. (2011) Relaying a war machine? In: Alliez, E. and Goffey, A. (eds.) *The Guattari Effect*. London: Continuum. pp. 134–155.

Terrific toll; blocks railways, roads; wires down; bridges out. (1938) *The Vermont Phoenix*, 23 September.

Walker, K. (2007) *After the Deluge*. New York, NY: Rizzoli.

Whitehead, A.N. (1978) *Process and Reality: An Essay in Cosmology*. New York, NY: The Free Press.

11 The cosmic flight of the *Aerocene Gemini*

Sasha Engelmann

27 August 2016. 7 a.m.
52°27'32.4"N 14°03'15.3"E
Schönfeld, Germany.[1]
Forecast: The current weather forecast looks very good (sunny). The wind speed[s] are not that high, so it might no fly so far.
We can only Do-It-Together (DIT)!
Plot your prediction on this map.
The transmission will include APRS position messages including inside and outside temperatures (the lifting power of the balloon), humidity, and air pressure.
Callsign: DL7AD-11
Frequency: 144.800 MHz AFSK1200
Packets: Low-duty.cycle APRS/SSDV images, Position packets, Log packets, Software error log packets (for debugging)
We run a special SSDV/APRS service which picks up the packets from the APRS-Igates and send them to Habhub.
Callsign: DL7AD-11
Frequency: 144.860 MHz 2GFSK9600
Packets: High-duty-cycle APRS/SSDV images
For decoding we run a specific Perl Script which work together with a TH-D72.
We would be interested in any ideas how to improve an open source solar flight predictor.
All images will be plotted on the SSDV page: http://ssdv.habhub.org/DL7AD[2]
Dear Ballooning Friends. . .[3]

A chase

The *Aerocene Gemini* launched into the air around 7:30 a.m., on a brilliantly sunny Saturday morning in Schönfeld, Germany. There was almost no wind. The two voluminous, aerostatic sculptures floated a dozen metres high, payloads trailing on ropes below. For some time, nothing happened. The two-part body hung there, like a pair of fragile creatures waiting for change, absorbing the sun's eager rays.[4] "*We are relearning how to float in the air*", Tomás Saraceno said.[5] Then, almost without warning, they caught a draft. A faster current. A line of flight. And they were off, joined together like the ancient twins Castor and Pollux, the *Dioscuri*, half immortal, bodies whose destinies are fatally joined, making their way steadily toward a line of tall trees in the distance.

This chapter follows the flight and fate of the *Aerocene Gemini*, tracing a cosmological and proto-aesthetic account of their journey. As they soar through the stratosphere, the *Gemini* will be imaginative companions for elaborating the micropolitics of creative experiments with atmosphere. Following Deleuze and Guattari (1994) I understand politics to be the art of tuning bodies to specific diagrams of the present and future, and macropolitics to be the striation or structuring of such diagrams. Micropolitics, always already intrinsic to politics, refers to the amplification of alternate attunements and dispositions, or "mutant existential virtualities" (Guattari, 1995: 120) toward present and future conditions. This chapter is a mythopoietic account enlivened by a Guattarian micropolitics as it is heralded in the proto-aesthetic textures of the Aerocene. It is also a written experiment conveying the drama of an unusual atmospheric event. In doing so, this chapter does not represent the journey of the *Gemini* as much as it collaborates with its retelling.

The *Aerocene Gemini* is a particular kind of airborne sculpture that gains buoyancy using only the sun and the air that it moves. Such a sculpture requires no helium or hydrogen, no combustion of fuel, and no propellers or turbines in order to fly. Implicit in the performance of Aerocene sculptures is the understanding that, "for things to interact they must be immersed in a kind of force field set up by the currents of the media that surround them" (Ingold, 2011: 93). Most Aerocene sculptures depend on several circumstantial factors in order to become airborne: solar irradiance, temperature, convection, weather, and the albedo of the Earth's surface. Crucially, they also depend on the different skills of actors who design, construct, launch, and follow them. The micropolitics of Aerocene sculptures is embodied in the solar-energetic relations these sculptures establish, and in the "ecology of practices" (Stengers, 2013) motivated by these relations.[6] Such ecologies of practice are not only functional. Rather, they "detach and deterritorialise a segment of the real" (Guattari, 1995: 131) that impresses practitioners with "different ways of seeing and making the world" (Guattari, 1995: 120).

What Guattari would term the "proto-aesthetic paradigm" of the *Aerocene Gemini* is the product of decades of research and experiment carried out not only in the sphere of art, but also between the Berlin-based studio of artist Tomás Saraceno and several scientific collaborators, including the Centre National des Études Spatiales (CNES), the citizen science organisation Public Lab, and the Massachusetts Institute of Technology's Earth, Atmospheric and Planetary Science department (MIT EAPS).[7] For Guattari the recognition of aesthetic relations between science, technology, art, and the social is made more urgent by numerous forms of toxicity and crisis (Guattari, 1995, 2000). Tomás Saraceno's practice is one of webbing between disciplines and discourses, yet these webs create distinct planes of consistency and attention in relation to planetary circumstances. The artist has met with international success in the last 15 years, in large part due to his abilities to think and work with the webs of a wide range of spider species, as well as webs modelled by scientists attempting to understand patterns of dark matter in the universe, and webs of technologies of sensing and communication.[8] At the same time, and over the length of his career, Saraceno has experimented with the aesthetic, social, and political textures of atmospheric journeys: those whose successes are dependent on the ways in which bodies like the *Aerocene Gemini* and the bodies of practitioners and atmospheric currents hold together.

Saraceno's work envisions different styles of atmospheric dwelling and moving, sensing and communicating that transcend scales, articulated in the body of work called *Cloud Cities*. In the three years during which I worked closely with Saraceno as part of the first doctoral thesis carried out in collaboration with his studio (see Engelmann, 2016), this body of work also became associated with two other titles: *Becoming Aerosolar*, and more recently, the *Aerocene*. *Aerocene* refers to a series of buoyant sun and air-powered sculptures (such as *Aerocene Gemini*) as well as a new planetary epoch: one in which novel collective investment in atmosphere is more widely performed. In Guattari's terms, this is to form "collective assemblages of enunciation" out of "fragmentary ventures" (Guattari, 1995: 120). Today, the Aerocene Community reaches internationally, and moves toward autonomous and heterogeneous governance. Recent activity has included awareness-raising and environmental sensing in Buenos Aires, and collaborations with low-tech sustainability and maker communities in Paris and elsewhere. In this sense, Aerocene brings the concerns of Cloud Cities together with a more urgent call for different planetary futures to be made and remade in acts of atmospheric sensing and publicity.

As studio assistant Kotryna Šlapšinskaitė related to me, the ascent was gentle. *Aerocene Gemini* approached the tall trees in the distance, and as the earthbound humans held their breaths, the sculptures barely cleared the treetops, payloads intact (Šlapšinskaitė, 2016). Then they disappeared into blue sky, dissolving into two particles whose presence had been so palpably felt an hour earlier as two membranous bodies hanging motionless over the field; two delicate creatures sharing a common filamentary web; two aerosols attracting the assembly of practitioners, technologies, cascades, predictions, and hopes in those particular cosmic conditions.

The chase began. A sense of excitement, of thrill, of adventure, and, equally, of suspense and trepidation. The *Dioscuri* were no longer immediately sensible, yet they communicated vital signs: they transmitted GPS location data, and readings of temperature, pressure, and humidity inside and outside their membranes. They captured and forwarded images of the view from their lofty aerial position to a live website where grounded practitioners could follow along. I was one of these grounded practitioners, constantly refreshing the SSDV site for new images. In this way my sensory realm was cosmologically extended: it was lured toward pulses of information from those distant, airborne twins, twins whose capacity to pulse such information depended on the "ecology of practices" (Stengers, 2013) that had unfolded at Studio Saraceno and on the early morning field. Guattari teaches us that micropolitics is always already related to macropolitics, and vice versa. The mutant-micropolitical ecology of practices of the *Aerocene Gemini* was constrained by legal frameworks: it had been organised to meet the guidelines of weight restriction for an aerostatic object launched in the airspace over Germany.[9] From the early morning onward, the *Aerocene Gemini* became a tiny red icon of a balloon on a map of continental Europe. Yet the conjoined sculptures were not reducible to the icon, since they continued to engender affective atmospheres, imaginative journeys, geo-political questions, and conversations played out in breathless proximity in the swelter of a hot summer day.

Flying using only the sun's rays and the wind, the *Aerocene Gemini* reached over 16,000 metres in altitude. The twins crossed the border of Germany to Poland, having caught a strong Southwesterly current. At this point, the images loading on the SSDV page were either neon rectangles, or impossibly granular. I later learned that this was not due to altitude or distance, simply to "software error" (Krahn, 2016). Where were they going? What were they seeing? Where would they land? Questions surfaced. Sensations of nervousness and worry intensified, caught up in that small red icon, hovering across the map. Was the flight too ambitious? Would the Gemini cross the border to Lithuania, perhaps even Russia? Would there be consequences of the *Aerocene Gemini* crossing these territorial borders?

The lost twins

A hopeful morning became a worried evening. The sun had set, therefore the sculptures were surely earthbound in Poland. But the signal was lost. It had faded altogether. The chase team still traversed the dark forests and great lakes of Poland near the province of Augustów. It was likely at that point that the sculptures had fallen into water, irretrievable. The chasers tried desperately to pick up the signal. A chance reception was the last remaining hope.

The far away searching for the sense of the sculpture, the search for a signal pulsing weakly through pine, landform, track, herds of elk, stream, stone, and fog, was relayed to those in Berlin via sporadic texts and calls. Bodies formed postures that curled into themselves, or slumped into grass. There were attempts to distract, to refocus, to worry a little less. We stared at the little red balloon icon on the map, marking where it had last been sounded. All of this punctuated by animated retellings and glossed-over eyes, stories imparting how the sculptures had lifted into the air, and the way they had hung, spectre-like, there, until the wind, as in myth, had carried them away.

Many things are lost. Many things find an upward draft out of the lives of creatures and never return again. But there is something particular about the loss of an entity that has woven so tightly into ecologies of practice that enable proto-aesthetic holding patterns (Guattari, 1995). These are ecologies of practice that inspire both individual and collective attachments as powerfully as they invoke collective dreams of how the world and its inhabitants might become. The loss of an entity that is not an entity at all, but a gravitational attractor, a filamentary web, or a "mutant existential virtuality" (Guattari, 1995: 120) is a loss that shakes at the filaments of these worlds. Such intensity of feeling is also an integral part of the micropolitics of such worlds.

The fates of the Gemini were lost in more ways than one. Scholars attest that the destinies of Castor and Pollux were recounted in the *Cypria*, an ancient poem that preceded the *Iliad*, but has long been lost. It is rumoured that one of the *Dioscuri* was mortal, the other immortal; and that upon Castor's fatal injury, Pollux offered him half his immortality, so that they could remain together, sharing their time between Elysium and Hades. In Homer's *Odyssey* is the following line: *One day both Dioscuri live, one day they are both dead* (Homer cited in Morelle, 1655: 25). In any affair, and in death itself, the Gemini are bonded: two bodies matching their ascent in cosmic adventures with wind and sun.

Circa 11:50 p.m: a signal emerges. Krell and Krahn hear the sculpture's unique call on a hand-held radio receiver. They know they are within 3 km. They drive down pitch-dark roads. The signal recedes; they retrace their steps. They turn down another road and the signal grows louder.

Circa 1:18 a.m: the *Gemini* are found. Krahn had exited the van with the radio antenna, and soon spotted the membrane, slumped over a bush in a field.

This news sparks collective elation. Calls and texts fly among phones at 2 a.m. Some feel intense relief, others pride, and others celebrate into the night. The sculpture is returned via car to Berlin, where it is unfolded at Studio Saraceno. Videos and data are downloaded. Damage is checked. One tiny pollution sensor fell off – probably somewhere over Poland. A few days later, friends of the *Aerocene* are summoned to hear a presentation of the event at Studio Saraceno. A dozen enthusiasts, in addition to many members of the studio, attend.

Nick Shapiro, a studio resident and collaborator, has brought his expertise on citizen science and open-source licencing to the Aerocene project. Shapiro says his work aims to "re-enchant devices" to allow non-academics to "creatively render" answers in their environments (Shapiro, 2016). Sven Steudte and Thomas Krahn present the APRS radio transmitters and camera-boards they invented for the *Aerocene Gemini*. Saraceno interjects often with comments and suggestions, then makes a short presentation of his own. Instead of highlighting the spectacular videos taken from the sculptures in the stratosphere, he speeds through dozens of photographs showing the long drive to the launch site, the unloading of cars, membranes spread on the field, many smiling faces in various poses, a sunny afternoon at the nearby lake, and a scene of waiting in his Berlin apartment. He presents the social-affective texture of the experiment: an equal part of the Aerocene's achievement. *What we are really doing here*, Saraceno says, *is relearning how to float in the air*.

We are (re)learning how to float in the air. This statement is simple and enigmatic. It is strange and alluring. It raises the question: when did "we" ever float? Perhaps as other forms of life, as bacterial, spore-like, metazoan, or amphibian creatures? Saraceno implies that Earth's human inhabitants have lost a collective sense of buoyancy, a sense that one can re-learn, perhaps, by engaging in experiments with solar and wind-powered drift. And in relearning this skill, this trait that fell away, we can float ourselves into more alluring futures. The directness of the statement – and the all-encompassing "we" – carries a tenor of heroism that is potentially problematic in relation to the proto-aesthetic and micropolitical.

Guattari warns of the micro-dogmas that are part and parcel of micropolitics, and that have powerful aesthetic allure. Yet this single articulation simultaneously manifests what is so compelling about Saraceno's practice for geographical thought and experiment, and more broadly, why such aesthetic practices reaching for the edges of earthly and atmospheric experience are so ethically and politically radical. The collective practice of relearning how to float in the air is one founded on sensory extension and amplification that does not remove hierarchy or difference among practitioners, objects, media, and materials, or vacate the macropolitics of community, as much as vibrate the threads among modes of being and becoming. Such sensory extension is ethical in its production of empathy (embodied and otherwise). And it is political in its conjuring of "enunciative territories" (Guattari, 1995) coalescing around atmospheric sensing: so many vibrating threads.

Art ruptures

As we have felt already, Guattari's thought is helpful in devising a vocabulary for such alter-adventures in practice, sensing, and politics. It is helpful in casting the proto-aesthetic tethers through which we can better grasp the flight of the *Aerocene Gemini*. A proto-aesthetic paradigm acknowledges that, "objects constitute themselves in transversal, vibratory position" (Guattari, 1995: 102). It attends to the aesthetics of "a world of strange mutually implicated beings cohering around objects and practices" (O'Sullivan, 2010: 260). In the unfolding experiment of two years of teaching creative practice together with Saraceno and members of his studio team, I began using the term "cosmological aesthetics" to better articulate the proto-aesthetic paradigm of the Aerocene. The word "cosmological" refers to the world-making function of "ecologies of practice" that weave, cohere, and endure. *Aesthetics* is understood as the play of sensation across such ecologies of practice and the matters and media they enrol. In this way aesthetics approaches *aesthesis*: a notion of sensitivity and feeling that is neither circumscribed by the human perceptual apparatus, nor by the cognitive faculties of perceiving subjects. Cosmological aesthetics proposes, first, that aesthetic experience can create tangible, sensible relations with contexts that are far removed, or much wider than, the particular conditions in which we experiment; and second, such adventures in aesthetics emerge from practices bound together by the forces of obligation, attachment, and – crucially – imagination.

Cosmological aesthetics resonates in the hand-held radio antenna, poised to receive a signal emanating through the thick darkness of the Polish night. Equally it is found in the rough packets of temperature, pressure, and image-data transmitting over vast expanses of horizon, atmosphere, and landscape in zeros and ones. And it is also found in the sleepy, strained, and sunburnt bodies of those who launched the sculpture at 7 a.m. on a field in Schönfeld, and waited all day for more news of its journey. It exists in all of these places at once because the sounds of the hand-held antenna, the flows of visual data, and the strains in the muscles of practitioners manifest the transmission and distribution of sense and sensing across spaces and scales, from a field in Germany to the stratosphere, yet bound together by a common imaginative lure. A lure that is partly localised in the *Aerocene Gemini* sculptures, but travelling far beyond them, in a body of atmospheric adventures that inspire different dispositions to matter, energy, and environment. Cosmological aesthetics accounts for the role of art and imagination in compelling such adventures in sensing: art is a cosmic force (Grosz, 2008; Deleuze and Guattari, 1994) that lures, detaches, and deterritorialises such ensembles of practitioners, devices, and media.

Taking cues from Guattari's *Chaosmosis*, the launch of the *Aerocene Gemini* is a force of rupture and a site where subjectivities crystallise. As aerodynamic aeromobility reaches astounding levels of growth, and the market for drone technology (both military and recreational) explodes, the launch of an aerostatic, solar-powered object, and the level of investment and commitment it fosters, is a glitch in this interface. The Aerocene also ruptures the space-times, scales, and social spheres of those invested in the project: compelling strange sleeping hours (e.g. working towards a launch at dawn), enlarging scales of thinking and experiment (e.g. how to track an entity with GPS over an altitude of 10,000 metres?), and producing novel social arrangements (e.g. the congregation of bodies, devices and materials around a delicate, lively membrane). To be sure, other forces also compel these arrangements: forces of economy, profession, friendship, activism, research, thrill, duty, status, obligation, risk, and love. However, the imaginative urgency, cosmicity, and refrain of these practices necessitate that we account for a moment of artistic rupture that has aesthetic, ethical, and political consequences.

Guattari asserts that art generates specific assemblages that present possibilities for being and becoming in contrast to the assemblages we find all around us (Guattari, 1995). Guattari might add that both art and processes of subjectivisation become modes of accessing different speeds and velocities of feeling, thinking, and becoming. Here we might remember Stengers' comments: "*the earth [must be] taken into account as a set of interdependent processes, capable of assemblages that are very different from the ones on which we depend*" (Stengers, 2011: 163). The role of the Aerocene as a rupture from current extractive and fuel-burning resource assemblages, and a window onto different, inventive assemblages of material collaboration is perhaps a key to our very survival on this planet (Guattari, 1992/1995; O'Sullivan, 2010). The ruptures created by the launch of an Aerocene sculpture fold into alternative geographic and political subjectivities.

Rather than an achievement in flight, the launch of the *Aerocene Gemini* is a feat of floating and sensing. This difference matters, because floating (or drifting) has a particular significance in relation to increasingly calculated, algorithmic mechanisms of control and surveillance of our planet's "elemental infrastructure" (McCormack, 2017), in which "a potential target appears as a drift" (Franke, 2016). Floating, or "quantum flows" for Deleuze and Guattari (1994), are inflections in striated space. Likewise, although it carries sensors of many kinds, the *Aerocene Gemini* does not (only) measure the atmosphere. The sculpture participates in a collective sensing that includes numerous "technical" readings but also, and equally, the data passing and digesting through bodies as the sculpture is launched, tracked, chased, sounded, and rediscovered as a faint echo on a radio antenna in the night.

Birch seeds

On 27 August, in the forests of Mostówka, Poland, Dr. Bronislaw Szerszynski was experimenting with some birch seeds. He had collected many over the course of the summer, which he spent, as usual, in a drafty family house, a house I had visited the summer before. This summer, he had invented a series of aerial experiments with the seeds: "They incidentally are a brilliant device . . . for revealing all the aerial eddies around the house and grounds. One just flew over our water pump and followed the tiny bound vortex leeward of the handle. A very elegant move too" (Szerszynski, 2016).

As this seed followed the vortex in the eddies of the forest, were the *Aerocene Gemini* passing overhead, following so many other flows, vortices, and eddies (*an elegant move too*)? Were the birch seeds and Gemini engaging in the same turbulence, resistance, and the same thermodynamic system? It is impossible to know. But here is another salient speculation: can we think of the *Aerocene Gemini* and the *Birch Seeds* in the same web of relations, a web that does not produce hierarchies in space or scale, but zones of proximity and synthesis?

On 27 August, Szerszynski was also following the *Aerocene Gemini*, watching the data and the SSDV image-stream as it passed over Poland and drifted to the North. Like those of us in Germany, he was extending himself cosmologically through the airborne sculpture and the practices and imaginaries enfolded within it, practices and imaginaries of which he was familiar (Szerszynski and Engelmann, 2015; Saraceno et al., 2015).[10] Whether or not the flight of a *Birch Seed* was contiguous with that of the *Aerocene Gemini* is less likely and less relevant than the notion that the Aerocene experiments in which Szerszynski and I participated inspired attentions and attunements to atmosphere that rippled in various ways in our respective lives and projects. It rippled in the presentations we gave at various conferences, in the "Dust Séance" we performed at the Haus der Kulturen der Welt with Tomás Saraceno (Saraceno et al., 2016), in research questions and conversations with many other scholars, and in aerial experiments with birch seeds in the forests of Mostówka.

Such micro-ripples of attentiveness – or shifts in atmospheric subjectivity – are aesthetically, ethically, and politically important. They shift material attunements and dispositions, as well as frames of reference, discipline, labour, and invention. They bring the textures of the world to the forefront of scholarship. They propose forms of observation that are ethical in their stance to the patterning of media, materials, and phenomena. They convey notions of sense and sensation that are not irreducibly human, nor even within the thresholds of human awareness. And crucially, these molecular modes of attention and attunement gather publics. Ultimately, the very possibility to tell a story of the Gemini and a birch seed in relation to a micropolitical investment in atmosphere implies the web of proximity and synthesis in which they are both entwined.

A hand to a photon to the stratosphere

It is important to emphasise that not all ecologies of practice, or the enunciative territories they perform, are equally potent in amplifying micropolitical tendencies. Some ecologies of practice are sticky and infectious, resilient and compelling, resonating in silhouettes and dreams, while others do not. Some practices mobilise the imagination and art as a cosmic force that obligates practitioners to each other and to non-humans, materials and technologies, in the same way that other cosmic forces give a spider web its orientation, or dance a photon around quercetrin in a leaf (Beresford-Kroeger and Thorne, 2013). For Guattari, as for Saraceno, the power of art as a force that compels, attracts, radiates, and lures is a fact animating the invention of practices that have cosmological scope.

In the context of an interview about the Aerocene for *Take On Art* magazine's special issue on Ecology, printed in February 2017, Saraceno invited Bronislaw Szerszynski and me to collaborate with him in answering a series of questions posed by art critic Stefanie Hessler. Hessler's first question was: "How would you describe the role of the artist; do artists have a certain responsibility toward the social and political?" (Hessler, 2017). Our collective response was to reframe the question with Guattari. We answered:

> how do we change mentalities, how do we reinvent social practices that would give back to humanity – if it ever had it – a sense of responsibility, not only for its own survival, but equally for the future of all life on the planet, for animal and vegetal species, likewise for incorporeal species such as music, the arts, cinema, the relation with time, love and compassion for others, the feeling of fusion at the heart of the cosmos?
>
> (Guattari, 1995: 119–120; cited in Hessler et al., 2017)

The aim in reframing the original question was not to evade it, but to make a Guattarian point that response-ability does not fall to artists alone, nor can art be instrumentalised for social and political change. Rather, the kinds of aesthetic and "artistic cartographies" (Guattari, 1995) that will engender responsibilities function at the level of subjectivities and mentalities that are not framed by art but are abroad in world. For the printed interview, and for this chapter, the word "responsibility" usefully folds together sensing and publicity: proto-aesthetics and micropolitics.

The social and political potential of the *Aerocene* lies in the act of proposing ecologies of practice that have a direct relation to the macropolitical structuring and striation of atmospheric space. It lies in offering many kinds of intervention and collaboration – from the highly technical to the speculative-philosophical, to the manual and labor-intensive, to the artistic and collective. If the *Aerocene*, like *Cloud Cities*, floats close to utopia, it is a porous and experimental utopia that can be (re)engineered and (re)mapped through different engagements. And through these engagements, we can contribute to an envisioning of alternative futures that are responsible to all forms of corporeal and "incorporeal species" on the planet, and that are urgently needed at a time of accelerating social, mental, and environmental toxicities (Guattari, 2000). This, more than the single cosmic flight of the *Aerocene Gemini*, is the micropolitics of the *Aerocene*.

Announcement: Next Aerocene Free Flight!

4 March 2017
Sunrise Time
Event is Weather Dependent
We will free fly Aerocene Sculptures from 52°27′32.4″N 14°03′15.3″E!
Join us!

Notes

1 The message from Studio Saraceno specified: "The location for this launch was selected because it falls outside the air traffic control zone in Berlin, necessary due to local air traffic regulations" (Studio Saraceno, personal communication, 2016).

2 SSDV stands for "Slow Scan Digital Video" and is a digital form of Slow Scan TeleVision. It can be used to transmit small images along with the telemetry data transmitted by an aerostat's payload during flight. This definition is borrowed from the UK High Altitude Society – "a loose collection of people who are interested in launching unmanned high altitude balloons into near space" (ukhas.org.uk).

3 Excerpts from an e-mail sent from Studio Saraceno on 26 August, 2016 to "friends of the Aerocene": in the message, people from many disciplines are invited to follow and track the Aerocene Gemini sculpture by accessing web sites and tracking information as it flew from Germany to an unknown location.

4 The best description of the *Aerocene Gemini* launch was offered by Kotryna Šlapšinskaitė, a studio member who was present at the event, and related the details to me later that day.

5 This comment was made three days later at an evening presentation at Studio Saraceno at which Saraceno, Daniel Schulz, Sven Steudte, Thomas X, and Nick Shapiro (Public Lab) presented the innovations and experience of this Aerocene Gemini launch.

6 In her article "Introductory Notes on an Ecology of Practices", Isabelle Stengers defines "ecology of practices" as "a tool for thinking" and enlarges in the following way: "The problem for each practice is how to foster its own force, make present what causes practitioners to think and feel and act. But it is a problem which may also produce an experimental togetherness among practices, a dynamics of pragmatic learning of what works and how" (Stengers, 2013: 195).

7 Aerocene is an open project initiated by artist Tomás Saraceno and developed by Aerocene Foundation.

At the same time, Aerocene is comprised of a dedicated and diverse global community of artists, geographers, philosophers, thinkers, speculative scientists, explorers, balloonists, technologists, and dreamers. The Aerocene project's primary collaborators and supporters are the Center for Art, Science & Technology (CAST) at the Massachusetts Institute of Technology (MIT), CNES (French National Space Agency), CCK Argentina, Public Lab, The Goethe Institute, Radioamateur, Freifunk, and IAK architecture-related Art Institute at Technische Universität Braunschweig, TBA21, among others. Anyone is invited to share, collaborate, and perform actions in the communal creation and development of the new Aerocene age.

8 Tomás Saraceno has collaborated extensively with scientists and academic institutions. Two important examples are: the invention of a laser-supported tomography method to scan and digitise spider webs, carried out in collaboration with engineers at TU

Darmstadt, and presented at the 18th International Congress of Arachnology; an ongoing collaboration with Dr. Leila Kinney, Dr. Lodovica Illari and Bill McKenna at MIT's Department of Earth, Atmospheric and Planetary Sciences on developing more nuanced models of atmospheric fluid dynamics, applications for monitoring of airborne pollution via aerosolar sculptures, and more sophisticated flight prediction for aerosolar vehicles.

9 Both Sven Steudte and Thomas Krahn are practiced at the art of launching and chasing balloons via APRS radio transmitters. Indeed there is a community of people who do this. It is legal to launch aerostats of a certain weight profile, and even to let them fly with no intention of finding them. However over a certain threshold, the aerostats must be equipped with more advanced transmitters (transponders) and be certified under different insurance schemes.

10 Indeed, Szerszynski's participation in an aerosolar sculpture workshop at the IAK, TU Braunchweig in November 2014 had inspired a draft paper we wrote together about "elemental dwelling" and aerosolar practices (see Szerszynski and Engelmann, 2015).

References

Beresford-Kroeger, D. and Thorne, K. (2013) Quantum violin. In: Turpin, E. (ed.) *Scapegoat Journal*, 5.

Deleuze, G. and Guattari, F. (1994) *What Is philosophy?* (trans. H. Tomlinson and G. Burchell). London: Verso. (Original work published 1991).

Engelmann, S. (2016) *The Cosmological Aesthetics of Tomás Saraceno's Atmospheric Experiments*. Unpublished doctoral thesis, University of Oxford.

Franke, A. (2016) *Nervous Systems Exhibition*. Berlin: Haus der Kulturen der Welt.

Grosz, E.A. (2008) *Chaos, Territory, Art: Deleuze and the Framing of the Earth*. New York, NY: Columbia University Press.

Guattari, F. (1995) *Chaosmosis: An Ethico-Aesthetic Paradigm* (trans. P. Bains and J. Pefanis). Sydney: Power Publications. (Original work published 1992).

Guattari, F. (2000) *The Three Ecologies*. London: Anthlone Press.

Hessler, S., Saraceno, T., Szerszynski, B. and Engelmann, S. (2017) Aerocene: A quasi-feasible utopia. Interview with Tomás Saraceno, Bronislaw Szerszynski and Sasha Engelmann. *Take on Art*, Special Issue on *Ecology*, 3(1), January–June.

Ingold, T. (2011) *Being Alive: Essays on Movement, Knowledge and Description*. Abingdon: Routledge.

Krahn. (2016) Personal communication.

McCormack, D. (2017) Elemental infrastructures for atmospheric media: On stratospheric variations, value and the commons. *Environment and Planning D: Society and Space*, 35(3): 418–437.

Morelle, M. (1655) *Tableaux du Temple des Muses*. (Plate 25). London: Warburg Institute.

O'Sullivan, S. (2010) Guattari's aesthetic paradigm: From the folding of the finite/infinite relation to schizoanalytic metamodelisation. *Deleuze Studies*, 4(2): 256–286.

Saraceno, T., Engelmann, S. and Szerszynski, B. (2015) Becoming aerosolar: Solar sculptures to cloud cities. In: Davis, H. and Turpin, E. (eds.) *Art in the Anthropocene: Encounters Among Aesthetics, Politics, Environments and Epistemologies*. London: Open Humanities Press. pp. 57–62.

Saraceno, T., Engelmann, S. and Szerszynski, B. (2016) *The Dust Séance [Public Performance as Part of Wisdom Techniques: Technosphere x Knowledge]* (K. Klingan, Head of Project). Berlin: Haus der Kulturen der Welt. Video of performance. Available online at: www.hkw.de/en/programm/projekte/veranstaltung/p_126262.php

Shapiro. (2016) Personal communication.

Šlapšinskaitė. (2016) Personal communication.
Stengers, I. (2011) *Thinking with Whitehead: A Free and Wild Creation of Concepts* (trans. Micheal Chase). Cambridge, MA: Harvard University Press.
Stengers, I. (2013) Introductory notes on an ecology of practices. *Cultural Studies Review*, 11(1): 183–196.
Szerszynski, B. (2016) Personal communication.
Szerszynski, B. and Engelmann, S. (2015) An elemental dwelling: The solar balloon and attuning to the powers of air. Paper presented at the symposium *Spaces of Attunement*, Cardiff University, 30–31 March.

Part III

Micropolitics

"But I believe that there is a collective, unformed search, from above and below, for another kind of politics. This is what I call 'micropolitics', and 'molecular revolution'. It begins with very immediate, daily, individual preoccupations, yet remains connected to what happens at the social level, and even, why not, at the cosmic level." (Guattari, 2009b: 138)

Guattari was forever searching for another kind of politics, for another way of going on in the world. A brief definition of micropolitics might be: "a politics of desire that questions all situations" (Deleuze and Guattari, 1986: 42). Elsewhere we are reminded that "every politics is simultaneously a macropolitics and a micropolitics" (2004: 235). Of course, such brevity does a disservice to the concept and was, arguably, much more developed in the solo work of Guattari, especially in his book *Molecular Revolution* which still awaits an unabridged translation.[1] It is for this reason, then, that the introduction to this third part of the book charts some of the ways micropolitics was theorised and the implications it has for thinking.[2]

Guattari writes of the intersection of different kinds of political struggles, outlining that "macro-political struggles . . . can be located at, for example, the electoral or union level, etc., and micropolitical struggles . . . can be situated at the same level, including that of the State, but which everywhere exceed social stratifications, institutional and legal limits" (2016: 52). Accordingly, "sometimes 'insignificant' events can set off considerable upheavals", although Guattari is keen to underline that it is never the case that one conditions the other but that the macro- and micropolitical are constantly interacting. As such, foregrounding the micropolitical, as we have chosen to do, could be seen to go against the ethos of the Guattarian project, a project which sought to overcome the dissociation between the two (see Rolnik, 2008: 12). However, as we hope to make clear, the reason for its prominence in the book is because the micropolitical's import (as much as the molecular or the minor) is invariably disqualified by dint of its *presumed size*. By resisting recourse to the epistemological shorthand of scale, we want to highlight how the micro need not be figured as a 'small' politics or, worse, a politics that is simply peripheral or inconsequential. Nothing could be further from the truth. Still further, we want to sketch how micropolitics is a liberation of politics writ large.

Micropolitics, for Guattari, is about disruption. It is a question of pragmatics, of "placing micropolitics everywhere", so as to trouble the "stereotyped relations of personal life, conjugal life, romantic life, and professional life, in which everything is guided by codes" (Guattari and Rolnik, 2008: 190). If there is a micropolitics to be practised, it is about not succumbing "to systems that coopt . . ., systems of neutralization, or processes of implosion or self-destruction"; or, in a more affirmative tenor, it consists of "apprehending how other assemblages of the production of life, the production of art, or the production of whatever you want might find their full expansion" (2008: 339). This is a micropolitics that "takes into account our own humble participation in the story; it is possible to work in the direction of dis-alienation, of a liberation of expression, of opening 'exit doors', if not 'lines of escape', from oppressive social stratifications" (Guattari, 2009b: 54–55).

In some ways, then, the micropolitical is most readily to be found around us. As Brian Massumi (2015: 70) has noted, while

> the 'micro' of micropolitical is not synonymous with small, and although the modulations that might be effected at that level can be widely distributed, there is no better place to start than the local context in which you live and work every day. Macropolitical positioning operates under the illusion that there is a neutral, higher-level vantage point.

The vocation of micropolitics is thus not about constructing or imposing global solutions; micropolitics is not programmatic. Indeed, it is quite the opposite. There is simply no guarantee to be found in micropolitics as "any praxis may or may not be of a repressive nature"; instead, "processual micropolitics can only – and should only – be found at each step" (Guattari, and Rolnik, 2008: 41). This means that micropolitical interventions or events can feel as if they are on the verge of taking shape: potentials that can be glimpsed but which may not come into focus. To return to Massumi, the goal is "not to overcome the incompleteness. It's to make it compelling. Compelling enough that you are moved to do it again, differently, bringing out another set of potentials, some more formed and focused, others that were clearly expressed before now backgrounded" (2015: 80).

The inspiration for Guattari's micropolitics, it has been suggested, was the work of Michel Foucault. Although the 'Foucault-Guattari' relation remains relatively under-examined (see Taga, 2014), Emily Apter (2018) has recently argued that Foucault's 'microphysics of power' served as the fulcrum of Guattari's 'micropolitics of desire'. Similarly, Stivale (2009: 16) has pointed to the inconsistencies between Guattari's interview with him, where he stated that he "was never influenced by Foucault's work . . . it was never of great importance" (Guattari, 2009b: 169), and yet presented a paper on Foucault (in the same year) where he drew connections between microphysics and micropolitics. Perhaps the key distinction for Guattari's micropolitics was its focus on the molecular.[3] As Apter notes, a "molecular micropolitics of desire is integral to the Guattarian vocabulary of anti-psychiatry, geophilosophy, chaosmosis, information theory and schizoanalytic cartography" (2018: 46). The conceptual associations of micropolitics with the molecular are clear in Guattari's writing, not just in the epigraph above, but also when he talks of "a 'molecular revolution' correlative to analytical practices and new micropolitics" (2009a: 203) or "a micropolitics of molecular transformations" (Guattari and Rolnik, 2008: 44). It can be tempting to read the micropolitical and macropolitical couplet as analogous to the molecular and the molar; to be sure, Guattari slips between these terms. But it is worth reiterating that for these couplets, there "is no distinctive opposition that depends on a logical principle of contradiction" (2008: 179). Although it may seem difficult, we need to change this logic. In the same way that physicists have had to admit that matter is both wave and particle (or, in Guattari's terms, "corpuscular and undulatory"), so must we accept that social struggles are both molar and molecular at the same time.

It is these connections between the micropolitical, the molecular, and the minor, which have interested social scientists in recent years (see Jellis and Gerlach, 2017). In the chapters that make up this part of the book, we find a sustained effort to think-with Guattari, and his insistence that we place micropolitics everywhere: in the invention of new micropolitical practices, in developing an analysis that makes more of a micropolitical register of intensities and the a-signifying, and of resisting an interpretive macropolitics that, in Massumi's words, "seeks to curtail the variety and exuberance of forms of life" (2015: 81). Ultimately, for Guattari, micropolitics held the promise of "soft subversions and imperceptible revolutions that will eventually change the face of the world, making it happier. Let's face it, it's long overdue" (Guattari, 2009b: 306).

Thomas Jellis and Joe Gerlach outline their approach to hitchhiking Guattari, where 'to hitch' takes on at least two senses. In this context, not only do they go for a ride with Guattari's ideas (his book, *Chaosmosis*, is in their backpack) but they also take these ideas out for a ride with others, along unexpected lines of flight. Exploring Guattari's enigmatic and minor notions of 'gentleness' and 'earthing', the chapter raises questions around the threshold for, and response to, micropolitical events. In so doing, they seek to reclaim hitchhiking as ethos, technique, and – above all – an effective practice of experimentation. As they note, hitchhiking entails a "shift in disposition, towards lines of movement and of affect, vectors which might point toward or be generative of micropolitical actions" (this volume: 180). The chapter tells the story of a series of hitchhikes in the south of the UK as part of a research-creation event organised by the SenseLab. As such, the chapter also animates accounts of commingling, or the activity of relating. Hitchhiking, with its speculative vectors, recombinant bodies, and transversal gentleness, is, then, as much a technique for existence as it is a technique for relation.

While Guattari never produced a dedicated study of the cinema, **Andrew Lapworth** argues that his approach to a-signifying cinema has been overlooked. Engaging with a series of recently translated texts by Guattari, the chapter foregrounds "a micropolitics of desire that creatively experiments with intensities of desire, exploring new associations of forces and bodies that disrupt established relations" (this volume: 187). Outlining the implications for Guattari's philosophy for contemporary understandings of a cinematic politics, Lapworth identifies two shifts, in terms of cinematic encounter and in analysis. In the former, there is a shift from structural representation to event of difference; in the latter, a shift from an interpretive politics to a micropolitics of a-signifying intensities. Such a micropolitics is not concerned, therefore, with the meaning or content of particular images but the way in which images can "capture and render sensible a-signifying processes that escape dominant semiologies" (this volume: 193). Lapworth also considers how micropolitical cinema is crucial in disrupting and remaking systems of valorisation, moving away from a political economy reading of commercial cinema to one where film can be considered in terms of ontological heterogenesis.

Nina Williams sketches out Guattari's long-standing interest in aesthetics as a way to reframe politics in art. Noting that his ethico-aesthetic paradigm – with its imperative to "attend to the immediate or nascent, material and sensory relations of art (not yet grasped or interpreted by the subject)" (this volume: 211) – is closely linked to his theorisation of politics – as something germinal, incipient, and emergent – Williams offers two propositions. The first is that art, like politics, is not limited to major (or molar) institutions. The second is that employing conventional means of valuing or engaging with art-encounters is inadequate. In other words, art cannot be reduced to utilitarian or instrumental logics. As she works through these propositions, it becomes clear that attending to the political content of art – art as a politics of the ready-made – can mean that we miss out on how art as a dynamic form is itself political. It is for this reason, then, that the onus is on a Guattarian aesthetics rather than art, one which is attentive to a non-discursive register of affects and sensations, of subjectifying forces that exceed institutionalised politics. Like other chapters in this section, the hope is for a politics that is inventive of new modes of relation.

Arun Saldanha's point of departure for thinking with Guattari is the rise of the Anthropocene as a term denoting a new phase of existence. In such an age, where talk of extinction is no longer the stuff of science fiction, he asks what kind of conceptual framework will "be adequate to this new, already catastrophic reality?" (this volume: 214). It is to Guattari that Saldanha turns, highlighting how important his terms – strata, IWC, machinic enslavement, and communism – are for confronting this crisis. If we linger for a moment on the first term that is examined, we might note how strata are caught up as part of Deleuze and Guattari's 'method', such that "RHIZOMATICS = SCHIZOANALYSIS = STRATOANAL YSIS = PRAGMATICS = MICROPOLITICS" (Deleuze and Guattari, 2004: 24). It is important to foreground here the resonances between strata and the molar – with their same tendency for structure, rigidity, and concreteness, although such a focus may well underline the conventional attempts to equate discussions of micropolitics with questions of scale. Crucially, for Saldanha, a stratoanalysis is incomplete without a more sustained engagement with IWC; in so doing, he presents a more pronounced Marxist propulsion of Guattari's thought. The Guattari detailed here is "at his best . . . when he demonstrates . . . the way capital works" (this volume: 217) and, through his theoretical eclecticism, has much to offer in terms of a renewed communism today.

Anja Kanngieser presents a fascinating and forensic taxonomy of sonic geographies in order to address the invisible forces of governance. As she notes: "Sound transverses events and states to bring into effect relational ecologies that reveal themselves to us in ways that require a particular sensitivity to apprehend" (this volume: 228). Crucially, this is not simply a call for more 'concerted listening' but an acknowledgement that sound decentres the human subject. Attending to the geographies of sound is to seek "chains of association across often incompatible and irreducible spaces and corporealities" (this volume: 230). To follow such threads, Kanngieser draws on Guattari's notion of transversality, which provides a means to trace how sound is used to make weapons of geophysical, atmospheric, and biological matter, but also to unpack disparate sonic territories and topologies. By charting the sonic technologies that make apparatuses of everyday discipline, the chapter also holds out hope for resisting, reshaping, and escaping such configurations. It is this micropolitical register, where sound is a 'commoning sensorial force', that holds the potential for building lines of evasion and of solidarity.

Notes

1 *The Molecular Revolution: Psychiatry and Politics* (Guattari, 1984) is the abridged version. Interestingly, a new French version came out in 2012 with Les Prairies Ordinaires, edited by Stéphane Naduad (Guattari, 2012).

2 As Andrew Goffey notes (in Guattari, 2016: 247n.3), "the correct orthography for 'micropolitics' and related terms is not clear. Guattari sometimes has 'micropolitiques' and sometimes 'micro-politiques'". As there is no obvious rationale for this difference, Goffey follows the convention for translating Deleuze and Guattari's work to render it as micropolitics.

3 Dosse (2009: 12) notes that "Prior to its association with Guattari, the 'molecular revolution' was Gramsci's creation".

References

Apter, E. (2018) *Unexceptional Politics: On Obstruction, Impasse, and the Impolitic*. London: Verso.

Deleuze, G. and Guattari, F. (1986) *Kafka: Towards a Minor Literature* (trans. D. Polan). Minneapolis, MN: University of Minnesota Press.

Deleuze, G. and Guattari, F. (2004) *A Thousand Plateaus: Capitalism and Schizophrenia* (trans. B. Massumi). London: Continuum.

Dosse, F. (2009) Introduction to Chaosophy. In: Guattari, F. (ed.) *Chaosophy: Texts and Interviews 1972–1977* (trans. D.L. Sweet, J. Becker and T. Adkins; ed. S. Lotringer). Los Angeles, CA: Semiotext(e).

Guattari, F. (1984) *Molecular Revolution: Psychiatry and Politics* (trans. R. Sheed). London: Penguin Books.

Guattari, F. (2009a) *Chaosophy: Texts and Interviews 1972–1977* (trans. D.L. Sweet, J. Becker and T. Adkins; ed. S. Lotringer). Los Angeles, CA: Semiotext(e).

Guattari, F. (2009b) *Soft Subversions: Texts and Interviews 1977–1985* (trans. C. Wiener and E. Wittman; ed. S. Lotringer). Los Angeles, CA: Semiotext(e).

Guattari, F. (2012) *La Révolution Moleculaire*. Paris: Les Prairies ordinaires.

Guattari, F. (2016) *Lines of Flight: For Another World of Possibilities* (trans. A. Goffey). London: Bloomsbury.
Guattari, F. and Rolnik, S. (2008) *Molecular Revolution in Brazil* (trans. K. Clapshow and B. Holmes). Los Angeles, CA: Semiotext(e).
Jellis, T. and Gerlach, J. (2017) Micropolitics and the minor. *Environment and Planning D: Society and Space*, 35(4): 563–567.
Massumi, B. (2015) *Politics of Affect*. Cambridge: Polity.
Rolnik, S. (2008) Preface to the 7th Brazilian edition. In: Guattari, F. and Rolnik, S. (eds.) *Molecular Revolution in Brazil* (trans. K. Clapshow and B. Holmes). Los Angeles, CA: Semiotext(e). pp. 9–13.
Stivale, C. (2009) Rethinking (with) Félix Guattari. In: Guattari, F. (ed.) *Soft Subversions: Texts and Interviews 1977–1985* (trans. C. Wiener and E. Wittman; ed. S. Lotringer). Los Angeles, CA: Semiotext(e). pp. 9–17.
Taga, S. (2014) Foucault et Guattari au croisement de la théorie du micropouvoir et de la psychothérapie institutionelle. In: Oulc'hen, H. (ed.) *Usages de Foucault*. Paris: PUF. pp. 99–107.

12 Hitchhiking Guattari

Thomas Jellis and Joe Gerlach

Thumbs up

Hitchhiking is both a neglected geographical problem (Cox, 1980; Day, 1980) and an easily caricatured vestige of heroically charged journeying.[1] Imagine, momentarily, Jack Kerouac on the road, in the motor-cycled retreat to solitude and the well-trodden path to 'finding oneself' in the fevered tumult of youth. We tend to think of hitchhiking as a counter-cultural practice; a borrowed ticket, from A to B to not-quite-sure-where. It is a mobile transaction that is repeatedly declared to be on the decline (Applebome, 1988; Moran, 2009; Woodcock 2011; Stromberg, 2015; Perkins, 2016), edging further away from littered grassy verges and worn asphalt margins. And for all the journalistic fervour which accompanies the 'end' of hitchhiking, there has only been sporadic academic interest in the topic (O'Regan, 2012; Laviolette, 2016; Mahood, 2016), and little in the way of sustained examination (although see Schlebecker, 1958; Chesters and Smith, 2001). How, then, to think hitchhiking *with* Félix Guattari; not just as a tactic for quotidian journeying, but as the reclamation of a technique for relational movement? How, in Guattari's (1995: 135) terms, can we work for hitchhiking's "liberation, that is, for its re-singularisation" as a fieldworking technology? More than this, though, how can we 'hitchhike' theory, such that it is something that we can take with us, but in a manner that it is allowed to pull us in different directions? Such a mode of theorising, to hitchhike theory, might consist of going along for the ride, without pre-supposing where these concepts might take you. In short, the encounter that we stage here between hitchhiking and Guattari is a point of departure for rethinking both fieldwork and the practice of theory in the social sciences, and specifically in Geography. Pushing against, then, the dangers, distrusts, and dubiety that linger in populist narratives of hitchhiking, the question becomes, why reclaim hitchhiking? To offer a tentative response from the outset, this process of reclamation entails cultivating an ethos toward what Guattari (1995, 2008) called a 'new gentleness', an affective and material openness to both the onflow of the world and an incessant valorisation of the co-productiveness of research-creation and micropolitical events.

In geographing a few days hitchhiking in the United Kingdom as part of a globally distributed ethico-aesthetic experiment, we call attention to the refrains, gestures, and affective atmospheres of research-creation; a, "speculative cartography that makes no pretence of providing a universal structural foundation" (Guattari, 1996a: 98). Put differently, this would constitute the sketching of existential choreo-cartographies where "subjectivity and sociality find new points of reference, new co-ordinates and possibilities of flight" (Zahm, 1994: 48) or, still further, "strategies for the constitution of new territories, other spaces of life and affect, a search for ways outside of territories seemingly without exits" (Guattari and Rolnik, 2008: 18). The intellectual milieu in which we attempted to rethink and reclaim hitchhiking as a fieldwork technology was that of the 'Society of Molecules', a global event composed of correlated local events (Manning and Massumi, 2014). Initiated by a research-creation collective in Montreal, *SenseLab*, the 'aim' (as far as one can state such a thing) was to 'respond to an issue of some kind'. Moreover, no constraint was placed on the content of the political-aesthetic interventions possible (see Manning and Plumb, 2009). This took place in mid-May 2009, across 17 'molecules' in 15 cities. Our 'molecule', already distributed, was drawn out across southern England, triangulated between Oxford-Chichester-London (see Abrahamsson et al., 2009; McCormack, 2013), its consistency formed around the intersections of the carto- and choreo-graphic, spinning out of our shared interests in dance, in maps, in lines, in rhythm, and in Guattari.[2] As a molecule, we responded to the 'call for participation' as an experiment in co-producing interventions with *ethico*-aesthetic reverberations, and in a micropolitical register. More specifically, for the two of us, this injunction propelled a sensibility towards enacting hitchhiking as a technique for geographical fieldwork, with unknown consequences. What follows is neither a 'roadside ethnography' nor a 'vagabond sociology' (see Purkis, 2012) but instead constitutes a series of exposures which seek to animate something of the staccato rhythms of movement, mutuality, and the madcap. It is this unpredictability, these 'accidental geographies' composed immanently of solitude and companionship, which we trace in vignettes. Alongside these interruptive tales, we examine Guattari's notions of gentleness and interpretation and, in so doing, make a case for reclaiming hitchhiking as a technology for relational movement, one which blurs distinctions between research, therapy, and adventure.

Molecules unbound

Barely a sign up, or a thumb extended. But there, on the slip-road, is a car, pulled up in anticipation. We exchange looks of surprise, then dash over. Where to? Southampton, superficially, at least. A couple of lines, meeting another. Others forming. Could we come along too? What are you doing? We're not sure. But we're part of a molecule. And you are too. We're interested in maps. Me too. A map person, in fact. On the move, conversation unfolds. A life-story; a life-line? Therapeutic. For us or them? For both? Could this be a small act of repair? We don't know. Too many questions in the space of immediacy.

But it just might be. A small act of repair, that is.

We're hoping to generate encounters. Different ways of meeting people, engaging with them, 'doing' research. We leave a postcard. Please draw us some lines. Perhaps your journey, or journeys. It's up to you. We've put an address and there's a stamp on it too. How about crossing the Channel, she suggests. Why not? These boys are on an adventure, she tells the security officer. But he's not having any of it. We're turned away, the line twists. We say thank you, thank you. What a start! We look for signs to the train station. We're following her advice. The train is slow and it doesn't feel right. We're not hitching. We read from Guattari's Chaosmosis. *Page 100. "Philosophy . . . generates its own register of creative constraints". Can we get a student rate? We're happy to work. No thanks. Please? Sorry. Are you sure? Sorry. We hit a blank. Perhaps we're not going to get across the water. Scratch-card fails. Enthusiasm drops. We leave a few lines in an email. We sign out and move on. Yellow chalk. On the floor, walls, pockets, hands. Traces of our trajectories. We draw a u-turn, an enfolding of those chalk traces, lecture theatre to tarmac. There are more to come. We look for places to stop and hitch. We look at the roads differently. We end up in the town-centre. We turn around. The signs aren't working. Nobody is stopping. A few waves. Some thumbs up. Several cars hurl abuse in our direction. We pack away the signs. The adventure was fun, but now let's head back. We keep our thumbs out. Another five minutes? Ok.*

Guattari's prose is exhilarating, crazed, insightful; it isn't easy-going for the most part (Gerlach and Jellis, 2015). He could be accused of writing heroically, of developing slogans, of making bold claims. For instance, when discussing his trip to Brazil with Suely Rolnik in 1982, Guattari writes that they had "invented a kind of explanatory machine that tries to penetrate different areas, different fields; to penetrate and at the same time invent, because, along the way, we had the impression of unleashing certain encounters, and perhaps even (why not) catalysing certain microevents" (Guattari and Rolnik, 2008: 427). He was also famously unable to settle on one idea – Deleuze described him as always onto the next thing – and he sometimes gave the "impression of being a kind of comet" (2008: 446), travelling not only to Brazil, but to Japan, Mexico, Poland, and the USA. And yet he resented occupying the position of an intellectual bringing "the latest word from Paris, the latest slogan", and instead encouraged the invention of different expressions and the formulation of relation-specific questions (2008: 435). Moreover, his ecosophical approach was not only an aesthetic or an analytic engagement, but an ethical one. For us, his invocation of a new gentleness, elsewhere translated as a new smoothness [*nouvelle douceur*], is suggestive of "new constellations of universe" (2008: 417), a proliferation of becomings.

Elsewhere, Guattari diagrams vectors for cultivating a type of fieldwork decoupled from an existential grid to which geography and geographers must ostensibly conform. This, in brief, entails an attentiveness to the refrain. Refrains, ritournelles, ritornellos affirmed by Derek McCormack as "affective blocks of space time" (2010: 213) in both rhythmic and plastic forms might be considered as techniques for both thinking through the processual natures of experience and for opening up enunciative territories through their diagrammatic and sensory functions. To this end, the refrains of, and for, hitchhiking are not given in advance, but emergent from the hitchhiking itself. That's to say hitchhiking doesn't move to a metronomic beat, yet is nonetheless generative of a certain kind of momentum, one that is central to acts of existence. Indeed, as Guattari suggests in unusually stark terms, "in life, one can only hold on to momentum. Subjectivity needs movement, directional vectors, ritournelles, rhythms and refrains that beat time to carry it along" (2009: 69). In other words, refrains can – though they may not – generate a certain 'expressive consistency' through repetition (see McCormack, 2013).

One refrain that permeates both the performances of hitchhiking and geographical fieldwork more generally is that of participation; notionally, at least. Despite its pervasiveness in thinking and doing research, the idea of participation or 'being participatory' is not without difficulty (Gerlach and Jellis, 2015). There is, albeit not obviously, a politics to participation. The mobilisation of hitchhiking here, however, is not to invoke an ethos of participation or to use participatory techniques as either a pedagogical method of how to do fieldwork, or as a substitute for ethics and ethical concerns. This means that hitchhiking is not about participation in overcoded, liberal terms. Participation through hitchhiking is more problematic insofar as it generated by and generative of processes, practices, and performances motored by actors that themselves are never fully formed and whose roles are never proscribed in advance. As Massumi remarks in this respect, "participation precedes recognition" (2002: 231).[3] And indeed Guattari (2015: 24) remarks, "when I participate in something, I don't wait to uncover its theoretical necessity". Participation, as with hitchhiking, then, has a 'speculative geography', "in which actors fold in and fold out of both being participant, and becoming eventful" (Gerlach, 2015: 281). This, then, is the heart of a Guattarian, eventful, geography in which we diagram participation in terms of singularisation, "a self-organising process that at its most basic level concerns bringing together ensembles of diverse components (material/semiotic; individual/collective)" (Genosko, 2002: 136). We might also think of a Guattarian geography as an attempt to develop "effective practices of experimentation" (Guattari, 2008: 24) – practices which might entail a shift in disposition, towards lines of movement and of affect, vectors which might point toward or be generative of micropolitical actions, or what we call alongside Guattari, a new gentleness.

New gentleness

Thumbs up, arms stretched, thumbs down, arms retracted, pause. The endless ballet of hitchhiking. Endless, to be sure, is pushing it, but lactic acid is its own infinity. Brakes screeching, a BMW double-backs on itself and the passenger door swings open: 'get in, let's go!' Then, before thoughts have the chance to settle, a line traced time and time again by Gareth the driver sets us on a course for Plymouth. Shared relief at getting an audacious ride soon turns to panic at 120 mph; this line could be cut short. The tattooed etches on the forearm point to somewhere else. Gareth, accelerating, then braking, and then shifting gear again little regard for the clutch, weaves his itinerant story of suicide, fights, police, and Eric

Clapton through the New Forest, Honiton, Exeter, and, all too soon, Plymouth. What to make of this journey? Madness, of a sort, but then madness, "with its halo of strangeness, reified in an alterity with no return, nevertheless inhabits our ordinary apprehension, without qualities, of the world" (Guattari, 2011a: 17). At such speeds it's difficult to comment on what coordinates were passed or created on the way, certainly not in any geometric sense. With Gareth and Guattari it's less a case of waypoints than it is of 'singularity points', "contingencies irreducible to serial generalisations" (2011b: 192); a fieldworking turning only on the conditions of unqualified potential and possibility.

Pause. Gather the materials need to hitch: destination boards, waterproofs, a modicum of determination, or at least a quantum of the inconstant joy that Spinoza identifies as hope. Early start, again. This time, with an objective: head home. Big breakfast and more Guattari for the road. Can't remember the page number. He suggests moving onwards, "don't hold back" (2011b: 195) is the clarion call from the maverick cartographer. We concur in obedience even if, as for Félix himself, "every orientation remains provisional" (2011b: 179). Having been dropped off after a short ride and offered a job by the unnameable driver paving residential patios, cash-in-hand, we walk to the Shell service station, just off the A38. It's a good location, or so we were told. Time passes slowly. Arms ache, thumbs are tense, and the remnants of yesterday's headache linger. We keep a certain distance between us, hoping to catch drivers' attention. Maybe 5 metres, more or less; a sort of synchronised ballet of thumbing and foot shaking. We take it in turns to take a quick rest, a momentary repose for orientation, "trying to see the trace of the real in continuity with the lines on the map" (Berardi, 2008: 5). We stretch our arms, moving them in different directions. Gestures, gestural lines, and directional refrains. A hitchhiker's' carto-choreography, or, "the process becoming processual" (Guattari, 1996b: 136). Still waiting for a lift, with not even the option of quitting in despair.

Eventually, we retreat to the petrol station itself; inhaling the fumes, we pack up our boards. A man hops out of a white van and takes an interest. We tell him we're seeking a ride – anywhere away from this spot. Once on board, racing towards the motorway, we listen in on xenophobic pronouncements. And yet he's driving to the airport to get out of the country. Liberal interventions soon evaporate when you're on the M5, tired, and wanting to be back on firmer existential territories. We leave him a postcard; he writes a few weeks later from a yacht in Thailand.

The temptation is to provide an explanation of such encounters, to read them and make sense of them. As Guattari notes, with "a little authority and much bluffing, one could always make all these details fit into the framework of basic interpretation" (Guattari, 2016: 232). It is a theme to which he returns again and again in his work, and particularly in *Schizoanalytic Cartographies*. Analysis "no longer resides in a derivation of interpretable signifying chains" (Guattari, 2012: 19) and should therefore extend – or to use his terms, enlarge – its means of intervention, so that it does not rest on the "interpretation of fantasms and the displacement of affects but endeavours to render both operative, to score them with a new range (in the musical sense)" (2012: 214). For instance, in one case which occupies him for some time, Guattari describes a singer who has had a recent bereavement and can no longer reach the same notes. It starts with a fairly conventional reading of the situation before Guattari offers a different interpretation of the condition. This one is much shorter, and notes that without the existence of the singing the patient might have lost other kinds of octaves in other registers, and culminates with a typical flourish of his:

> That is all one can say about it! Certain long marked out pathways . . . undergo a pragmatic transformation. Should these facts be recorded in the column of debts and losses? Nothing is less sure!
>
> (Guattari, 2012: 25)

So if interpretation is not what we have in mind, what are we to *do* with this material? Again, we find ourselves drawn to Guattari's fondness for the speculative (Jellis, 2014). For him, an interpretation is always one among many and its strength rests on if it 'works' – that is to say, if the case takes on a different consistency. With all that said, what of Gareth and Simon, their lines of flight, their finite vectors, their commentaries on life?

The hitchhikers' mode of existence is often one of anticipation, always on the ready, or on the cusp of readiness. That edginess persists, even once car doors are slammed shut; relaxation does not come into it. There is a unwritten obligation to keep your conversational end up (Moss, 2009): to nod, murmur assent, or appear attentive. You are not only thrust into someone else's journey, but also their life, however portentous that might sound. People, it turns out, have stories to tell. And what was striking was quite how much people wanted to tell you. To be sure, this may be due to the allure of storytelling, where the narrator is able to reinvent a wished-for life from scratch. One imagines that similar strategies may be found across other forms of travelling, from ocean liner cruises to gap year backpacking. But it is also because there is someone there who will listen.

This theme of the hitchhiker as a careful listener is not new. Grace Small, the 'grandmother' hitchhiker, described herself as a counsellor to lost and lonely drivers. Recalling one journey, Small writes that she could not figure out the driver. By the time they reached their destination, however, the driver's, "troubles just poured out. The years totalled many that he had not talked about his unhappy childhood" (Small and Davis, 1990: 191). But perhaps it is less what our drivers said, the *content* of our discussions, and rather that they were willing to say it at all. Although it has often been remarked that there is spirit of mutuality when it comes to hitchhiking – that there is, in fact, often extraordinary hospitality – we are often hamstrung as researchers by the kinds of locations in which interviews or encounters ought to happen. Some of this is premised on common sense or even safety. But it might well limit the spaces in which encounters can happen. The fleeting nature of the connection and the intimacy of the moving car may well be a catalyst for confession. In other words, hitchhiking is a technique for relational movement – one which could sweep you up and take you along unimagined trajectories.

Mutant coordinates

Nobody is stopping; just a handful of people wave. Some passengers imitate us, their own thumbs up, grinning at us as they speed on. Intermittently, our now well-rehearsed refrain of thumbs up is met with middle-fingers. A car skids into the lay-by but the virtual promise of the lift soon gives way to profanity. We brace ourselves for further insult as another car swerves over, but this time the driver waves his arm at us. We run over, the window wide-open. We struggle for conversation but leave a postcard all the same. He suggests we stop at some services. They're not services though: it's another petrol station. He drives off. We get the thumbs out. The noise washes over. The rhythm of the traffic; the wind as the cars speed past. There are fewer waves, and even fewer thumbs up. Several cars pull over. They need petrol, not hitchhikers. Time slows. We draw some chalk lines during the interludes. They are squiggles. Some split, others turn on themselves. When will they let you pass? This could be the end of everything.

What to make, in conclusion, of these ethico-aesthetic, or micropolitical, interventions? If they don't add up straightforwardly to what we call geo-graphing, then perhaps hitchhiking has some bearing on Guattari's notion of 'earthing', a perilous and transversal movement between subjectivation and intensification of affects, bodies, and materials through which "something else becomes possible", specifically that of the "emergence of coefficients of processual creativity" (Guattari, 1995: 82). Hitchhiking as a technique of relational movement turns on the generation of, 'mutant coordinates', proto-locative points that disrupt the staid geometries of Euclid and instead affirm the molecular vectors of everyday mobilities – in other words, the "unthinkable qualities of being" (Guattari, 1995: 106). What then, of a new gentleness, or a new smoothness? We suggest that this is a way into thinking and practising the micropolitical (see Jellis and Gerlach, 2017), which is among other things, an "insistence on the pragmatics of experience/experiment" (McCormack, 2010: 213). Such an insistence reaffirms research-creation as what Brian Massumi calls a 'commingling': the "activity of relating" (2011: 103). Hitchhiking, with its speculative vectors, recombinant bodies, and its transversal gentleness, is as much a technique of existence as a technique of relation.

Many years later, driving to work, you narrowly avoid somebody walking in the road. As you pass, you see a hand dangling by the side, thumb pointing outwards. As you speed on by, you recall your own spells of hitchhiking. At the next possible opportunity, you double back, and then turn again, then pull up on the verge. The hitchhiker is nowhere to be seen. A moment or so later, he emerges from a bend and you beckon him over. Where are you going? The danger, distrust, and dubiety that you have been writing of, in various draft forms, comes bubbling up. An uneasiness about it all, about letting a stranger join you on your journey. He's grateful beyond expression. You ask where he's going: Oxford.

Acknowledgements

We are immensely grateful to all the drivers who have stopped to give us a ride around south-west England over the last few years; thank you for sharing your journeys with us. Versions of this paper were presented at the Siobhan Davies Dance Studio in 2009, in the 'For Félix' session at the AAG 2012 in New York, and at the Nottingham Contemporary art gallery in 2015, for an event organised by Andrew Goffey at the Centre for Critical Theory. Thank you to the others in our molecule – especially Sarah Rubidge and Derek McCormack – for your encouragement. And thank you to J.D. for your comments on an earlier draft of this.

Notes

1 The discussion of hitchhiking in geography has been not only been brief but, perhaps more surprisingly, undertaken by physical geographers.
2 Local groupings were referred to as 'molecules', "a reference to Deleuze and Guattari's concept of the 'minor' as 'molecular'" (Manning and Massumi, 2014: 106). These molecules consisted of between 3 and 10 individuals.
3 The radicalness of this idea is always on the cusp of dissipating if participation simply comes back to questions of subject positions (which have to do with claims and recognition).

References

Abrahamsson, S., Clarke, G., Henry, D., Hung, J., Gerlach, J., Gunduz, Z., Jannides, C., Jellis, T., McCormack, D., Rubidge, S., Stones, A. and Wilford, A. (2009) How to devise ethico-aesthetic practices by moving between cartography and choreography? *Inflexions*, 3. Tangents.

Applebome, P. (1988) On the road, 1988: New dangers and decline in hitchhiking. *New York Times*, 3 April.

Berardi (Bifo), F. (2008) *Félix Guattari: Thought, Friendship and Visionary Cartography* (trans. G. Mecchia and C.J. Stivale). Basingstoke: Palgrave Macmillan.

Chesters, G. and Smith, D. (2001) The neglected art of hitch-hiking: Risk, trust and sustainability. *Sociological Research Online*, 6(3).

Cox, N.J. (1980) On the theory of hitch-hiking as a spatial stochastic process. *Professional Geographer*, 32(1): 51–54.

Day, M.J. (1980) Comments arising from 'On the Theory of Hitch-Hiking as a Spatial Stochastic Process' by Nicholas J. Cox. *Professional Geographers*, 32(4): 480–482.

Deleuze, G. (2007) Letter to Uno: How Félix and I worked together. In: Deleuze, G. *Two Regimes of Madness: Texts and Interviews 1975–1995* (trans. A. Hodges and M. Taorima; ed. D. Lapoujade). Los Angeles, CA: Semiotext(e).

Genosko, G. (2002) *Félix Guattari: An Aberrant Introduction*. London: Continuum.

Gerlach, J. (2015) Editing worlds: Participatory mapping and a minor geopolitics. *Transactions of the Institute of British Geographers*, 40: 273–286.

Gerlach, J. and Jellis, T. (2015) Guattari: Impractical philosophy. *Dialogues in Human Geography*, 5(2): 131–148.

Guattari, F. (1995) *Chaomosis: An Ethico-Aesthetic Paradigm* (trans. P. Bains and J. Pefanis). Sydney: Power Publications.

Guattari, F. (1996a) Regimes, pathways, subjects (trans. B. Massumi). In: Genosko, G. (ed.) *The Guattari Reader*. Oxford: Blackwell Publishers. pp. 95–108.

Guattari, F. (1996b) Institutional practice and politics. In: Genosko, G. (ed.) *The Guattari Reader*. Oxford: Blackwell Publishers. pp. 121–138.

Guattari, F. (2008) *The Three Ecologies* (trans. I. Pindar and P. Sutton). London: Continuum.

Guattari, F. (2009) *Soft Subversions: Texts and Interviews 1977–1985* (trans. C. Weiner and E. Wittman). London: The MIT Press.

Guattari, F. (2011a) Schizo Chaosmosis (trans. A. Goffey). In, Alliez, E. and Goffy, A. (eds.) *The Guattari Effect*. London: Continuum.

Guattari, F. (2011b) *The Machinic Unconscious: Essays in Schizoanalysis* (trans. T. Adkins). Los Angeles, CA: Semiotext(e).

Guattari, F. (2012) *Schizoanalytic Cartographies* (trans. A. Goffey). London: Bloomsbury.

Guattari, F. (2015) *Machinic Eros: Writings on Japan* (trans. G. Genosko and T. Adams). Minneapolis, MN: Univocal Press.

Guattari, F. (2016) *Lines of Flight: For Another World of Possibilities* (trans. A. Goffey). London: Bloomsbury.

Guattari, F. and Rolnik, S. (2008) *Molecular Revolution in Brazil* (trans. K. Clapshow and B. Holmes). London: The MIT Press.

Jellis, T. (2014) Book review: *Schizoanalytic Cartographies* by Félix Guattari. *Society and Space*. Available online: http://societyandspace.org/2014/07/19/schizoanalytic-cartographies-by-felix-guattari-reviewed-by-thomas-jellis/

Jellis, T. and Gerlach, J. (2017) Micropolitics and the minor. *Environment and Planning D: Society and Space*, 35(4): 563–567.

Laviolette, P. (2016) Why did the anthropologist cross the road? HitchHiking as a stochastic modality of travel. *Ethnos*, 81(3): 379–401.

McCormack, D.P. (2010) Thinking in transition: The affirmative refrain of experience/experiment. In: Anderson, B. and Harrison, P. (eds.) *Taking Place: Non-Representational Theories and Geography*. London: Ashgate. pp. 201–220.

McCormack, D.P. (2013) *Refrains for Moving Bodies: Experience and Experiment in Affective Spaces*. Durham, NC: Duke University Press.

Mahood, L. (2016) Thumb wars: Hitchhiking, Canadian youth rituals and risk in the twentieth century. *Journal of Social History*, 49(3): 647–670.

Manning, E. and Massumi, B. (2014) *Thought in the Act: Passages in the Ecology of Experience*. Minneapolis, MN: University of Minnesota Press.

Manning, E. and Plumb, L. (eds.) (2009) Micropolitics: Exploring ethico-aesthetics. *Inflexions*, 3. Tangents. Available online: www.inflexions.org/issues.html#no3

Massumi, B. (2002) *Parables for the Virtual*. Durham, NC: Duke University Press.

Massumi, B. (2011) *Semblance and Event: Activist Philosophy and the Occurrent Arts*. Cambridge, MA: The MIT Press.

Moran, J. (2009) A guide to hitchhiking's decline. *The Guardian*, 5 June.

Moss, S. (2009) What killed hitchhiking? *The Guardian*, 29 May.

O'Regan, M. (2012) Alternative mobility cultures and the resurgence of hitch-hiking. In: Fullagar, S., Markwell, K. and Wilson, E. (eds.) *Slow Tourism: Experiences and Mobilities*. Bristol: Channel View.

Perkins, A. (2016) The death of hitchhiking is a modern tragedy. *The Guardian*, 21 September.

Purkis, J. (2012) The hitchhiker as theorist: Rethinking sociology and anthropology from an anarchist perspective. In: Kinna, R. (ed.) *The Continuum Companion to Anarchism*. London: Continuum. pp. 140–161.

Schlebecker, J.T. (1958) An informal history of hitchhiking. *The Historian*, 20(3): 305–327.

Small, G. and Davis, R.B. (1990) *The Hitchhiking Grandmother*. Forest Grove, OR: Pilgrim Way Press.

Stromberg, J. (2015) The forgotten art of hitchhiking – and why it disappeared. *Vox*, 10 June.

Sykes, S. and Sykes T. (eds.) (2005) *No Such Thing as a Free Ride? A Collection of Hitch-Hiking Tales*. London: Cassell Illustrated.

Woodcock, S. (2011) The lost art of hitchhiking. *The Guardian*, 2 August.

Zahm, O. (1994) Félix Guattari et l'art contemporain. *Chimères*, 23: np.

13 Guattari and the micropolitics of cinema

The desiring-machines of Satoshi Kon

Andrew Lapworth

Introduction: cinema and the micropolitics of desire

One of the central preoccupations defining Guattari's (2009a: 142) engagement with politics in both writing and activism was the attempt to rethink the stakes of contemporary political struggle in terms of the "liberation of *desire*" from dominant systems and institutions of *signification*. A crucial aspect of this struggle for Guattari was the need to invent a different thinking of desire beyond the "personological" understandings typically found in philosophy and the social sciences, which reduce it to subjectivist realms of identification and unconscious representations. In contrast, Guattari conceives desire not as a property held by an individual subject, but instead as material processes of individuation that bring subjects, objects, and their relations into being.[1] Desire therefore takes on a preindividual and transformative accent in Guattari's thought through its capacity to produce a "rupture in the economy of the 'already-there' [. . .] activating another world of possibilities" (Guattari, 2015a: 36–37). This leads Guattari (2016a: 74–75) in his book *Lines of Flight* to distinguish between two possible senses of politics in relation to desire: a *micropolitics of desire* that creatively experiments with intensities of desire, exploring new associations of forces and bodies that disrupt established relations; and a *molar politics of signification* oriented towards the capture and codification of these desiring intensities according to the recognisable forms and dominant value systems of the present (see also Guattari, 1984: 85–86). Psychoanalysis, Guattari argues, clearly operates on this latter terrain of a politics of signification through its attempts to classify legitimate and illegitimate forms of desire, and to proliferate normalised models of subjectivity and the unconscious that are compatible with the capitalist order (Watson, 2008).[2] While the critique of psychoanalysis is a focus of Guattari's early work, Lazzarato (2014: 112) notes how in his later writings it becomes just one of many "social machines of signification" (including the mass media, education, the law, finance, the family) that impose a semiotic modelling of identities, roles, and functions on bodies. Guattari's point here is that one of the main ways in which power operates in contemporary society is through the constitution of dominant signifying semiologies that control the images we have of ourselves, and of what we might become (Elliot, 2012).

If artistic modes of expression are privileged throughout Guattari's work as a means for escaping these repressive signifying semiologies, it is because they appeal to what he terms *a-signifying semiotic registers* that bypass language and meaning to act directly on bodies as intensities (of colour, movement, rhythm, time) that can only be sensed (Guattari, 2011).[3] These a-signifying intensities, as Genosko (2012: 250) writes, "come flush with sensibilities not yet entangled in dominant modelisations of identity and social relations", and which instead create new potentials of thinking and feeling. Of the arts, Guattari (2016b: 54) always had a special fascination with *cinema* as an "extraordinary instrument for the production of subjectivity". In a series of essays published in the late 1970s and early 1980s, Guattari argued that the cinema is a "revolutionary machine" for capturing and expressing desire, but one which is just as capable of inciting transformative processes of becoming as it is reproducing impoverished identities and models of behaviour: "one can expect the best or the worst from cinema" (Guattari, 2009b: 267). Indeed, and as he argues in the essay *Cinema of Desire*, the history of cinema can be imagined as a site of ongoing political struggle between signifying semiologies that homogenise and personalise desire (through clichéd plots and clearly individuated, stock characters), and an a-signifying semiotics that participates in the liberation of desire from "the territorialising alienations and pre-set coordinates" of national, familial, personal, humanist, and transcendent values (Guattari, 1984: 98).

However, these a-signifying registers and potentials of the cinematic encounter have largely been ignored in conventional approaches to theorising the politics of film, which instead tend to think cinema through representational logics of signification and meaning (linguistics) and the encounter through the interpretative semiologies of the subject/object split (psychoanalysis). Indeed, these a-signifying elements have even been downplayed in contemporary engagements with Guattari in the film literature, which have mainly focused on the potential 'utility' and 'applicability' of his (and Deleuze's) concept of "the minor" for theorising postcolonial and emerging national cinemas (e.g. Martin-Jones, 2006; Marshall, 2009). The problem with such piecemeal applications of specific concepts, as Gerlach and Jellis (2015: 132) highlight, is that they leave us with a rather limited sense of the "conceptual-empirical potential" of Guattari's philosophy for contemporary thought, which in the work on 'minor cinema' has often narrowly focused the question of politics on the subversion of narrative structures and themes or the affirmation of alternate identities. Rather than attempting to apply concepts from elsewhere in his work to cinema, in this chapter I engage directly with Guattari's explicit discussions of the relations between film, politics, and desire in a series of recently translated writings (Guattari, 2009b, 2016b). In the following section, I explore the implications of Guattari's philosophy for contemporary understandings of a cinematic politics by foregrounding two important theoretical shifts operative in these essays: from the cinematic encounter as a structural representation of the same (psychoanalysis) to the encounter as a *machinic event* of difference; and from an interpretative politics of signification (linguistics) to an experimental 'micropolitics' of *a-signifying intensities*. At stake in Guattari's work, I claim, is a more transformative sense of cinema as what he terms a "desiring-machine", which generates a-signifying processes that disrupt existing forms of representational coding, and opens us instead to "new collective arrangements of desire" (Guattari, 2009b: 241). Finally, I explore this potential through an engagement with the expressions of a machinic unconscious in Satoshi Kon's, 2006 anime, *Paprika*, a film which forcefully highlights cinema's capacity to think desire as a process of connection and individuation[4] (rather than through forms of identification or representation).

Machinic thinking: Guattari's a-signifying cinema

> One goes to the cinema to suspend the usual modes of communication for a while.
> (Guattari, 2009b: 258)

Unlike some of his philosophical contemporaries (Badiou, Baudrillard, Deleuze, Rancière), Guattari never produced a dedicated study of the cinema. Yet the essays he did write (most of which appear in a section on 'Cinemachines' in the *Chaosophy* volume (Guattari, 2009b)) reveal a perceptive and original approach to thinking the relations of film and politics. Despite a few recent attempts to explore the implications of Guattari's philosophy for contemporary film theory (see Genosko, 2009; MacCormack, 2005; Herzog, 2008), his thinking on the politics of cinema has largely been overshadowed by the more sustained engagements of his philosophical collaborator, Gilles Deleuze. Where Guattari does appear in the film literature there has been a tendency to simply conflate his thought with that of Deleuze's (usually to the detriment of Guattari, whose influence on concepts such as 'minor cinema' is reduced or elided), and thus missing some important differences between their respective approaches. Deleuze (2005a, 2005b) was always clear that his work on cinema should be read as a resolutely philosophical project, concerned above all with addressing the new problematics for thinking movement and time that the cinematic encounter raises for philosophy. For the most part, and with some modifications, Deleuze's philosophy of the cinema drew heavily on the key terms, taxonomies, and auteurist focus defining post-war French film theory (and especially the writings of Andre Bazin and Serge Daney in the film-journal *Cahiers du Cinema* (Dosse, 2010)). Guattari's engagement with film, by contrast, was far more undisciplined, dispensing with most of the terminology and taxonomic categories of conventional film theory, and interested as much in the actual pragmatics of making politically engaged cinema as it was with the philosophical analysis of it.[5] Relatedly, then, we also find in Guattari's (1984: 98) writing a greater attention to what he terms the "transversal connections" that cinema can establish with, for example, the affective and perceptual capacities of different audiences, new social and political movements (e.g. documentary cinema and the anti-psychiatry movement), alternate systems of cinematic production and financing, and so on.[6] Therefore, and while Deleuze (2005b) in *Cinema II* offers a precise theoretical characterisation of what political (or 'minor') cinema *is*, it is Guattari who provides a more pragmatic account of the material processes and transversal relations that define *how* it might be created (Elliot, 2012).

Despite these differences, what Guattari shared with Deleuze was a belief in the inadequacy of dominant psychoanalytical and linguistic approaches – bound as they are to subject-centred and representational modes of thought – for thinking the transformative potentials of the cinematic encounter. Central to both Guattari and Deleuze's accounts is the claim that cinema thinks for itself – it does not need the analyst or theorist to bring thought to it, and it certainly doesn't aspire to simply reproduce the clichéd logics and structures of human thought (Clarke and Doel, 2016; Lapworth, 2016). Cinema is instead a non-human, "machinic intelligence" capable of expressing new powers and potentialities of thought through the different images and signs it creates (Epstein, 2014: 67). What Guattari (2009b) argues is most valuable about the cinema, then, is that it turns us away from the dominant images and thoughts that we already have: while watching a film we encounter affects, perceptions, and thoughts that are not ours, and which confront us with what we have yet to think (Rushton, 2012). In the essay *The Poor Man's Couch*, Guattari (2009b) argues that psychoanalytic engagements with cinema have often tried to constrain and absorb these becomings of thought by reducing film to a representational 'object' that can be interpreted according to the pre-given structures and meanings of the psychoanalytic unconscious. Indeed, he notes how this literature has typically understood cinema as a mass form of the "psychoanalytic cure", but one that is even more potent (and thus potentially more repressive) in its effects because it exceeds the linguistic focus of the psychoanalytic 'transference' procedure to also engage with various non-linguistic semiotic registers and processes of subjectivation (such as rhythms, durations, intensities) (Guattari, 2009b: 262). Guattari particularly denounces the co-option of "commercial cinema" (i.e. Hollywood) into a privileged instrument of contemporary social control, one that chains heterogeneous forces and intensities of desire to a signifying grid that fixes narrative possibilities, crystallises recognisable forms of identity, and reproduces the behavioural norms required by the capitalist order (Lazzarato, 2014).

While much of cinema is "undeniably personological, Oedipal, reactionary", Guattari (2009b: 267) argues that it isn't necessarily so in the same way that psychoanalysis is.[7] Psychoanalysis is patently unable to think the unconscious processes and becomings involved in cinema because it remains tied to a restrictive conception of the unconscious which reduces it to the psychic interiority of the constituted subject (Freud) or to the structural field of the linguistic signifier (Lacan) (Sauvagnargues, 2016). What is at stake in the cinema is a radically different thinking of the unconscious, one in which the psychoanalytic "theatre of representation" (in which already-constituted forms and identities are repeated) is substituted for the individuating "factory of production" (through which these forms become-other) (Deleuze and Guattari, 2004: 25). Importantly, to grasp these individuating potentials of cinema demands a reorientation in thinking that Guattari (2015b) first outlined in his earliest work, *Psychoanalysis and Transversality*, from a notion of *structure*, which thinks the encounter (cinematic or otherwise) through the dialectical coordinates of subject-object relations, towards a concept of the *machine*, which thinks the encounter as an event that gives rise to new constellations of bodies, desire, images, and affects. Throughout his work, Guattari (1995: 108) frequently distinguished his sense of the 'machinic' from common-sense understandings of the 'mechanical', with the latter referring to the more limited sense of a structural relationship between discrete set of parts which function together to perform a pre-defined set of actions. The 'machinic', by contrast, takes on a more transformative meaning in his work through its links to "procedures that deterritorialise its elements, functions, and relations of alterity" (Guattari, 1995: 37; Antonioli, 2012). Following Guattari, the machine thus describes systems of intensive connection by which heterogeneous components encounter and affect each other, with each encounter having the possibility of redistributing the material flows and relations between assembled elements (MacCormack, 2005).

The cinema might therefore be reimagined as a form of *machinic encounter* between the deterritorialising intensities that compose the filmic assemblage of body-brain-screen. This in turn has important implications for how we might theorise the politics of the cinema, which is no longer a matter of what an image *is* (its representational contents or themes) but rather *how it works* to potentially modify existing arrangements of desire (Sauvagnargues, 2016). As Guattari (2009b: 264–265) puts it: "a machine treats you like a machine, and the essential thing is not what it says, but the sort of vertigo of abolition that the fact of being 'machinized' provides for you". The political question here thus becomes whether the cinematic machine is oriented around predictable identities and meanings that cause intensities to follow familiar trajectories; or, alternatively, whether it creates modes of expression that reorganise flows between components in different directions, generating a rupture in our existing habits that opens towards new horizons of possibility for thought and experience (Hynes, 2013; O'Sullivan, 2010).

An important part of cinema's machinic capacity to break expectations and rupture dominant orders of sense is its creation of what Guattari (2009a) calls 'a-signifying intensities'. Therefore, the second shift in Guattari's thinking on the politics of film that I want to highlight is from a focus on *signifying semiologies* towards a thought of *a-signifying semiotics*. By drawing attention to the a-signifying semiotic registers of cinema, Guattari attempts to strike back against a prevalent (Saussurean) linguistic structuralism that reduces film to a matter of human language and systems of signification. As he argues in the essay *The Cinema of Desire*:

> It is vitally important to expose and insist on the independence of an asignifying semiotics. It is this, in fact, that will allow us to understand what permits cinema to escape the semiologies of signification and to participate in collective assemblages of desire.
>
> (Guattari, 2009a: 241)

'Semiologies of signification', as noted earlier, are based on dominant social forms of encoding that constitute a stable, transcendent signifier which "systematizes, hierarchizes, and structures" heterogeneous matters of expression into representational chains of meaning (Watson, 2008: n.p.). A-signifying semiotics, in contrast, "break the effects of signification and interpretation, thwart the system of dominant redundancies, accelerate the most 'innovative', 'constructive', 'rhizomatic' components" (Guattari, 2009b: 307–308). These a-signifying processes operate at the level of material intensity, acting immanently on bodies without recourse to meaning or signification. Cinema is populated by a multiplicity of different a-signifying elements, including light and shadow, sensations of colour (Godard's (1969: 7) formula: "It's not blood, it's red"), non-linguistic sounds, music, movements, gestures of bodies, experiences of duration, and rhythms of spatial editing. Quoting the film theorist Christian Metz, Guattari (2009b: 243) writes that this a-signifying "polyvocality" means film is irreducible to the perspectives of representation or language, "since the semiotic alloy that composes its matter of expression is not fixed and is itself open to multiple systems of external intensity that go in different directions".[8] Instead of the system of double articulation operative in linguistics (signifier/signified), Guattari argues that cinema puts into play '*n*' material components of expression that move beyond signifying semiologies and the personological individuations (i.e. 'normal' and model characters) and dominant values (adaptation and cooperation with authority) they enshrine (Genosko, 2012).

The a-signifying intensities produced by the cinema thus lie beyond the reach of an "impotentializing reflexive consciousness" (Guattari, 1984: 137). They are instead linked to material events of thought and affect, "acting directly on the nervous system, rather than passing through the detour of the brain" (Deleuze, 2005c: 43). Thought materially encounters these a-signifying elements as the bearers of new possible 'Universes of reference' currently unknown to us, comprising forces and becomings that exceed the conventional images we have of ourselves and our relations to others. At their best, and as Lazzarato (2014: 111) notes, encounters with cinema have the potential to act as events of "disindividuation", which suspend the habitual coordinates linking ordinary situations, images, and movements in predictable patterns, and force us to cross over existential thresholds into new territories of experience.

So, how might this attention to the a-signifying registers of the filmic encounter offer us a different way of thinking about the politics of cinema? Key here, I argue, is how Guattari's work attempts to unsettle many of the conventional terms and criteria through which we would judge the political valence of film, which in the psychoanalytic and linguistic film literature often centres around issues of narrative or psychological contents and themes (Guattari, 2016b). Through the concept of "a-signifying semiotics", Guattari attempts to untether a thinking of politics from these representational questions of 'content', instead opening thought to intensive forces and processes of 'individuation' (Massumi, 2011: 54). Therefore, what defines the politics of film is not simply the question of its subject matter:

> Cinema is political whatever its subject [. . .] one can make a film having life in a religious convent as its theme that puts the revolutionary libido in

> motion; one can make a film in defence of revolution that is fascist from the point of view of the economy of desire.
>
> (Guattari, 2009b: 245–246)

What defines the 'micropolitics' of cinema, Guattari argues, is thus not the meanings or contents of specific images, but rather the extent to which these images capture and render sensible a-signifying processes that escape dominant semiologies. The images and signs of cinema can thus operate as "a singularity, a rupture in sense, a cut, the detachment of a machinic component from the flux of components", involving the production of a new kind of affective rhythm which, for a moment at least, can open other creative directions in the dynamic production of subjectivity (Guattari, 1995: 18). Of course, and as Maria Hynes (2013) notes, Guattari was careful to avoid a naïve celebration of this machinic process, seeing the logic of detachability at the heart of capitalist modes of subjectivation in which affective singularities and intensities are torn from semiotic fields to become constrained into repetitively drawn associations and predictable forms. Against the identificatory closure of capitalist forms of subjectivation, a micropolitical cinema is instead oriented towards the "resingularisation of subjectivity [. . .] and the overthrowing of dominant redundancies" (Guattari, 1995: 20–21). For Guattari (2009b: 246), "every filmic production, every sequence, every frame" takes sides in this micropolitical struggle between signifying semiologies that impose roles and functions on bodies and a-signifying ruptures that unleash creative becomings; between a mechanical reproduction of the same and the machinic production of difference; and between a "conservative economy of desire" and the possibility of a "revolutionary breakthrough" that invents new potential forms of life.

The 'dream-machines' of Satoshi Kon's *Paprika*

Few contemporary directors, I would argue, have as deep an affinity with the spirit of Guattari's thought as the Japanese filmmaker, Satoshi Kon, whose work similarly takes as its focus issues of human-machine relations and becomings, the connections of contemporary social and mental ecologies, the non-linearity of time and space, and the transversal scrambling of conventional genres. *Paprika* (2006), Kon's last anime film before his sudden death in 2010 at the age of 46, is perhaps the most resonant with Guattari's ideas through its explicit critique of psychoanalytical repression and its exploration of the machinic process of desire and the unconscious. Based on a novel of the same name by the Japanese author Yasutaka Tsutsui (1993), *Paprika* is set in a near-future Tokyo in which the psychiatrist Dr. Atsuko Chiba and her team are working on a new technological device called the 'DC Mini' which allows analysts to enter their patient's dreams as a form of psychotherapy. Despite still being in development, Chiba secretly uses the device in private therapy sessions outside the research facility, during which she appears in her patient's dreams as her lively alter-ego, Paprika. However, early in the film three of the dream-machine prototypes are stolen and subsequently used by the thief to invade the dreams of specific individuals, which leaves them in a comatose state from which they cannot awake. As more people are targeted, their dreams and psyches converge to form an ever more powerful and dangerous dream that increasingly dissolves any line separating the real world and the dream-world, and threatens to draw all that exists into the abyss of a collective delirium that none can escape.[9]

Following Guattari, what interests me here about *Paprika* is less the (molar) politics of its narrative contents and themes (of which there are many: gender and violence; technology and subjectivity; psychoanalysis and social control), and more the ways in which the film creatively engages with a-signifying processes and intensities to generate new possibilities for thinking subjectivity and the unconscious. What I particularly want to highlight is how the film's expression of the collective affects and virtual forces of *dreams* replaces a psychoanalytical conception of the unconscious as an individualised entity bound to personological representations and meaning, with a thought of the unconscious as a "desiring-machine" predicated on transindividual connectivity and becoming (Guattari, 2009b). Of course, cinematic dream states have usually been framed as the interpretative object *par excellence* of psychoanalytical approaches to film, which read them as symbols of the repressed desires and pathologies of the dreaming subject (Powell, 2007). In the essay *Balance-Sheet for Desiring-Machines*, Deleuze and Guattari (2009: 101) note that psychoanalysis foregrounds what they term the "Oedipal pole" of dreams – or, the "dream-theatre" – where the dream becomes an object of molar interpretation in which representational narratives and symbolism prevail over a-signifying sequences and material intensities ('it's your father', 'it's your mother', 'it's you!'). Despite the Oedipal colonisation of dreams by psychoanalysis, Deleuze and Guattari (2004: 348) still see in them the potential for transformative becomings, arguing that "at the heart of dreams themselves – as with fantasy and delirium – desiring-machines function as indices of deterritorialisation". They thus distinguish a second, "deterritorialising pole" of dreams – the "dream-machine" – in which what matters "is desiring-production, machinic operation, the establishment of connections, the vanishing points or those of the deterritorialisation of the libido being engulfed in the nonhuman molecular element, the circulation of flows, the injection of intensities" (Deleuze and Guattari, 2009: 101). Therefore, and while the psychoanalytical dream-theatre 'reterritorialises' by fixing their significance on imaginary and structural representations, this machinic pole follows the deterritorialising forces and intensities of dreams that break up the symbolic chain of (Oedipal) associations.

Following Deleuze and Guattari, psychoanalysis thus deprives dreams of their micropolitical significance in two main ways: it lacks sensitivity to the a-signifying processes and non-human intensities at work in dreams; and it fails to recognise the dream as a machine for connecting with forces of the outside world (Guattari, 2009b). This reterritorialisation of dreams is also the case in much conventional Hollywood cinema, where dream sequences are usually bracketed off from the central plot through techniques of deframings, dissolves, or special effects that clearly signal to the viewer that what they are watching is a dream (Deamer, 2016). The purpose of doing this is usually to help orientate the viewer to the direction of the film's narrative flow by explicitly demarcating the line between reality and fantasy, and by linking the contents of the dream to the psychology and actions of clearly individuated subjects. Part of what makes *Paprika* such a disorienting experience, then, is the way it breaks with many of these conventional cinematic grammars and techniques through which dreams are usually expressed on-screen. As the film progresses, *Paprika's* 'dream-world' and 'real-world' increasingly become entangled, reaching a point of indiscernibility in the second half of the film where our ability to recognise what is and is not real is suspended. One of the key techniques through this experience of indiscernibility is generated is through Kon's use of *non-linear processes of spatial editing*, in which the movements of characters and objects in the film create new and unexpected connections between different images and spaces. Early in the film these disorienting interruptions in spatial continuity are a defining feature of events in the 'dream-world', such as in the film's precredit sequence where the main characters charge at a frenetic pace through multiple dream-spaces that are loosely connected through a series of match cuts. However, later in the film these non-linear processes and images overspill their containment in dreams and begin to materially transform the spatial fabric and relations of 'real-world' Tokyo. A clear example of this is the 'Dream Parade' of marching objects which appears in a burst of sound and colour at various points in the film to signal the crossing of a threshold between dream and reality. As the thief's deranged collective dreams grows ever more powerful, the parade suddenly appears marching through the streets of Tokyo, transforming stunned onlookers into an array of fantastical objects (Buddha statues, robots, wind-up toys) that join the hypnotic march. What is especially affective about this sequence is its jarring use of both sound (the parade marches to the pulsating technopop soundtrack of composer Susumu Hirasawa) and intensities of colour, which contrast disconcertingly with the almost photorealistic urban backgrounds for which Kon is famed (Osmond, 2009). Here we shift from an emphasis on the intensive differences *between* specific sequences or frames (as in the 'dream-montage' that opened the film), to the image *itself* as site of difference in which a multiplicity of actual forms and virtual forces encounter and transform one another.

Through these cinematic experiments with a-signifying intensities of colour, sound, and non-linear editing, *Paprika* thus expresses a different conception of dreams as no longer simply the private psychological state of an individuated subject, but rather as a transindividual process that connects heterogeneous signs, bodies, ideas, and affects in unforeseen ways. Desire and dreams, the film suggests, are thus not mere unconscious representations of reality as psychoanalysis would have us believe, but are instead machinic forces implicated in the material production of different subjects and worlds. What is interesting here about Kon's *Paprika* is how its anti-Oedipal expressions of the dreams and the unconscious contrast sharply with the dominant Oedipal framing of desire and sexuality usually found in Japanese anime and manga (Brown, 2006).[10] This Oedipalisation of desire in Japanese popular culture is something Guattari (2015a: 14) himself highlights in *Machinic Eros: Writings on Japan*, when he warns against "a capitalistic *infantilism* with its vibrating zones of collective hysteria, such as the syndrome of puerile cute culture (*kawaii*), and the reading-drug of manga comics". Such popular forms of visual culture, Guattari argues, thus often retreat into infantilised representations of desiring-subjectivity that are defined by incestuous libidinal fixations on childhood and infantile attachments to cultural traditions (Asada, 1989). But this is not an irreversible situation, and Guattari points in his text to several Japanese artists and filmmakers (such as Yayoi Kusama, Min Tanaka, Keiichi Tahara, Mitsuo Sato) whose works allow us to distinguish between "an unconscious shackled to the past, to archaic fixations and investments, and a forward looking, machinic conception" (Genosko, 2002: 132). Kon's *Paprika*, I argue, clearly expresses this latter conception through its attempts to think desire as a machinic process of connection that exceeds and transforms personological modes of identity and molar forms of interindividual relation.

At stake in the cinematic encounter is thus also the possibility for a different thinking of subjectivity beyond its typical transcendent and 'egological' articulations. Throughout his work, Guattari stressed that there are multiple possible modalities of subjectivation yet to be imagined and produced, and maybe the dominant model of psychoanalytic subjectivity – with its obsession with familial representations and its face turned towards the fixations of the past – is neither the most productive or relevant for us today (Guattari, 1995: 12, 1996). Instead of a discrete and autonomous individual, subjectivity for Guattari emerges amidst a multiplicity of semiotic and affective processes of which he argues cinema, in particular, "can show how they are orchestrated, 'machinated', and infinitely manipulated" (Guattari, 2009b: 263). It is this capacity to produce new singularising processes that can subvert dominant individualised modelisations of subjectivity that defines for Guattari the cinematic encounter's micropolitical potential. In *Paprika*, this disruption of an image of the individualised subject is clear in the way in which it presents the characters of Dr. Chiba (the psychoanalyst) and Paprika (her dreamworld alter-ego), which in the first half of the film are easily distinguishable as occupying different 'worlds' (real/imaginary), but as these thresholds dissolve later in the film and Chiba/Paprika increasingly share the same frame they become more like the actual and virtual elements of a collective subjectivity. Here, the image of a stable or 'true' self dissipates into a preindividual field of intensities, movements, and encounters whose expression depends on a logic different to that of the personological enunciation of the human subject: "it reaches the point where one no longer knows who is speaking or who is who" (Guattari, 2009b: 264).

Conclusion

> We are witnessing the death of German, Japanese and Italian cinema. It is an incredible disaster, the death of a species such as this, comparable to the disappearance of certain species of birds and mammals.
>
> (Guattari, 2011: 50)

In an interview on contemporary art only a few months before his death, Guattari (2011: 50) made a plea for social scientists and philosophers to extend our understanding of contemporary ecological crisis to include the degradation and loss of various "incorporeal species", such as cinema (Dewsbury, 2015). Guattari's (2011: 21) worry, of course, was not that films would stop being made, but was rather about the movie industry becoming increasingly co-opted to serve a capitalist logic of "homogenesis", which proliferates reductive forms of subjectivity and desire, and leads to the "loss of the colours, the flavours, and the timbres of different universes of reference". It is in these later writings that we see Guattari's thinking on micropolitics and social transformation increasingly become reframed around the problem of the revaluation (or 'resingularisation') of *value* (Guattari, 1995, 2008). The issue with capitalism is that it valorises only a certain form of value-production (exchange-value) which subsumes all other "existential riches and sign-qualities" to the generalised equivalence of capital. But Guattari (1995: 55) notes there are also "machinic values" and "values of desire" that are values of subjective creation and becoming. These are modalities of value, he argues, that do not have an immediate effect on exchange values, "and which, for this reason, actually deserve to be financed" (Guattari, 2015a: 31). It is this question that comes to define the micropolitical stakes of encounters with art and cinema in Guattari's final writings: "will its innovations be instrumentalised and subsumed in a new capitalist axiomatic, or can it produce an autonomous machinic surplus-value" (Zepke, 2011: 207).

Whereas the dominant political economy of commercial cinema valorises only the exchange-value potential of film (as box-office returns and DVD sales), Guattari (2015a: 31) argues that what is at stake in micropolitical cinema is the "heterogenesis of systems of valorisation", producing new machinic values that offer an enrichment of potentialities. What defines the 'micropolitical' potential of film is thus its orientation towards this logic of ontological heterogenesis, involving the enrichment and transformation of habitual processes of subject production. To grasp these transformative potentials of film, this chapter highlighted how Guattari's philosophy shifts thought beyond conventional (psychoanalytic and linguistic) understandings of the cinematic image as a 'representation', towards a theorisation of the image as a material process of 'individuation'. When watching a film, then, "we are not in the presence of a passively representative image, but of a vector of subjectivation" (Guattari, 1995: 25). As we saw in the encounter with the dream-world *Paprika*, a crucial part of cinema's individuating capacities is its production of a-signifying intensities that escape the snares of established systems of representation and signification. The a-signifying intensities of cinema intervene directly in our relations with the external world, interrupting existing habits of response by opening new possible connections of thought with the different forces, bodies, and materialities that weave the immanent "semiotic net" of life (Guattari, 1995: 60). And even if, as Guattari notes (2009b: 266), our social world is increasingly "contaminated by dominant representations", there is always the possibility for a "minimal aperture" to emerge out of the encounter with cinema, and through which we might transform the conventional images we have of ourselves and our ways of thinking and perceiving the world around us.

Notes

1 Hence Deleuze and Guattari's (2004) preference in *Anti-Oedipus* for the term 'desiring-production', which highlights its connection to the material production of social reality, rather than to a private or individualised realm of mentalities and unconscious representations.

2 Guattari (1984: 168) elaborates on this connection between psychoanalysis and the capitalist social order in *Molecular Revolution*, when he writes: "For the psychoanalyst, it has now become a crucially important question of power: all expressions of desire must be made to come under the control of the same interpretative language. This is a way of making deviant individuals of all kinds submit to the laws of the ruling power, and it is this that the psychoanalyst specialises in".

3 Guattari notes that it is not just art that produces such a-signifying semiotics, and he also mentions computing and artificial intelligence, mathematical equations, scientific formulae, money and finance, and aesthetic practices more generally as generative of their own a-signifying processes. However, he argues that art's thinking surpasses that of politics, philosophy, and science in this regard because these other domains "lack the existential impact of art [. . .] for grasping the creative potentiality at the root of sensible finitude", and expressing these forces towards the construction of new subjective and social realities (Guattari, 1995: 134; see also Zepke, 2011).

4 The concept of individuation is drawn here from the philosophy of Gilbert Simondon (2009) and its uptake in the thought of Guattari (1995; Deleuze and Guattari, 2004) where it refers to a mode of thought that takes as primary not the already-constituted individual or subject, but rather preindividual forces and material processes of emergence. In thinking the image as a process of individuation, Guattari (1995, 2009a) attempts to shift the political status of cinema beyond the representational economy of interpretation and signification of conventional philosophy and film theory (Sauvagnargues, 2016). As an individuating encounter, cinema for Guattari connects thought to the a-signifying intensities and machinic relations that make it creative.

5 As attested by Guattari's (2016b) (unrealised) attempt between 1980 and 1987 to direct an experimental Sci-Fi film, entitled *A Love of UIQ*, based on a screenplay he wrote with the assistance of the American filmmaker Robert Kramer. In the preamble to the film, Guattari set out his aim for the film as attempting to harness cinema's extraordinary capacities to produce subjectivity as a means to reconfigure existing patterns of collective unconscious desire. See also the commentary on Guattari's script by Maglioni and Thomson (2016).

6 Here, I am not suggesting that Deleuze was ignorant of questions relating to technical development, audience reception, industrial and commercial processes, political censorship, and their effects on cinema (indeed, these themes do appear periodically in the Cinema books). But that he tends to bracket these broader themes and questions in order to say something philosophically precise about the cinematic image and its relations to movement, time, and thought.

7 Guattari (2009b) gives the examples of *Badlands* (Malick, 1973) and *Eraserhead* (Lynch, 1977) as commercially successful films that elaborate a non-normative thinking of desire and the unconscious that attacks the dominant and reactionary values of both conventional Hollywood cinema and the field of psychoanalysis.

8 Guattari was thus especially fascinated by the technological and aesthetic developments associated with cinema (e.g. 'the talkies', colour film, television, CGI, 3D), which he thought could be both opportunities for enriching collective modes of expression as well as instruments for repressive systems of social control. These technologies and processes are thus not revolutionary in and of themselves – what matters is whether they work to connect us in new ways to other groups and practices, or whether they simply reproduce already established relations and patterns of thought. In his 1973 essay Cinema of Desire, for example, Guattari specifically highlights the Super

8 camera as opening the potential for a molecularization and democratisation of visual technology by facilitating new possibilities for making cinema beyond the dominant studio system of finance and production, and outside the boundaries of conventional narrative cinema.

9 Readers may notice some similarities here with the plot of Christopher Nolan's, 2010 blockbuster, *Inception*. Nolan credits *Paprika* as an important source of inspiration for his own film, and in fact copies sequences from Kon's anime almost frame-by-frame, including a sequence where Ellen Page's character Ariadne (a dream detective herself modelled on *Paprika*'s titular character) shatters the boundaries of the dream world like glass. While Satoshi Kon's films may not be that well known in the West (outside fans of anime), his plots and images have inspired several other American filmmakers, including Darren Aronofsky whose psycho-thriller *Black Swan* (2010) is an adaptation of Kon's first anime feature, *Perfect Blue* (1997). Aronofsky's *Requiem for a Dream* (2000) also takes inspiration for certain scenes from Kon's film.

10 A famous example of this Oedipalisation of desire can be found in the anime TV series *Neon Genesis Evangelion*, in which the relationship between the two central protagonists – Shinji Ikari and Rei Ayanami – is narratively and visually framed in terms of Shinji's desire for his deceased mother, who he believes Rei strongly resembles.

References

Antonioli, M. (2012) The literary machine. *Contemporary French and Francophone Studies*, 16(2): 157–170.

Aronofsky, D. (2000) *Requiem for a Dream*. Santa Monica, CA: Artisan Entertainment.

Aronofsky, D. (2010) *Black Swan*. Los Angeles, CA: Cross Creek Pictures.

Asada, A. (1989) Infantile capitalism and Japan's postmodernism: A fairy tale. In: Miyoshi, M. and Haratoonian, H.D. (eds.) *Japan and Postmodernism*. Durham, NC: Duke University Press. pp. 271–278.

Brown, S. (2006) Screening anime. In: Brown, S. (ed.) *Cinema Anime: Critical Engagements with Japanese Animation*. Basingstoke: Palgrave Macmillan. pp. 1–22.

Clarke, D.B. and Doel, M.A. (2016) Cinematicity: City and cinema after Deleuze. *Journal of Urban Cultural Studies*, 3(1): 3–11.

Deamer, D. (2016) *Deleuze's Cinema Books: Three Introductions to the Taxonomy of Images*. Edinburgh: Edinburgh University Press.

Deleuze, G. (2005a) *Cinema I: The Movement-Image* (trans. Hugh Tomlinson and Barbara Habberjam). London: Continuum.

Deleuze, G. (2005b) *Cinema II: The Time-Image* (trans. Hugh Tomlinson and Robert Galeta). London: Continuum.

Deleuze, G. (2005c) *Francis Bacon: The Logic of Sensation* (trans. Daniel W. Smith). London: Continuum.

Deleuze, G. and Guattari, F. (2004) *Anti-Oedipus: Capitalism and Schizophrenia* (trans. R. Hurley, M. Seem and H.R. Lane). London: Continuum.

Deleuze, G. and Guattari. F. (2009) Balance-sheet for desiring-machines. In: Guattari, F. (ed.) *Chaosophy: Texts and Interviews 1972–1977*. Los Angeles, CA: Semiotext(e). pp. 90–115.

Dewsbury, J-D. (2015) Guattari's resingularisation of existence: Pooling uncertainties. *Dialogues in Human Geography*, 5(2): 155–161.

Dosse, F. (2010) *Gilles Deleuze and Félix Guattari: Intersecting Lives* (trans. D. Glassman). New York, NY: Columbia University Press.

Elliot, P. (2012) *Guattari Reframed*. New York, NY: I. B. Tauris.

Epstein, J. (2014) *The Intelligence of a Machine* (trans. C. Wall-Romana). Minneapolis, MN: Univocal Press.

Genosko, G. (2002) *Félix Guattari: An Aberrant Introduction*. London: Continuum.

Genosko, G. (2009) *Félix Guattari: A Critical Introduction*. London: Pluto Press.

Genosko, G. (2012) Félix Guattari. In: Colman, F. (ed.) *Film Theory and Philosophy*. London: Routledge. pp. 243–252.

Gerlach, J. and Jellis, T. (2015) Guattari: Impractical philosophy. *Dialogues in Human Geography*, 5(2): 131–148.

Godard, J-L. (1969) Let's talk about Pierrot: Interview with Jean-Luc Godard. In: Whitehead, P. (ed.) *Pierrot Le Fou: A Film by Jean-Luc Godard*. London: Lorimer Publishing. pp. 5–19.

Guattari, F. (1984) *Molecular Revolution: Psychiatry and Politics* (trans. R. Sheed). New York, NY: Peregrine Books.

Guattari, F. (1995) *Chaosmosis* (trans. P. Bains and J. Pefanis). Sydney: Power Publications.

Guattari, F. (1996) Subjectivities: For better and for worse. In: Genosko, G. (ed.) *The Guattari Reader*. Oxford: Blackwell Publishers. pp. 193–203.

Guattari, F. (2008) *The Three Ecologies* (trans. I. Pindar and P. Sutton). London: Continuum.

Guattari, F. (2009a) *Soft Subversions: Texts and Interviews 1977–1985* (trans. C. Wiener and E. Wittman). Los Angeles, CA: Semiotext(e).

Guattari, F. (2009b) *Chaosophy: Texts and Interviews 1972–1977* (trans. D.L. Sweet, J. Becker and T. Adkins). Los Angeles, CA: Semiotext(e).

Guattari, F. (2011) On contemporary art. In: Alliez, E. and Goffey, A. (eds.) *The Guattari Effect*. London: Continuum. pp. 40–53.

Guattari, F. (2015a) *Machinic Eros: Writings on Japan*. Minneapolis, MN: Univocal Press.

Guattari, F. (2015b) *Psychoanalysis and Transversality: Texts and Interviews 1955–1971* (trans. A. Hodges). Los Angeles, CA: Semiotext(e).

Guattari, F. (2016a) *Lines of Flight: For Another World of Possibilities* (trans. A. Goffey). London: Bloomsbury.

Guattari, F. (2016b) *A Love of UIQ* (trans. S. Maglioni and G. Thomson). Minneapolis, MN: Univocal Press.

Herzog, A. (2008) Suspended gestures: Schizoanalysis, affect, and the face in cinema. In: Buchanan, I. and MacCormack, P. (eds.) *Deleuze and the Schizoanalysis of Cinema*. London: Continuum. pp. 63–74.

Hynes, M. (2013) The ethico-aesthetics of life: Guattari and the problem of bioethics. *Environment and Planning A*, 45(8): 1929–1943.

Kon, S. (1997) *Perfect Blue*. Tokyo: Madhouse.

Kon, S. (2006) *Paprika*. Tokyo: Madhouse.

Lapworth, A. (2016) Cinema, thought, immanence: Contemplating signs and empty spaces in the films of Ozu. *Journal of Urban Cultural Studies*, 3(1): 13–31.

Lazzarato, M. (2014) *Signs and Machines: Capitalism and the Production of Subjectivity* (trans. J.D. Jordan). Los Angeles, CA: Semiotext(e).

Lynch, D. (1977) *Eraserhead*. Los Angeles, CA: American Film Institute.

MacCormack, P. (2005) A cinema of desire: Cinesexuality and Guattari's a-signifying cinema. *Women: A Cultural Review*, 16(3): 340–355.

Maglioni, S. and Thomson, G. (2016) UIQ: Towards an infra-quark cinema (or an unmaking of). In: Guattari, F. (ed.) *A Love of UIQ*. Minneapolis, MN: Univocal Press. pp. 13–44.

Malick, T. (1973) *Badlands*. Burbank, CA: Warner Bros Studios.

Marshall, B. (2009) Cinemas of minor Frenchness. In: Buchanan, I. and MacCormack, P. (eds.) *Deleuze and the Schizoanalysis of Cinema*. London: Continuum. pp. 89–101.

Martin-Jones, D. (2006) *Deleuze, Cinema, and National Identity: Narrative Time in National Contexts*. Edinburgh: Edinburgh University Press.

Massumi, B. (2011) *Semblance and Event: Activity Philosophy and the Occurent Arts*. Cambridge, MA: The MIT Press.

Nolan, C. (2010) *Inception*. Burbank, CA: Warner Bros Studios.

Osmond, A. (2009) *Satoshi Kon: The Illusionist*. Berkeley, CA: Stone Bridge Press.

O'Sullivan, S. (2010) Guattari's aesthetic paradigm: From the folding of the finite/infinite relation to shizoanalytic metamodelisation. *Deleuze Studies*, 4(2): 256–286.

Powell, A. (2007) *Deleuze, Altered States, and Film*. Edinburgh: Edinburgh University Press.

Rushton, R. (2012) *Cinema After Deleuze*. London: Continuum.

Sauvagnargues, A. (2016) *Artmachines: Deleuze, Guattari, Simondon* (trans. S. Verderber and E. Holland). Edinburgh: Edinburgh University Press.

Simondon, G. (2009) The position of the problem of ontogenesis. *Parrhesia*, 7(1): 4-16.

Tsutsui, Y. (2013[1993]) *Paprika* (trans. A. Driver). New York, NY: Vintage Books.

Watson, J. (2008) Schizoanalysis as metamodelling. *Fibreculture Journal*, 12. Available online at: http://twelve.fibreculturejournal.org/fcj-077-schizoanalysis-as-metamodeling/

Zepke, S. (2011) From aesthetic autonomy to autonomist aesthetics: art and life in Guattari. In: Alliez, E. and Goffey, A. (eds.) *The Guattari Effect*. London: Continuum. pp. 205–219.

14 Reframing politics in art

From representational subjects to aesthetic subjectification

Nina Williams

When we consider the politics of art – from Dadaism to Graffiti – we might think of the forms, motifs, and meanings expressing radical statements about how society is or should be organised. This is a representational politics, arising out of the conceptual intentions of the artist or object, because it is concerned with communicating a message. As a starting point, this chapter wants to open up the ways we think and frame politics in art, starting with the view that there are many kinds of politics at work in an aesthetic encounter, some of which are not expressed at this representational level as social commentary or the making more visible of already recognised problems. Whilst I don't want to deny the potential of political art in this sense, what I see as problematic is that in approaching art in this way, it directs the art in one direction – the conceptual impetus of the object – and restricts it to this model of why something should be deemed important. Instead, in this chapter, I explore a politics[1] emergent in Guattari's 'ethico-aesthetic paradigm', which is about opening up to and expressing other ways of becoming in the world which cannot be reduced to utilitarian or instrumental logics. Guattari's conceptualisation of aesthetics (elaborated most explicitly in *Chaosmosis*: *An Ethico-aesthetic Paradigm* (1995)) is significant for drawing attention to the emergent and transformative qualities of an aesthetic encounter, which foregrounds a politics that is inventive of new modes of relation, rather than communication.

Politics in art

Zepke notes that contemporary art has "tended to abandon sensation in favour of discursive representations seeking to intervene into the realm of the 'political'" (2009: 192). Yet as Massumi (2011: 54) recognises, "artistic practices that explicitly attempt to be political often fail at it, because they construe being political as having political content, when what counts is the dynamic form. An art practice can be aesthetically political, inventive of new life potential, of new potential forms of life, and have no overtly political content". The analysis of art in the social sciences has followed this route criticised by Zepke and Massumi, it subsumes art to a politics of the ready-made, rather than as the transformative "production of a sensation that takes us beyond the 'lived experience' of a phenomenological flesh" (Zepke, 2009: 176).

If the kind of politics engaged in art is most often that of the dominant real, which as Guattari reminds us is a politics of consciousness (2015b: 182),[2] aesthetics is precisely about a pre-conscious assemblage "deployed in a non-discursive register" (Guattari, 2015a: 53). This is by no means a-political, rather it is about reframing the politics of art by attending to a 'non-discursive register', which is where it develops alternative avenues for thought and action in the world, or simply generates an affect or sensation.

My approach in this chapter is to draw on the minor aesthetics of Guattari. I do so in order to pose a provocation to the way art is evaluated as a form of political expression, by exploring how art is an active aesthetic process oriented at the production of modes of existence. By the minor I mean both everyday encounters with art – the visit to a gallery, for instance – but also the more visceral, ineffable, and speculative moments of aesthetics – the feeling between an artwork and spectator, the movements between a body and materials, or the apparition of an idea.

There are two propositions which are crucial to this argument: the first is that art is not limited to major institutional spaces. In the next section I argue for a shift from a thinking of 'art' that continues to be invested in the molar logic of subject-object, and towards an engagement with the *process* of aesthetics. Aesthetics here denotes not just major artistic institutions, individuals, or objects, but also what we might term the 'minor registers of aesthetics', such as the sensations produced in the artistic encounter or the materiality of colour on the canvas. To elaborate what is at stake in this shift I turn to Guattari who offers an engaged and rich conceptual vocabulary for making these processes tangible. This first proposition is necessary to the question of how we evaluate the arts because it aims to open up a space for minor forms of arts practice outside of their institutional form. This is important because the art encounter in this sense propagates stages of subjectification, problematising stagnant modes of being and inventing new modes of life. In this way, to be seized by these potentialities of art and aesthetics requires us to recognise that the world is not fixed, and that we do not know the full extent of the forces that constitute and transform it.

The second proposition is that the use of certain kinds of politics as a means of valuing or engaging with art-encounters may not be adequate. Instead, as I explore in the final section of the chapter, an ethics of art opens up what we mean by politics to attend to those subjectifying forces that exceed politics in its institutionalised form. Key to this, and something I want to highlight more forcefully in relation to both propositions, is that aesthetics is not limited to the intended representational content, communicated through the subject and creator of the artistic object, but is part of a wider system of sensation, meaning and (co-)constitution, which transforms subjectivity in the event of its encounter. Indeed, Guattari rejects a binary role of artist and spectator, content and meaning, and instead presents the spectator as a co-creator of aesthetic enunciation (1995: 14). It is this rejection which is crucial for reframing the relations between art and politics, which sets Guattari apart from others who have theorised aesthetics.

One way of understanding this difference is in the different emphasis of politics in art between Guattari's (1995) "ethico-aesthetic paradigm" and Rancière's

(2008) "Politics of Aesthetics". Rancière importantly alerts us to a "politics of aesthetics that predates artistic intentions and strategies" and thus amplifies a political impetus in art irrespective "of the specific message such-and-such an artist intends to convey" (2010: 141). This has proved a productive point of reference for contemporary geographic research on aesthetics (Dixon, 2009; Yusoff, 2012; Millner, 2015), where an emphasis on the political intensity of artistic products has highlighted the transformative capacity of art to alter social perceptions and structures. There are two aspects of Rancière's approach, however, which do not coincide with the propositions outlined in this chapter: one is his fidelity to politics as the means of transformation (which is understood as a much more interruptive force than for Guattari), and the second is his attention to primarily human institutions and regimes as the agents of that transformation. Thus, Rancière's aesthetic politics remains tied to groups or subjects, and the creation of "specific forms of 'common sense'" (2010: 141), whereas for Guattari there is no subject of art. The ethico-aesthetic paradigm is not concerned with molar institutions (the gallery space, the funding body, etc.) or a logic of subject and object, but is attuned to the "prophetic" potential of minor aesthetic practice (O'Sullivan, 2005). In this way, it implies an openness to the *sensory* rather than analytic or semantic modes through which we recognise politics *and* art. It is not setting up aesthetics as being about how to judge something as good or bad, but as something that can propagate new modes of subjectivity. It is about new modes of relation which cut across the categories of subject and object and are about the aesthetic shaping of a new mode of existence. Guattari alerts us to the fact that by valuing the politics of art in terms of subjects and groups, like Rancière does,[3] politics remains bound to a closed linguistic apparatus determining who or what has the agency for transformation – the radical subject – and overlooks a minor, imperceptible political force which does not neatly reside within the individual or group.

Minor aesthetics

By emphasising minor aesthetics, I am highlighting the sensory relations of art over the conceptual intentions, which takes us from the signifying qualities *within* an artwork to the a-signifying process of aesthetics. Rather than connoting meaning and communicating a message, the potential of aesthetics is in "rupturing with forms and significations circulating trivially in the social field" (Guattari, 2000: 52). This reading of aesthetics stems from the Greek term *aesthesis*, implying a sensory mode of perception. Following Papenburg and Zarzycka (2013: 3), this presents a "move away from the understanding of aesthetics as being concerned with 'beauty' and question[s] the universalising claims that this concept implies . . . and underlines the significance of cultural valuing of certain sense impressions and the dismissal of others".[4] Further to challenging art's visual-centrism, or even for questioning its underlying cultural values, as implied by Zarzycka and Papenburg, the notion of aisthesis is also significant for introducing sensation as the opposite of intellect and thought: in short, a sensory unconscious.

This is precisely the pre-linguistic or a-signifying register of aesthetics amplified in Guattari's work.

In highlighting the sensory dimensions of art-encounters, aesthetics can no longer be understood solely in relation to institutionalised art where particular practices are validated according to a pre-defined conception of what constitutes art or politics. As Zepke recognises, a focus on institutionalised art implies "a concept-function defining a proposition about an object in the world – *it's art!* – but this *analysis* nevertheless conforms to the form of recognition (the true and the false) by which all information is communicated" (2009: 182). As Guattari discusses, it is the major institutional orientations of an event that limit the potential of real transformation:

> Wherever a desire to create or an inclination to really live springs up, wherever something is happening, be it in the sciences or the arts, one encounters a rejection of contemporary systems of organization and hierarchization. Scientific progress and aesthetic or cultural mutations never proceed from authoritative means. As soon as some general headquarters tries to legislate the visual arts, literature, science, etc., research and creation come to a dead halt.
>
> (2009b: 28)

In this interview, Guattari makes clear that the problem of evaluating manifestations in art according to the orthodox systems of organisation is that it deadens the nascent impulse of aesthetics. To engender rather than deaden aesthetic mutations, it is apparent that a more open approach to understanding artistic practice – one which is not concerned with a legislation of or for the arts – is necessary.

Minor aesthetics is, therefore, not about a set of transcendent values; it moves away from a representational politics to recognise that aesthetic capacities can reveal their potential outside of orthodox artistic frameworks. As Guattari (1995: 103) states:

> It might be better here to speak of a proto-aesthetic paradigm, to emphasise that we are not referring to institutionalised art, to its works manifested in the social field but to a dimension of creation in a nascent state, perpetually in advance of itself, its power of emergence subsuming the contingency of hazards and activities that bring immaterial Universes into being.

In this extract Guattari presents an aesthetics that is not limited to institutionalised art; instead he wants to open up and make active these nascent states of creation, which the social sciences currently fail to challenge within the normative linguistic or even embodied frameworks. Guattari's aesthetics is a productive point of reference for the social sciences because he reconfigures the conceptual starting point of our debates on the spaces and subjects of aesthetics away from the institutionalised artistic object and towards the process. This reconfiguration parallels a shift from major (institutionalised) to minor art; it introduces an alternative perspective on aesthetics than that of the institutional gallery space or the intentional subject.

A move from institutionalised art to minor aesthetics is by no means a rejection of the fixed figures of artistic production. As Massumi (2014: 39) notes, it is not sufficient to emphasise the "creative surpassing" of art since "each gives affordance to each other". Rather, the point of the shift from institutionalised art to minor aesthetics is to "inhabit the in-between of their processual interlacing" (Massumi, 2014: 39). That is, to encounter an aesthetic force that runs alongside the "already expressed" categories of the artistic object. In similar vein, it is important to note that the institutionalisation and representationalism associated here with major art is not inferior – the gallery, for instance, is very much an enabling space in artistic contexts – nor do I want to set up a binary between major art and minor aesthetic processes. Rather, the aim, to paraphrase Deleuze and Guattari (1987: 428), is to not take the primacy of the institutionalised arts for granted, so that minor aesthetics is not portrayed as a pre-artistic or para-artistic or sub-artistic agency.

To be clear, in emphasising the sensory qualities of art, I do not mean to ignore the material qualities of an artistic object. On the contrary, as Sauvagnargues recognises, a minor approach "makes it possible to think expression in art without abandoning its mimetic responsibility" (2016: 73). Maria Hynes (2013) makes explicit a notion of minor aesthetics directly through an institutionalised model – that of Marina Abramović's art – to develop an exploration of the experience of an aesthetic encounter. The experiences Hynes foregrounds, whether they are situated within traditional art institutions (as with Hynes' illustrative examples) or otherwise, are better understood through the aesthetic processes that constitute the encounter, rather than through an analysis of the object of the artwork or the intention of the subject as the creator. This is because "the artwork thinks life as an encounter with force, an encounter that works at the level of sensation and thus before the organisation required by human perception" (Hynes, 2013: 1937). My aim in this chapter is to bring a notion of minor aesthetic – working at the level of sensation – to the fore in social scientific research on the art encounter in order to reframe what we think of as political in art. Guattari's ethico-aesthetic paradigm is invaluable in order to endorse this minor aesthetic as a starting point for politics: a politics that is inventive of new modes of life in that it "experiments liberally with the production of subjectivities" (Hynes, 2013: 1940). Such experimentation necessarily resists the constraints of institutional categorisation, acting as a "force for seizing the creative potentiality at the root of sensible finitude – 'before' it is applied to works, philosophical concepts, scientific functions and mental and social objects" (Guattari, 1995: 112).

This approach endorses less exclusive modes of art and goes beyond the centredness of the individual subject as the bearer of creative capacities. Rather than placing emphasis on the agent that makes the art or the object that speaks of (and to) the world (both of which render the force of art as within either of the end points of the subject-object dualism), Guattari forces us to think about the immanent aesthetic process between these end points, and thus to a force which is neither applied by the individual or finished in the product. The importance of Guattari's ethico-aesthetic paradigm is thus that it steers a broader conceptualisation of art towards the new modes and potentials of subjectivity that precede and cut across a subject. It is an actualisation of difference and heterogeneity (Guattari, 1995: 115). What I want to emphasise as key to Guattari's reading of aesthetics is, firstly, its transformative nature, and, secondly, the way that this rests upon a pre-subjective process. Both components are significant for alerting us to the value of aesthetic registers beyond a representational politics.

Two components of aesthetics

The first component of aesthetics – that it is a transformative process – means that it requires a break from normative codes, embracing an awareness of the fragility of life and in that the potential for creative transformation. This creative transformation is based on novelty and heterogeneity – that is the ability of aesthetics to produce new and diverse associations in the world – and is realised through its potential to engender new subjectivities. We can interpret this as minor transformation, not because it is less significant, but because it operates foremost at the molecular level of bodies and matter. Rather than a radical or eventful interruption, it is a transformation of experience through which new modes of being emerge. Guattari is interested in how aesthetic modalities, perhaps a piece of art but not necessarily, can produce new ways of experiencing the world. What is important is not the artworks themselves, but their capacity to create these affective experiences, how the artwork might transform our dispositions. Taking this idea outside of major art, we might consider how the various shades of green detected by the human eye shape a certain sensibility whilst walking, coupling, and decoupling a body within an environment. The transformative aspect of Guattari's aesthetics, then, takes us closer to the creative root of transversal relations, which cut across the individual, the collective, and the environment, as part of a challenge to normative forms of signification and expression.

The second component of Guattari's conceptualisation of aesthetics as a pre-subjective process speaks directly to a more molecular and material understanding of agency that we see emerging in contemporary social sciences of art. An emphasis on the pre-subjective process of aesthetics could be about the colour in a Kandinsky painting or the reverberation of sound in an installation. As O'Sullivan recognises, as a *proto*-aesthetic paradigm "the distinctions of subject-object have yet to be fixed and reified, a world of strange mutually implicated beings cohering around objects and practices" (2010: 260). This pre-individual quality of aesthetics described by O'Sullivan is what I mean when calling for attention to the minor as the nascent, pre-articulated, sensory encounter with aesthetics. With both components, that it is transformative and pre-subjective, we see the privileged position Guattari holds for aesthetics in the process of subjectification.

Aesthetic subjectification

O'Sullivan recognises that Guattari's work on subjectivity is "a call to creativity, a call to become actively involved in various strategies and practices that will allow us to produce/ transform, and perhaps even go beyond, our habitual selves" (2006: 1). We see here how, in its relation to subjectification, aesthetics arises as a prophetic force rather than a discourse of beauty expressed at the level of the signifier. Guattari's thought enables us to recognise the potential of aesthetics as a process that stimulates new modes of subjectivity whereby we engage in the world differently – beyond our habitual selves. Crucially, it is a process targeted at subjectification rather than the political intent or communicative message of the artistic subject in an artistic object. As such, it is precisely this foregrounding of the aesthetic throughout his work that implicates a generative rather than representational politics of art. Generative because, as Brunner et al. recognise, for Guattari "aesthetics itself shapes a mode of existence that accounts for the transversal relations between subject and object, and between corporeal and incorporeal forces, that together make up the real" (2013: 43).

The potential of aesthetic practices in the process of subjectification is apparent in Guattari's discussion of Keiichi Tahara's photography. As part of a number of visits Guattari made to Japan during the 1980s,[5] Guattari met with Keiichi Tahara and became interested in the methods of lighting and framing that are central in his portraits and which Guattari suggests bring the face out of "pre-established montages of signification" (2015a: 56, 68). Guattari writes of Tahara's photographic method that "it is no longer a question of denoting an identity or of connoting a message. We are no longer in the register of identifications and mediatized communications. It is by immediate transference without any hesitation that these deterritorialized bodies are given to us" (2015a: 62). We can see in Figure 14.1 how a body is deterritorialised: it is split into three points – one 'subject' and two reflections – which amplifies an ego to the point of disintegration. Whilst Guattari does not explicitly engage this portrait, I am drawn to it here because of the way it mutates the subject, intimating the invention of divergent subjectivities in Guattari's aesthetic paradigm. Coupled with a seemingly dismembered hand, "shadows cast by the frames", and an "intrusion of surface singularities" (streaks, scratches), we are presented with a "machinic or machined subjectivity produced by this complex *dispositif*, which is the most intimate that can be sensed without ever becoming familiar" (Guattari, 2015a: 67–68). It never becomes familiar because it breaks from the 'objective denotation' of portraiture, turning itself upon the identity of the subject and conventional tools of representation. The potential in this disruption can be set against a representational politics in order to affirm the a-signifying and minor registers of aesthetics that reframe the politics of art.

Given its capacity to transform and generate heterogeneous modes of life, that is, to engender "unprecedented, unforeseen, and unthinkable qualities of being" (Guattari, 1995: 106), Guattari's conceptualisation of aesthetics is presented as an ethics. This is an ethics that is absolutely political because it is about inventing new forms of life. Throughout his life and work, Guattari has consistently fought against the repression of difference under capitalism, which sees subjectivity

Figure 14.1 Laura Betti in Keiichi Tahara's portrait collection 1978–1987 © Keiichi Tahara

"held in a constant state of lack, debt, procrastination and so forth. Immanence is captured by a transcendent apparatus and, as such, subjectivity is standardised" (O'Sullivan, 2010: 261). Against this homogenising effect, it is artistic practices and techniques that take a privileged role in overcoming the stagnant states of subjectivity within capitalism. The ethical dimension is therefore not tied to a subject who responds to an event but is located in the intensity of the aesthetic impulse, where something is produced differently. This is significant for a social-scientific frame because it emphasises aesthetic, subjective heterogeneity over and above the pre-established politics of subjects or groups. The ethico-aesthetic paradigm thus necessitates a shift in attention to aspects of life that are not already in our common purview and as such is about producing new forms of thinking.

Reframing ethico-political value

To be clear, Guattari's work on aesthetics is not depoliticising aesthetics by emphasising its ethical dimension:

> The new aesthetic paradigm has ethico-political implications because to speak of creation is to speak of the responsibility of the creative instance

> with regard to the thing created, inflection of the state of things, bifurcation beyond pre-established schemas, once again taking into account the fate of alterity in its extreme modalities.
>
> (1995: 107)

However, the kind of politics that comes to mind in relation to art tends to be that of the common sense – where assumptions (about where it takes place, say, or about agency) have already been made. And this understanding does not typically regard a 'creative instance' over the thing created, or go beyond 'pre-established schemas'. Such an approach reinforces dogmatic conceptions of both politics and art – those that are concerned with the 'already expressed' frameworks of thought – meaning that we overlook the minor sensory registers of aesthetics. Guattari's ethico-aesthetic paradigm invites us to rethink assumptions about why something is important and more precisely develop conceptual work in relation to the event at hand.

Extending social science's ethico-political valorisation of aesthetics beyond a representational logic is necessary to endorse art's capacity to generate new forms of life. If, as Hynes suggests, ethico-aesthetics seeks "neither to secure against life's mutations, nor [is it] content to merely reflect on what is given to it", the implication is not to reject modes of art which enliven the political but to place research "at the point of life's openness and creativity" (2013: 1931–1932). Far from being a moral ethics "whereby the dominant, cognitive model of the subject is positioned at the centre of deliberation over ways of life" (Hynes, 2013: 1933),[6] Guattari presents us with a political ethics concerned with inflections of life outside of its institutional capture. The ethics is that of becoming-minor, through the aesthetic potential available in art and elsewhere.

As Massumi (2015: 11) reminds us, "the ethical value of an action is what it brings out in the situation, for its transformation, how it breaks sociality open. Ethics is about how we inhabit uncertainty together. It's not about judging each other right or wrong". By emphasising uncertainty and not foreclosing what counts as politics in art, the ethico-aesthetic paradigm becomes "a mode of critical inquiry that is situated at the point of life's openness – at that point when both subject and object remain open questions" (Hynes, 2013: 1933). Following Massumi, the task of reframing ethico-political value demands that the social sciences don't "attach positive or negative values to actions based on a characterization or classification of them according to a pre-set system of judgment. It means assessing what kind of potential they tap into and express" (2015: 11). What this means for researching art is that attention is directed to the eventfulness of aesthetic instances or the immanence of an art encounter rather than the communicational content. The point is to accept that our representational understanding of an event, artistic or otherwise, is always predetermined by institutional frameworks and expectations. Guattari's thought enables us to think an ethico-politics of art that is not constrained by that. The potential of aesthetics emerges in the excess over a molar logic of subject-object, where new relations – that are neither subject nor object – are assembled in the encounter with art. Whilst they might be ephemeral (and thus seemingly subtle), they are not trivial (lesser, marginal, prior to) but generate heterogeneity which resists pre-established coding. The aesthetic assemblages are "not the invention of a new medium of communication but *the invention of a new kind of relation with what is communicated*" (Guattari, 2009a: 383) – an ethico-aesthetic of becoming-minor.

My argument therefore, is that we need to radically reframe the concepts used to valorise what is political in art in order to avoid homogenous social narrative and open new avenues for apprehending the aesthetic encounter. The aesthetic task is not about the subject but about exploring the conditions of becoming-other in *any* practice, where a new affect or perception in turn alters a feeling or perspective on something. This is not limited to the alternative ways of sensing produced – what art communicates – but is about how the affective or sensory impulse of aesthetics acts as a subjectifying force before what is communicated. This is a radical proposition because it invites us to re-evaluate the kind of practices we deem important in the first instance, in order to accept the potential of transformative shifts that emerge out of the present. This requires that we set aside the standard barometers of success and progression, not only to celebrate the upwards motions but to pay closer attention to the shifting tides out of which a new politics becomes possible.

Notes

1 This type of politics is akin to Deleuze and Guattarian theorisation of micropolitics, which as Massumi (2009, 2011, 2015) has consistently revealed, is a powerful tool for rethinking the potential of immanence and becoming. The concept, understood as "a question of reconnection processually with what's germinal in your living, with the conditions of emergence of the situations you live through" (Massumi, 2015: 79), resonates with an ethico-aesthetic imperative to attend to the immediate or nascent, material, and sensory relations of art (not yet grasped or interpreted by the subject).

2 This departure from a politics of the dominant real is connected to Guattari's attempts to deprivilege signifying semiotic registers and disengage the production of subjectivity from the conscious individual. Instead, and importantly for the aims of this chapter, Guattari wants to make evident a political force which acts at a level of subjectification before the conscious individual has made sense of or communicated it. The ethico-aesthetic paradigm emerges as an important approach for propagating such a politics, then, because it is about "existential absorption, a pre-egoic, pre-identifactory agglomeration" rather than a "discursive scheme" (Guattari, 1995: 79).

3 Whilst Guattari does not explicitly comment on Rancière's work on aesthetics, in making this distinction I am following Lazzarato (2014) who argues that contrary to Guattari, Rancière establishes a politics that is tied to linguistic discourses whereby the political act begins with conscious thought and action. Politics in this sense relies upon humanist assumptions about who or what has the agency for transformation – the radical subject – and overlooks a minor political force which does not neatly reside within the individual, and which Guattari is able to address through his attention to the process of subjectification. It is important to note, however, that Rancière (2016: 34) in recent work has begun to acknowledge this criticism of an 'internalist' method by moving away from a solely discursive mode of political analysis.

4 Against Kant's aesthetics of the sublime, Zepke argues that "Deleuze and Guattari read the sublime through a Nietzschean filter that removes its romanticism, making of sensation the overcoming of the self in an emergence of a new life – and even of a living

nature – that is utterly inhuman" (2009: 186). In this sense, aesthetics subsumes commentaries on beauty and the sublime in the Kantian sense, but only alongside an immanent, and inhuman force that precedes cognitive recognition.

5 These visits and exchanges are documented in *Machnic Eros: Writings in Japan* (2015a), a collection of Guattari's writings edited by Gary Genosko and Jay Hetrick. Further details are provided in Dario Lolli's (2015) review of the book for *Theory, Culture, Society*.

6 Hynes (2013) draws several distinctions between bioethics and Guattari's ethico-aesthetic paradigm, which provides a valuable context for thinking ethics in the Guattarian sense.

References

Brunner, C. Nigro, R. and Raunig, G. (2013) Post-media activism, social ecology and eco-art. *Third Text*, 27(1): 10–16.

Deleuze, G. and Guattari, F. (1987) *A Thousand Plateaus: Capitalism and Schizophrenia* Minneapolis, MN: University of Minnesota Press.

Dixon, D. (2009) Creating the semi-living: On politics, aesthetics and the more than-human. *Transactions of the Institute of British Geographers*, 34(4): 411–425.

Guattari, F. (1995) *Chaosmosis: An Ethico-Aesthetic Paradigm*. Bloomington, IN: Indiana University Press.

Guattari, F. (2000) *The Three Ecologies*. London: Continuum.

Guattari, F. (2009a) *Chaosophy: Texts and Interviews 1972–1977*. London: Semiotext(e).

Guattari, F. (2009b) *Soft Subversions: Texts and Interviews 1977–1985*. London: Semiotext(e).

Guattari, F. (2015a) *Machinic Eros: Writings on Japan*. Minneapolis, MN: Univocal Press.

Guattari, F. (2015b) *Lines of Flight: For Another World of Possibilities*. London: Bloomsbury.

Hynes, M. (2013) The ethico-aesthetics of life: Guattari and the problem of bioethics. *Environment and Planning A*, 45(8): 1929–1943.

Lazzarato, M. (2014) *Signs and Machines: Capitalism and the Production of Subjectivity*. London: Semiotext(e).

Massumi, B. (2009) Technical mentality revisited: Brian Massumi on Gilbert Simondon. *Parrhesia*, 7: 36–45.

Massumi, B. (2011) *Semblance and Event: Activist Philosophy and the Occurrent Arts*. Cambridge, MA: The MIT Press.

Massumi, B. (2014) *What Animals Teach Us About Politics*. Durham, NC: Duke University Press.

Massumi, B. (2015) *The Politics of Affect*. Cambridge: Polity Press.

Millner, N. (2015) Activist pedagogies through Rancière's aesthetic lens. In Hawkins, H. and Straughan, E. (eds.) *Geographical Aesthetics: Imagining Space, Staging Encounters*. Farnham: Ashgate Press.

O'Sullivan, S. (2005) Notes towards a minor art practice. *Drain*, 5. Available online at: http://drainmag.com/index_nov.htm

O'Sullivan, S. (2006) *Art Encounters Deleuze and Guattari: Thought Beyond Representation* London: Palgrave Macmillan.

O'Sullivan, S. (2010) Guattari's aesthetic paradigm: From the folding of the finite/infinite relation to schizoanalytic metamodelization. *Deleuze Studies*, 4(2): 256–286.

Papenburg, B. and Zarzycka, M. (2013) Introduction. In Papenburg, B. and Zarzycka, M. (eds.) *Carnal Aesthetics: Transgressive Imagery and Feminist Politics*. London: I.B. Tauris.

Rancière, J. (2008) *The Politics of Aesthetics: The Distribution of the Sensible*. London: Continuum.

Rancière, J. (2010) *Dissensus: On Politics an Aesthetics*. London: Continuum.

Rancière, J. (2016) *The Method of Equality: Interviews with Laurent Jeanpierre and Dork Zabunyan*. Cambridge: Polity Press.

Sauvagnargues, A. (2016) *Artmachines: Deleuze, Guattari, Simondon* (trans. S. Verderber and E.W. Holland). Edinburgh: Edinburgh University Press.

Yusoff, K. (2012) Aesthetics of loss: Biodiversity, banal violence and biotic subjects. *Transactions of the Institute of British Geographers*, 37(4): 578–592.

Zepke, S. (2009) Deleuze, Guattari and contemporary art. In: Holland, E., Smith, D. and Stivale, C. (eds.) *Gilles Deleuze: Image and Text*. London: Continuum. pp. 176–197.

15 Communist stratoanalysis

Arun Saldanha

It has been officially proposed to the International Geological Congress that the planet has extremely recently entered a new phase in its existence called the Anthropocene. The Anthropocene will be the name for the geological age in which the human species has affected the earth to such an extent its traces will forever (or until the planet gets gobbled up by the exploding sun) remain discernible. It is no wonder the press and the arts and humanities have suddenly become so intrigued with earth science. By presenting an unfathomably ancient ground for existence and basis for future speculation geologists provincialise human temporal cycles and eschatologies, including those of the great monotheistic and imperial traditions, with a confidence no postmodernism or actor-network theory ever achieved. Earth science alerts the humanities the course of human society has come to a critical threshold. The possibility of human extinction is no longer only the stuff of dystopian science fiction. What conceptual framework will be adequate to this new, already catastrophic reality? Many point out the Anthropocene concept itself erases the social particularities under the vague and old-fashioned essentialism of *anthropos* (Moore, 2016). It is not humanity as such but industrial capitalism as developed from European colonialism, heterosexism, racism, and indeed anthropocentrism that has burnt through deposits, depleted rivers, wrecked ecosystems, and heated the planet's atmosphere. What concepts can describe the various levels of uneven organisation which the Anthropocene names, not simply for the sake of description, for science, but in order to formulate collective responses which could halt the catastrophic course of capitalism – that is, for politics?

Guattari was one of the few post-war theorists, especially in France, who had a keen interest in environmentalism, and he folded its more-than-human scope into his eclectic vision of social change. This chapter will argue many of his concepts were constructed for confronting exactly the kind of capitalist crisis the Anthropocene entails. We will focus on strata, Integrated World Capitalism, machinic enslavement, and communism.

I

Stratoanalysis is only one of the many names Deleuze and Guattari give to their critical method: "RHIZOMATICS = SCHIZOANALYSIS = STRATOANALYSIS =

PRAGMATICS = MICROPOLITICS" (1987: 22). By conceptualising change as layered, stratoanalysis offers a four-dimensional spatiality commensurate with planetary processes. For conceptualising the more-than-human temporal and spatial scales of the Anthropocene, the concept of stratification proves very useful. Of course, strata have long been basic metaphors in sociology, especially in Marxism. Elaborating Freud, Guattari also uses it as metaphor for the layers of the unconscious and of social machines. In his programmatic essay "Machinic propositions", Guattari (1984: 150–151) prefigures his later obsession with typologising socio-ontological parameters, suggesting social machines are made up of flows, strata, and "object-species totalities". Strata are composed of "knots of intensities", resonances, or redundancies in a quasi-cybernetic sense, depending on the kind of society (palaeolithic, archaic, capitalist, fascist, revolutionary) they structure. But Guattari also experiments with the strata concept to embed socio-psychical machines within a deeper, earthlier reality. "The heaped-up strata form a kind of humus, or what one might call a system of soups. Behind life there is a biological soup, beyond the biological soup a physico-chemical soup and so on" (1984: 130). At its most general, stratification is an imprisonment inside a level of material organisation. What Guattari is interested in from the beginning is how to bring about a deterritorialisation whereby a process can become autonomous from the strata which have hitherto structured it. As *A Thousand Plateaus* will put it:

> Strata are acts of capture, they are like "black holes" or occlusions striving to seize whatever comes within their reach. They operate by coding and territorialization upon the earth; they proceed simultaneously by code and by territoriality. The strata are judgments of God; stratification in general is the entire system of the judgment of God (but the earth, or the body without organs, constantly eludes that judgment, flees and becomes destratified, decoded, deterritorialized).
>
> (Deleuze and Guattari 1987: 40)

Deleuze and Guattari's "system of strata" is a unique contribution to thinking the earth without anthropocentrism yet from the vantage point of a species which has irretrievably changed its milieu and life chances. Against theological or theosophical notions of a cosmic evolution reaching its pinnacle in man Deleuze and Guattari stress that strata are not stages. No stratum is superior or more perfect than another. Astrophysics is not more fundamental than embryology or aesthetics. In concluding we should not think Deleuze and Guattari seek to vilify stability or naturalise anarchy. Though they privilege the liberation of desire away from strata and back into the immanence of the earth they also point out that intentional destratification usually ends in failure, or worse, fascism or psychosis.

> Every undertaking of destratification (for example, going beyond the organism, plunging into a becoming) must therefore observe concrete rules of extreme caution: a too-sudden destratification may be suicidal, or turn cancerous. In other words, it will sometimes end in chaos, the void and destruction,

> and sometimes lock us back into the strata, which become more rigid still, losing their degree of diversity, differentiation, and mobility.
>
> (1987: 503)

Insofar as it presents the constructedness of human social formations at timescales from the geological to the cybernetic, stratoanalysis is immediately political. His and Deleuze's concept of strata is revolutionary in providing a geological critique of all questions moral, but they also understands God's judgements are not so easily overturned.

The Anthropocene is a global reality which of all philosophical works *A Thousand Plateaus* comes closest to providing an ontological framework for. In fact, the Anthropocene can be called a profoundly Deleuze-Guattarian epoch (see Saldanha and Stark, 2016). And the necessity for revolutionary politics to plant itself into earthly phyla has become even more obvious now than in Guattari's lifetime. Atmospheric and oceanic strata respond in excessive and unpredictable ways to a particularly violent appropriation of water, oil, metals, sand, salt, and so on. As is being understood much better now, relentless extraction has been fundamental to the expansion of capitalism (see Malm, 2016). But now the hubris of the efforts of the anthropic stratum to see itself as separate and above the other strata (man masters nature) comes crashing down as the increase in atmospheric carbon dioxide and the corresponding higher retention of solar heat is making the effects of all inter-stratal processes (like earth quakes and tsunamis) ever more catastrophic. In merely two centuries man finds himself not at a Promethean pinnacle, but more enslaved and befuddled than ever before by the geophysical and agro-ecological strata which he thought he had conquered.

As seen in the speeding up of extraction through suburban sprawl, fracking, tourism, and smart cities, and in the incapacity of governments and publics to conjure alternatives, capitalism everywhere continues to hold onto a disastrous course. Given his commitment to both environmentalism and the communist tradition Guattari would wholeheartedly agree with many of today's climate justice activists but insist on the critique of the dominant mode of production (which is also a production of subjects). Since it is the uncontrollable self-organising machinism of capital that has been madly ploughing through the machinic phyla and keeping humans addicted to oil, humanity's objective can only be to dismantle it.

II

Stratoanalysis would be incomplete without a precise understanding of capital as the agent of destratification which has irrevocably altered the composition of the earth's strata and machinic phyla. From the 1960s until his death, Guattari was engaged in developing an understanding of capitalism which could serve as a conduit for political intervention. While theorists who befriended him like Franco 'Bifo' Berardi and Maurizio Lazzarato have always grasped this Marxist, even Leninist, dimension and continue developing it themselves (Berardi, 2008: chapter 3; Lazzarato, 2014: introduction; see also Sibertin-Blanc, 2016; Rauning,

2016), many readers of Guattari adeptly remove it. This is not only intellectually dishonest but undermines the constructive potential of his thinking in times neoliberal capitalism is giving way to even more reactionary regimes across the world. Like most poststructuralist thinkers, Guattari offers a creative fidelity to Marxian concept-creation even if it might be unrecognisable to traditional Marxists. Most fundamentally, he follows Marx closely in arguing for the unprecedented intensity and scope of deterritorialisation that *money* harbours, and that any revolutionary politics must engage with the capitalist system as a convergent megamachine. It is not just that anything on earth and beyond can be commodified and speculated on through a general quantitative equivalency, but that money under capitalism becomes fully autonomous and sweeps (integrates) entire ecosystems, populations, and daily routines into its homogenising orbit. Governments and individual capitalists and consumers have only limited control over capital's voracious and monstrous semiotic, what Deleuze and Guattari in plateau 13 of *A Thousand Plateaus* call its 'axiomatic'. The result is Integrated World Capitalism (IWC). As in calculus, capital's integration is logically counterbalanced by its differentiation, the stupefying kaleidoscopic stream of goods, images, cities, and experiences we inhabit as consumers. More than surplus value deriving from the unpaid work of wage-labourers, Guattari argues that wealth in IWC comes from money's own self-expansion through technological and consumerist means.

Geographers know well that IWC is not at all homogeneous, and could never be. Space and bodies become organised into highly uneven chunks of nation-states, rich and poor, management and manual labour, so-called communist and so-called democratic, racial formations and genders and varying levels of fitness. This organisation is what plateau 9 calls 'segmentarity'. The value created by labour for capital is therefore not only monetary but machinic. By working we also mold our desires and spaces into identities which reproduces the capitalist system. Guattari is at his best, as Bifo and Lazzarato argue, when he demonstrates that the way capital works in itself brings forth a veritable "machinic enslavement", especially through the twin process of miniaturisation and expansion of the *dispositions* of biopower which media technologies and institutions have made possible since the 1950s.

> Capitalism seizes individuals from the inside. Alienation by means of images and ideas is only an aspect of a general system of enslavement of their fundamental modes of semiotization, both individual and collective. Individuals are 'equipped' with modes of perception or normalization of desire just as they are with factories, schools, and territories. The expansion of the division of labor to planetary levels implies, on the part of global Capitalism, not only an attempt to integrate productive forces of every social category, but moreover a permanent recomposition, a reinvention of this collective workforce. The ideal of Capital is no longer to bother with individuals endowed with passions, capable of ambiguity, hesitation and refusal as well as enthusiasm, but exclusively human robots.
>
> (Guattari, 2009: 262)

Implicitly deepening Marx's notion of commodity fetishism (1992), Guattari claims throughout his oeuvre that capitalism's taken-for-grantedness is materially based on consumers bathing ever more in an irresistible and inexhaustible universe of signs. If we link this ever-densifying machinic enslavement back to the Anthropocene, the catastrophic nature of capitalism becomes still more apparent (see Parikka, 2015). The more the planet gets trapped in capital's course, the more humans are blinded to their exploitation through a fantasy realm spun from ubiquitous screens and libidinal prompts, and the more, also, ecosystems are exploited to sustain these technological addictions.

Guattari's reliance on world-systems theory is seldom noted. Building on Marx and Braudel but further inspired by anticolonial struggles, world-systems theory, as formulated by Samir Amin (1976), who is apparently Guattari's main inspiration, and Immanuel Wallerstein (1979) who became its most influential proponent, analyses the long-term interconnections underpinning the shifts in global and regional hegemony. The dominance of a core or centre is contingent on the provision of raw material, labour and consumers from peripheral states, usually colonies or ex-colonies. In the 1970s and 1980s a semiperiphery arose comprising states like Argentina, Saudi Arabia, and Taiwan, while China and India are now evolving from peripheral to central status. There is but one capitalist world-system, which "tolerates, in fact requires, a certain peripheral polymorphy, to the extent that it is not saturated, to the extent that it actively repels its own limits" (Deleuze and Guattari, 1987: 436). Far from proving free trade brings wellbeing to all, the collapse of state communism, Asian growth, and the resurgence of protectionism à la Trump all simply illustrate capital's continued quest for exploitable landscapes and bodies. The only explicit criticism Deleuze and Guattari make of world-systems theory is that it forgets the stratification within the West: ghettoes, refugees, bums, white trash, people "abandoned to erratic work (subcontracting, temporary work, or work in the underground economy)" (1987: 469). As many have argued, and as was foreseen by Guattari, the neoliberal paradigm demolished the old imperial idea of universal welfare but would not necessarily get rid of new forms of authoritarianism and nationalism. US hegemony started to appear doomed already in the mid-1970s (Deleuze and Guattari, 1987: 566n.23) and the Soviet Union was waiting to implode. Capital's axiomatic creativity is exactly what allows it to start rearranging the North/South and East/West polarities globally in the name of the sacrosanct market. Evidently capital has no special fondness for white people.

In *Anti-Oedipus* (1983) Deleuze and Guattari showed great interest in a question fundamental to both comparative history and, at the beginning of capitalism's adventure, Hegel: why did *Europe* become dominant? "The only universal history is the history of contingency. Let us return to this eminently contingent question that modern historians know how to ask: why Europe, why not China?" (1983: 224). While shipping in other civilisations supported commerce and despotic territoriality, Western Europeans were the only to develop the mad desire to cross the oceans for discovery's sake and subsequent territorial expansion. Capitalism proper requires a singular kind of adventurous zeal to inhabit the new conjunction of capital and labour flows. Closely following Marx and Braudel, Deleuze and Guattari argue that the flows of commodities, technology, industry, money, and armies were not allowed to "stream together" (1987: 452) in previous social formations (like the Roman Empire, the trade network of Arabs, or the Ottoman world economy). Capitalism only appears where and when labour is deterritorialised from serfdom and village, and capital from land and market square. The two flows are "conjugated" through the comparability granted by a uniform system of prices (including wages) and exchange rates. Hence conjugation is what capitalist flow is all about. The machinic entangling of money, raw materials, tools, invention, and labour power in eighteenth-century Lancashire is the archetypical conjugation. This new highly lucrative form of exploitation rapidly diffused by conjugating (deterritorialising) one market after another, so that Marx and Engels would in five decades talk of one economic world-system. Globalisation was an old dream of emperors and holy men across the world, but it was only industrial wage-labour, financed through land-grabs, genocide, slavery, and accumulated mercantile wealth, that could kick-start the runaway process.

Unlike the territorial morphology of the antiquities and feudalisms (whether of Western Europe, Mexico, or Japan), capital strives to absorb and overcode all political entities of the world. Now "we can depict an enormous, so-called stateless, monetary mass that circulates through foreign exchange and across borders, eluding control by the States, forming a multinational ecumenical organization, constituting a de facto supranational power untouched by governmental decisions" (Deleuze and Guattari, 1987: 453). However, states have always been crucial for providing infrastructure and laying down axioms conducive to capital flow, especially through war and technology. For Marx (1992), of course, the "so-called primitive accumulation" by states on the basis of dispossessing colonised populations was in hindsight a necessary territorialising project for capitalism to take off. Deleuze and Guattari push this further: "there is a violence *that necessarily operates through the State*, precedes the capitalist mode of production, constitutes the 'primitive accumulation', and makes possible the capitalist mode of production itself" (1987: 447). A long passage in plateau 13 locates the origins of geographical inequality in the extortion practices of ancient empires (1987: 437–447): agriculture produces a flow of surplus which has to be siphoned and distributed, creating wealth for those who capture it. A state is in fact nothing but what plateau 13 christens an *apparatus of capture*. The despot invents a new kind of arithmetic, *money*, in order to seize rent and tribute, and from the sixteenth century, profits from trade and manufacturing. For thousands of years, therefore, the state machine has involved a "general space of comparison and a mobile center of appropriation" (1987: 444). Deleuze and Guattari make the startling claim that any sedentary society is *already urban*, stockpiling by exploiting many peripherals at once (1987: 440). In retrospect, but only in retrospect, the state, the city, and capitalism were and are tangible in nomadic and agricultural societies as transcendental *limits* in both the mathematical and geographic sense. These limits were actively staved off as undesirable by the nomads and savages. It is as if they presaged the global catastrophes of the 21st century.

When in early modern Atlantic Europe peasants were dispossessed and capital was available from long-distance trade and slave labour, as described by generations of Marxists, the 'space of comparison' of the state enabled a mode of homogenising populations and markets. In the process, exploitation became untied from the body of the despot, but the spatial logic of social stratification and inequality (central wealth, poor peripheries) remained in place. While for Deleuze and Guattari money originally propped the state and concomitant caste distinctions, capitalist axiomatics dramatically transforms the essence of production and exchange. Without understanding capital flow, spatial divisions of labour, and the complex desires sustaining them, the differentiation into rich and poor tends to be explained in moral instead of systemic terms: the greed for power and possession of individuals instead of the outcome of a distributive logic. Deleuze and Guattari do not locate the basis of inequality in the use of money or debt as an instrument of power but in *circulation itself*. With every new wave of financialisation and computerisation, circulation becomes less "human", less constrained by consciousness. Capitalists and states no doubt enrich themselves without scruples through their privileged connection to capital. But with numbers running into quadrillions, algorithms rewriting themselves continuously, and futures folded into the present in ever more complex ways, capital becomes increasingly unmappable by any one power or intelligence.

The concept of IWC ultimately breaks with the binary of centre and periphery. The global economic system is a network of networks with shifting *gradients* of economic power. Capital flows become viscous in finance hubs like Frankfurt, for example, and manufacturing hubs like Ciudad Juarez. They thereby bypass countless other places. Capitalist world-space is far too uneven, volatile, and cramped to fit a bipolarity. As Doreen Massey (1995) showed definitively, it is the differential connectedness of a place, not its prior centrality or peripherality that determines its capacity to attract and steer further flows. This happens at all scales, and the differences far exceed what the human sciences usually pick up on. World-systems theory is on the right track with its geohistories of European dominance and Global South poverty over the *longue durée*, but the centre-periphery polarity prevents it from examining inequalities on all strata. Meanwhile it is fair to say Deleuze and Guattari have no theory of location and spatial division of labour as such – how is machinic surplus *re*territorialised in this place and not another? They could be criticised for the tendency common to philosophers of abstracting the geographies of money and commodity chains into a homogeneous becoming-planetary as if place no longer matters. But because they accept the premises of world-systems theory, there can at least be the start of a philosophy of globalisation attentive to the differences from which globalisation ensues.

III

Another theoretical innovation in what we could call 'Guattarian political economy' is its argument that the capitalist axiomatic commands not only production and circulation but also subjection or subjectification within specific segments of society, corresponding roughly to Foucault (2003) calls 'biopower'. Commodity fetishism, the retrieval of archaic affective registers to bestow an addictive allure to products and services, became ever more neurotically geared to work, compete, and enjoy. Ultimately the spatio-temporal rhythms of human bodies are orchestrated by the movements of capital. Guattari does not shy away from calling this situation *machinic enslavement*, a concept running through his own oeuvre, the co-authored *Capitalism and Schizophrenia* project, and commentaries on the continent, but one which is almost absent in Anglophone reception. Guattari was quite prophetic about the extent bodies are today commanded 24/7 to generate value and the distinctions between work and leisure collapses (see Crary, 2014). Bodies are literally suspended in flows of images and data, fantasies of holidaying and sexual omnipotence, mixed with perpetual anxiety of falling behind. They produce value simply by being alive. Guattari differs fundamentally from Foucault, however, in continuing to think in a communist vein about what exactly constitutes the collective movements of revolution. With subjectification there is necessarily the potentiality of *subjectivation* which rides the social megamachine's lines of flight towards a new socius.

Hence *Anti-Oedipus* refines and does not abandon the Marxist critique of alienation. The book's central question is one which has driven political philosophy and psychoanalysis from the beginning: why do people desire their own enslavement? More specifically, how does excess capital "buy" the docility of billions of bodies? If inequalities within and between countries are cemented ideologically, for example, through the representations on television and blogs, they are reinforced much more directly through bodily practice and the senses. Whether rich or poor, all people love handling cash, seeing returns on their investments, fondling new gadgets. Rather than money "representing" shit as psychoanalysts suggested, money and shit are interchangeable flows within the same libidinal economy (Deleuze and Guattari, 1983: 28–29, 302). "It is at the level of flows, the monetary flows included, and not at the level of ideology, that the integration of desire is achieved" (1983: 239). But this does not mean ideology doesn't exist. Some readers falsely conclude that Deleuzian affect and the Guattarian machinic seek to overcome the Althusserian theory of ideology. By underscoring the non-representational and pathological nature of capitalist biopower Deleuze and Guattari in fact offer a strongly antihumanist concept of ideology in which it is no longer the distortion of an otherwise stable social reality but becomes the desirous embracing of the social's cumulative potentialities (see Saldanha, 2015). This perspective on ideology as machinic enslavement yields a richer appreciation of how socialising processes work than is often available in Marxism. Physically slotted into machines for production and consumption like cars, offices, call centres, theme parks, casinos, and Facebook, bodies have no illusion they are being abused for making money for an invisible oligarchy. It is just that habit, peer pressure, and sheer enjoyment make them actively hostile to exploring how to interrupt this situation and discover its lines of flight. Unlike the pessimism of the Frankfurt School, Guattari and Deleuze place a qualified hope in the many subtle molecular ways in which desire for radical change lurks even beneath the most insipid of suburban or derelict spaces.

The deterritorialisation and scrambling of traditional cultural codes *alienates* to the extent there is a societal condition Guattari says is psychotic. For Lacan (1993) psychosis is the untying of what holds signification together in a subject whose desire is then no longer barred by paternal prohibition. It is the most horrific experience possible. Guattari airlifts the term 'schizophrenia' (the mental disorder prone to psychoses) from psychiatry to mean capitalism's generalised and intrinsic breakdown of meaning. When identities and history lose footing, delirium and hallucination take over. There are artistic and intellectual movements which deliberately play with this psychedelic and psychotic aspect of accelerating capitalism, from Baudelaire to Artaud to postmodernism. Schizophrenia is the limit condition against which capitalism as system continually brushes. In both Marxism and Deleuze and Guattari this "limit" is always both geographical and psychocultural.

The irreverent revalidation of schizos in *Anti-Oedipus* can be distinguished from antipsychiatry and Romanticism which tend to see in madness a liberating vehicle. It should instead be appreciated in light of Guattari's critiques in the 1960s of the focus in mainstream psychoanalytical practice on neurosis, ego, and disciplining. Schizoanalysis, both the practice and the concept, has led to many misunderstandings, but it should be remembered that it is a methodology for diagnosing capitalism as system, which manifests itself in the clinical setting but also in art or science. Guattari continued refining the term, culminating in his later years in the befuddling *Schizoanalytic Cartographies* (2013). It makes no sense to analyse the geography of money without analysing the strata of the unconscious and everyday spaces through which money can flow and bodies are enslaved. What *Capitalism and Schizophrenia* argues is that only a certain madness can be structurally adequate to the spiralling, inhuman speeds of money-as-capital. The bottom-line of our mode of production is not reason and freedom but the desperate attempt to control the uncontrollable. No wonder, therefore, it unleashed juggernauts like world war, famine, extreme weather, rising sea levels, and mass extinction. It is not difficult to draw the logical conclusion: avoiding further catastrophic madness requires another mode of production.

IV

To end this chapter let us return to Guattari's relation to communism. The catastrophic nature of the Anthropocene would have certainly spurred Guattari, if he were to have known the concept, to argue it intrinsically entails a rigorous return to the communist project. But a return is a reinvention. While Marx and Lenin thought in terms of proletarian consciousness and historical stages, Guattari envisages capital as an all-engulfing psychotic process which subsumes the earth itself and can only be derailed through marshalling its lines of flight through a multiplicity of molecular revolutions. The neoliberal consensus of the last few decades had been based on the astonishingly irrational belief that markets can themselves amend the nefarious tendencies of capitalist accumulation. Even more irrational is now the racist nationalism that is appearing in country after country in the wake of the contradictions in neoliberal globalisation around migration and violence. To imagine a different society is not at all utopian, it is the rational thing to do. It is a kind of *realism* because it understands capitalism is self-destructive. History is accelerating and coming to a threshold, a boiling point, and it is scary how few dare to imagine, to recall a well-known quip, the end of capitalism instead of the end of the world.

> Utopia, today, is to believe that current societies will be able to continue along on their merry little way without major upheavals. Social modes of organization that prevail today on earth *are not holding up*, literally and figuratively. History is gripped by crazy parameters: demography, energy, the technological-scientific explosion, pollution, the arms race. . . . The earth is deterritorializing itself at top speed. The true utopians are the conservatives of all shapes and sizes who would like for this 'to hold up all the same', to return to yesterday and the day before yesterday. What is terrifying is our lack of collective imagination in a world that has reached such a boiling point, our myopia before all the 'molecular revolutions' which keep pulling the rug out from under us, at an accelerated pace.
>
> (Guattari, 2009: 307)

In the quote above "molecular revolution" refers to the rapid multifarious societal changes which destabilise life, but usually it is Guattari's term for a radical politics which continues the mass uprisings of 1968 and learns from the ideological dead-end that was centrally planned communism. His book *Molecular Revolution* was published the same year as *Anti-Oedipus*, 1972, and likewise a condensation of the intense political debates that occurred in Paris in the wake of the attempt at mass revolution of May 1968 and the intellectual confluence of structuralism, psychoanalysis, radical art, and the far left, especially Maoism. Guattari's main question was how to prevent protest movement becoming diffuse or co-opted by liberal-bourgeois society. By 1979 most of the political fervour of the sixties had dissipated. That year Guattari proposed a "Plan for the Planet" which called for revolutions in the plural that would enact change at the levels of the body, sexuality, the household, institutional racism, and the media in addition to the factory and the school, and without a vanguard party. However,

> it is impossible to make a clear-cut distinction between the fringe ideas that can be recuperated and those that lead don the slippery slope to authentic 'molecular revolutions'. The borderline remains fluid, and fluctuates both in time and place. The essential difference is whether, in the final analysis, a given phenomenon – however broad its implications – is one that remains on the margins of the socius, or whether it poses a fundamental threat to it. In this sense, what characterizes the 'molecular' is that the lines of flight combine with the objective lines of deterritorialization of the system to create an irrepressible aspiration of new spaces of freedom.
>
> (Guattari, 1984: 269, translation modified)

Guattari criticised both Maoist and other leftwing fringe groups (*groupuscules*) and the orthodoxy of the official French Communist Party for trying to jealously emulate industrial and consumer capitalism by means of ostensibly scientific state control.

> A communist society must be designed not with reference to consumption, but to the desire and the goals of mankind. [. . .] There can be social planning in terms of organising production – though there still remain a lot of unanswered questions – but it cannot claim to be able to give *a priori* answers in terms of the desire objectives of individuals and subject groups.
>
> (1984: 43)

Guattari corresponded with his Italian friend Antonio Negri, imprisoned on spurious links to far-left terrorism, about radical politics. Published just before Gorbachev came to power, the book based on their conversations shows Negri inviting Guattari to rethink the future of "communism", a term that otherwise appears seldom in the latter's work (Guattari and Negri, 1990). While the book's vocabulary and style is Guattari's, the focus on this term probably comes from Negri since of the two he uses it regularly. Semiotext(e) retitled the translation *Communists Like Us* and published it in 1990, that is, just while the great experiment of state socialism was ending. Against the triumphalism of liberal ideologues, the term 'communism' was given an entirely new subversive quality. To identify as communist was no longer to be tethered to determinism, nostalgia, and the state, but to construct "new lines of alliance, new spaces of liberty", as the subtitle has it.

Communism for Guattari and Negri had to be rescued from itself. The Soviet system was imploding under its own bureaucratic, military, and ideological black holes, as well as its continued reliance on capitalist modes of surveillance: "we have in place one vast machine, extending over the planet an enslavement of all mankind" (1990: 26). In line with the *operaismo* movement with which Negri is associated, Guattari in this book holds that communism must entail the abolition of work, or at least, of wage-labour. Guattari and Negri are not the only theorists on the left to suggest the development of collective productive power makes the end of work as humans have known it increasingly possible, even if such radical change is fundamentally precluded by the liberal-capitalist insistence on growth, utility, and scarcity (see, for example, Frase, 2016). The immense problems associated with the perpetual growth model, not in the least in the form of the nuclear arms race, neuroses, and climate catastrophe, make switching to a mode of production focused on creativity instead of exploitation a no-brainer.

> We envisage an imaginative, creative process at once singular and collective, sweeping the world with a great wave of refusal and of hope. Communism is nothing other than a call to life [. . .] Preventing catastrophe will require a collective mobilization of freedom [. . .] Extermination or communism is the choice – but this communism must be more than just the sharing of

> wealth (who wants all this shit?) – it must inaugurate a whole way of working together.
>
> (Guattari and Negri, 1990: 27–28)

The machinic enslavement of bodies into globally integrated capitalism and the Anthropocene has since Guattari's death attained another dangerous impetus only embryonic in his time. The resurgence of nationalist, racist, and totalitarian tendencies in George W.H. Bush and Nicolas Sarkozy, then the repackaged fascism of Marine Le Pen, Donald Trump, Nigel Farage, and Jair Bolsonaro are based on a mediatic appeal which *Anti-Oedipus* can serve as a guide to understand. The desire the masses harness towards such obscene figures of liberal democracy is strictly coterminous with their machinic integration into global neoliberal consumerism. The new fascists thrive thanks to the crises, but raise alarm about minorities and terrorism instead of capital and class exploitation. As many point out, their mass appeal is to a significant extent the fault of the mainstream left having abandoned its traditional battles to imagine and build alternatives to capitalism. Amazingly, Guattari had in the later 1980s already heard of Trump's predatory ways, though the fascist streak would develop later. Guattari understood the real-estate mogul could only lay destruction to American society because he could spin what he did as the only way out of crisis and because he encountered so little effective institutional resistance.

> In the field of social ecology, men like Donald Trump are permitted to proliferate freely, like another species of algae, taking over entire districts of New York and Atlantic City; he "redevelops" by raising rents, thereby driving out tens of thousands of poor families, most of whom are condemned to homelessness, becoming the equivalent of the dead fish of environmental ecology.
>
> (2008: 21)

The Three Ecologies (2008, French edition 1989) argues that the crises of capitalism are felt across three ecologies, three scales, three kinds of being: the social (mostly, exploitation and conflict), the mental (addictions, anxieties, depression, a schizophrenia particular to the capitalist socius), and the environmental (from plastic and water pollution to endangered species to resource depletion and, of course, global warming, then already a hot topic). From his first days in office Trump has set the agenda for wrecking havoc on all three ecological fronts for no other reason than his own narcissistic greed and that of his crony friends. But Guattari the communist reminds us: in democracy, *we* on the left are to blame for permitting such fascism to proliferate. With every year getting hotter, life for the world's urban poor becoming ever more precarious, and another mass shooting in the US, Guattari's multifarious post-psychoanalytical environmental thought or "ecosophy" becomes relevant to elaborate. Communism's "call for life" derives its propulsion not simply from a utopian promise, then, but first of all from sketching maps of the multiple crises we find ourselves in and grasping the mass desire for alternatives which those in power are perpetually trying to constrain.

V

The aim of this chapter has been modest yet eminently urgent: to retrieve the heterodox Marxist inflection of Guattari's thought. Capitalism only holds together through deterritorialisation happening at dizzying speeds through many strata at once. In hindsight, the Anthropocene has been an inevitable result of the exponentially increasing scope and depth of the profit-driven axiomatic since about 1800. What is needed is an earthly communism, a *geocommunism* which can turn the deterritorialisation inherent in the global system against itself and thereby create forms of life not produced for profit. It is true Guattari, like Marx, does not provide much practical guidance on age-old strategic conundrums. Memories of the failure of May '68 are still rehearsed in thinking through the antiglobalisation movement, Occupy Wall Street, and the various mass protests and riots of the last few years. On questions of political organising – how can Trump's onslaught be stopped? how can the social justice principles of the European Union be saved from below? how can an internationalism prevent the vitriolic anti-migration rhetoric? – there is much discussion to be had. Most actually existing communists have been skeptical about the quasi-anarchist ideas of postmodern Marxisms such as Guattari's. But precisely through his eclecticism, he can inspire new debates on the classic questions of communism like movement leadership, the law, the withering-away of the state, whether to industrialise, national divisions, patriarchy, and so on. Stratoanalysis and schizoanalysis can inspire communists to investigate not just liberal democracy's but their own complicity with the megamachines imperilling the planetary and socio-psychical ecology. How the many molecular revolutions will amalgamate into a system change is structurally impossible to foresee. But that fundamental change is necessary for the survival of "civilization as we know it" is becoming dreadfully obvious.

References

Amin, S. (1976) *Unequal Development: An Essay on the Social Formations of Peripheral Capitalism* (trans. B. Pearce). New York, NY: Monthly Review Press.

Berardi (Bifo), F. (2008) *Félix Guattari: Thought, Friendship and Visionary Cartography* (trans. G. Mecchia and C.J. Stivale). Basingstoke: Palgrave Macmillan.

Crary, J. (2014) *24/7: Late Capitalism and the Ends of Sleep*. London: Verso.

Deleuze, G. and Guattari, F. (1983) *Anti-Oedipus: Capitalism and Schizophrenia* (trans. R. Hurly, M. Seem and H.R. Lane). Minneapolis, MN: University of Minnesota Press.

Deleuze, G. and Guattari, F. (1987) *A Thousand Plateaus: Capitalism and Schizophrenia* (trans. B. Massumi). Minneapolis, MN: University of Minnesota Press.

Foucault, M. (2003) *Society Must Be Defended* (trans. D. Macey). London: Picador.

Frase, P. (2016) *Four Futures: Life After Capitalism*. London: Verso.

Guattari, F. (1984) *Molecular Revolution: Psychiatry and Politics* (trans. R. Sheed). London: Penguin Books.

Guattari, F. (2008) *The Three Ecologies* (trans. I. Pindar and P. Sutton). London: Continuum.

Guattari, F. (2009) *Soft Subversions: Texts and Interviews 1983–1985* (ed. S. Lotingen; trans. C. Wiener and E. Wittman). New York, NY: Semiotext(e).

Guattari, F. (2013) *Schizoanalytical Cartographies* (trans. A. Goffey). London: Continuum.

Guattari, F. and Negri, A. (1990) *Communists Like Us: New Spaces of Liberty, New Lines of Alliance* (trans. M. Ryan). New York, NY: Semiotext(e).

Lacan, J. (1993) *The Psychoses* (trans. R. Grigg). London: Routledge.

Lazzarato, M. (2014) *Signs and Machines: Capitalism and the Production of Subjectivity* (trans. J.D. Jordan). New York, NY: Semiotext(e).

Malm, A. (2016) *Fossil Capital: The Rise of Steam Power and the Roots of Global Warming*. London: Verso.

Marx, K. (1992) *Capital: A Critique of Political Economy*, Vol. 1. London: Penguin Books.

Massey, D. (1995) *Spatial Divisions of Labor: Social Structures and the Geography of Production*, second edition. London: Routledge.

Moore, J. (ed.) (2016) *Anthropocene or Capitalocene? Nature, History, and the Crisis of Capitalism*. Oakland, CA: PM Press.

Parikka, J. (2015) *A Geology of Media*. Minneapolis, MN: University of Minnesota Press.

Rauning, G. (2016) *Dividuum: Machinic Capitalism and Molecular Revolution* (trans. A. Derieg). New York, NY: Semiotext(e).

Saldanha, A. (2015) Guattari, ideology, and the ironies of theoretical (im)practice. *Dialogues in Human Geography*, 5(2): 149–154.

Saldanha, A. and Stark, H. (eds.) (2016) A new earth: Deleuze and Guattari in the Anthropocene. Special issue of *Deleuze Studies*, 10(4).

Sibertin-Blanc, G. (2016) *State and Politics: Deleuze and Guattari and Marx*. New York, NY: Semiotext(e).

Wallerstein, I. (1979) *The Capitalist World-Economy*. Cambridge: Cambridge University Press.

16 Transversal geo-politics

The violence of sound

Anja Kanngieser

On a website called 'The Electronic Intifada' are a series of recordings made by Palestinian blogger Rana Baker during the 2012 Israeli attacks on Gaza. The four recordings, taken in total over 8 minutes between 8:50 p.m. and 3:50 a.m. on the 20th and 21st of November (a few days preceding the ceasefire), document deployments of missiles from Apache helicopters and live ammunition fired from the ground. The recordings themselves are low fidelity with static overlaying the relays of explosions and retorts. This static fuzz lends the recordings a peculiar stillness that suspends in tension life within the intermittent echoes of warfare woven through. This discomfort is heightened by the large absence of animal or human voices. Occasional words and squeals punctuate the drone of planes, sometimes the inquisitive piping of a child, at times in the far off distance an amplified vocal distortion or perhaps a television or radio can be discerned but the words are muddied and incomprehensible. In the recording 'Apaches and apache strikes right now in Gaza Nov 20 9.37pm' an infant's wail is syncopated sharply. Moments where the recordist brushes her hand over the microphone (or perhaps it is some kind of signal interference) recall a situated embodiment that for the large part the recordings deny. One senses for a few seconds that she is sitting there, silently, listening, allowing us bear witness to what she hears.

These recordings remind us how profoundly the sounds of a place can evoke its conditions. Most importantly for geo-politics scholars, I argue, they compel us to critically address the largely overlooked 'invisible', atmospheric forces of governance, and their role in determining everyday experiences of space and place (Feigenbaum and Kanngieser, 2015). By listening to the ambiances or atmospheres of a place we get a sense of the complex and shifting terrains that constitute it, the sudden flashes of activity, the unremarkable meanderings and stutters, the compositions of organic and inorganic matter. Listening closely allows us to hear for sonic topologies – those continuous tones and harmonics that hum throughout moments as they articulate themselves. Sound transverses events and states to bring into effect relational ecologies that reveal themselves to us in ways that require a particular sensitivity to apprehend. Through concerted listening we are able to encounter sound as a way of 'knowing', as an acoustemology. Put another way, through emphasising the temporal, active, and collective dimensions of sound, as Jean-Paul Thibaud (2011) stresses, we are able to study and to document the unfolding of an atmosphere.

Félix Guattari's work, specifically his conceptualisation of transversality, provides a framework for unpacking the disparate territories and themes of these sonic topologies. For Guattari, the concept of transversality provided a way to "think the interactions between ecosystems, the mecanosphere, and social and individual universes of reference" (1989: 135). From its inception, his idea of the transversal was tied to investigations of power and the formation of subjects (subjectivation). Through the early experiments in psychoanalytical and political organisation, to later explorations of creativity and aesthetics, transversality signalled for Guattari a means to think about the crossing over between different identities, structures, and hierarchies. Subjectivation, for Guattari, is thus

> based on transversal practices that allow the collective emergence and entanglement of existential territories and universes of value. Transversality as a field of expression provides the milieu for a creative emergence from disparate forces. As a 'dimension to overcome two impasses,' it is not a mere connecting device but a practice for novelty to emerge.
>
> (Brunner and Rhoades, 2010: np)

Following Guattari's commitment to experimentation, this chapter uses a transversal technique to move between sites, times, machines, and nation-states and/or privatised entities – *it becomes itself an instance of transversal expression*. Transversality is used to map out an 'assemblage of vulnerabilities' – bodies and forms of life vulnerable to sonic governance. By naming these vulnerabilities, and casting light on these assemblages, it is my intention to highlight the necessity for a more concerted focus on sound within geo-political research.

Sound has several unique affordances that make it relevant to a transversal style of investigation. Firstly, the processual character of the transversal – something that cuts across scales and levels – is fundamental to the way in which sound, as a phenomenological and as a vibrational-kinetic entity, works. Transversality always affects that which it crosses through; it does not just 'connect'; it changes things; it brings into the world novel relations; it shifts paradigms and builds new formulations. These may be significant or not, temporary or not, evident or not. Second, sound as transversal subjectivates; sound is productive of place, space, and identity. (Nancy, 2007: 17). These sonic processes are not indicative of spatial or temporal permanence; they are highly contingent and eventful – sounds require space and air for their form, they "take shape on different scales of space" just as they do different temporal scales (Roads, 2001: 39). The communication of auditory trauma is thus always fraught. In the moment of its activity, sound can cause acute physical and psychological violence, which may resonate well after the vibration itself has diminished (see Revill, 2014: 3). This, however, is in excess of its possible articulation, leaving the narration of sonic assault as impartial and unseen.

Thirdly, transversality reveals the vulnerabilities that sound governance capitalises upon because it shows how bodies are asymmetrically affected within common conditions. Put another way, transversal modes let us shift between the general to the particular, while retaining a view to the collective. Sound has a tendency toward immersion: when it is not heard it still penetrates the proximate skin and bodily cavities, organs, and cells. As a tool of warfare, this asymmetry to sound exacerbated by intensity of volume or frequency, makes it unevenly harmful. Sound enacts an elementary affective force influencing how we relate to the world and the materials that comprise it (Ingold, 2011). It is highly promiscuous; it goes into space and time where bodies cannot (LaBelle, 2010) and it envelops that which it comes into contact with, creating a collective effect (though it may not be experienced in the same way throughout that collectivity). At the same time, as seen with holosonics, sound can be highly specific when directed, affecting certain kinds of bodies more intensively than others and fostering awareness of precise relations of subject and object.

Finally, sound decentres human subjects. Everything makes sound and is touched by sound, regardless of how imperceptible to human sensibilities this may be. Attending to the geographies of sound requires seeking chains of association across often incompatible and irreducible spaces and corporealities, working in a motion of propagation like sound itself. This movement reveals surprising or unconventional connections, offering a productive means for reconfiguring how we understand the interrelations between biotic and abiotic subjects. This helps us to conceive a more nuanced, and less humanistic, view of ecosystem function. In sum, transversality situates sound in an always-political dynamic delimiting how we conceive of political agency (Kanngieser, forthcoming) into a more 'cosmopolitical' (Stengers, 2005) approach. This has consequences for how we consider sound to generate knowledge of environments, the power structures that it illuminates and participates in, the intersections of state, military, and corporate interests, and so forth. Vital to this are different ways of listening or attenuating in order to become more sensitive to these structures and dynamics. In what follows, this chapter, through maintaining a transversal momentum, provides a pathway to intervention into such ways of attending to the complex and shifting interchanges of governance and subjectivation – namely by exploring how sound is weaponised across geophysical, atmospheric, and biological scales.

From the air to the earth

> In modern warfare, mechanical and metallic, the element of sight is almost zero. The sense, the significance, and the expressiveness of noise, however are infinite.
>
> (Russollo, 1986: 49–50)

> All one has left is a resonance chamber well on the way to forming a black hole.
>
> (Deleuze and Guattari, 2004: 379)

It came from above: air strikes. The sound of helicopters chopping the sky with their blades overhead, the repetitive whips undercut by cracking drones oscillating as they moved from one location to the next, hovering then peeling away. Atmospheres of war, made in the air, etch out zones of militarisation as much vertical as horizontal. The sound of missiles, machines, infrastructural collapse are carried through the air in waves. The waveforms of sound make it almost impossible to escape as they are felt as tactile vibration as much as they are heard. Sound penetrates space and matter.

Writing on the affective force of sound, Steven Goodman explores how sound warfare impacts on the ways "populations feel – not just their individualized, subjective, personal emotions, but more their collective moods or affects" (2009: xiv). For Goodman "sound contributes to an immersive atmosphere or ambiance of fear and dread"; the vibration of sonic weapons, however, "threatens not just the traumatized emotional disposition and physiology of the population, but also the very structure of the built environment" (2009: xiv). Defined under the rubric of nonlethal or 'less-lethal' (and psy-op)[1] warfare, not intended to inflict injury or death, sound devices thoroughly destabilise the integrity of infrastructure – geological, material, social-political – and can, and do, cause massive amounts of physical pain and damage, even death (Altmann, 2001; Arkin, 1997), depending on the parameters of their deployment.

In the 2014 report by the Who Profits Research Centre on the use of nonlethal or less-lethal weapons, a section is dedicated to a device nicknamed 'the scream'. First used in Bil'in in 2005 during a protest against the Israeli West Bank barrier, the Israeli manufactured anti-riot system 'SHOPHAR'[2] consisted of 36 horns stacked upon each other mounted onto a jeep (Businessweek, 2014). Several years later a different iteration of the scream appeared on the West Bank: an LRAD (long-range acoustic device) produced by the American Technology Corporation – a "high-intensity directional acoustic hailer designed for long-range communication and issuing powerful warning tones" with a reach of up to 3,000 metres depending on the model (LRAD nd). Used to both establish exclusionary zones, and police civil demonstrations, Volcler (2013)[3] reported effects included temporary loss of hearing, nausea, and dizziness (Altmann, 2008).

The past years have seen increasing use of diversionary devices such as stun grenades or 'flash bangs' in occupied were used territories, with reports particularly focusing on instances of violence against civilians. Comprising a mix of mercury and magnesium powder, and with peak sound pressure levels between 130 and 190 dB at 1.5 metres (well above safety limits), stun hand grenades such as the M84 are designed to distract noncombatant targets through the combination of pyrotechnics and sound (Department of the Army, 2009: A–11). In 2014 three incidents wherein stun grenades outside of political and religious demonstrations received coverage in English-language media – the employment of grenades in combination with tear gas against school children in al-Khalil (ISM, 2014), the firing of grenades into a wedding march in al-Issawiya (Ma'an, 2014), and the attack by Israeli police on the al-Aqsa mosque, in which grenades were detonated inside the mosque itself (Lewis, 2014). While stun grenades are designed to be discharged within confined and enclosed spaces, this is with the regulation that they do not come within 1.5 metres of human bodies. Alongside ongoing concerns around the ways in which war recasts and shapes everyday geographies into escalating spaces of terror, such incidents demonstrate the ambiguous position that less-than-lethal sound devices hold. The disorientating effects of flash bang grenades have been widely documented – causing severe dizziness, temporary deafness and blindness; and the devices have led to acute injury and death (Bogue et al., 2014; Gurr, 1997; Wright, 2001). Nonetheless, they are designated less-than-lethal status.

The calculated use of the significant affective, emotional, psychological, and physical violence of sonic weapons is perhaps most markedly captured by the production of sonic booms through shock waves caused by jets travelling beyond the speed of sound. Shortly following the enaction of the Israeli Disengagement Plan from Gaza in 2005, it was reported that IDF jets were flying at extremely high speed and low altitude over the densely populated Strip, creating what were described as "sound bombs" or thunderclaps over residential areas several times a night. Felt as "a wall of air, painful on the ears", sonic booms give the effect of aerial bombardment; the force of the boom cracked edifices, blew doors off buildings, and shattered glass (McGreal, 2005). This intense sonic physicality was accompanied further by physical and psychological harm including nose-bleeds, hearing loss, symptoms of anxiety and hypertension, exhaustion, and, according to the Palestinian health ministry, increased miscarriages and heart problems, with the most severe impacts on children and infants. Medical human rights groups argued that, given the ubiquity of the sonic reverberation making it impossible to shield from, the tactic amounted to "collective penalties" defined as illegal by customary international humanitarian law and the Geneva Conventions (International Committee of the Red Cross, 2014).

As from the air, sonic technologies constrain and control life beneath the soils of Gaza. Taken from Naval methods for finding oil, and used to guard prisons, security installations, sea ports, government facilities, airports, and banks, microphones are placed 1.5 metres within the ground to measure low frequency and infrasonic energy travelling through the Earth's layers. This information is processed by "intrusion recognition algorithms" to help ascertain the type of activity causing the sound – be it walking, digging, vehicular, and so forth (Elpam, 2014). Elpam Electronics who engineer these geophones claim that the system can ascertain minute movements up to 10 metres below surface level. With a plan to bury hundreds of sensors around the Gaza border, the company is working on the means to differentiate human from other subterranean sounds.

From extreme volume to below human thresholds of hearing "sonic warfare is as much about the logistics of imperceptions (unsound) as it is perception" (Goodman, 2009: 9). The affective resonances of sound are not limited to human bodies, nor to biotic bodies at all. Sound not only emanates from and impacts upon bodies but also vibrates through environments and objects; in a text on non-cochlear sound, Will Scrimshaw emphasises the nature of sonic affects and signals in excess of their human audibility or perceptibility (2013: 28). This kind of "sonic materialism" (Cox, 2011) considers sound beyond its attributed phenomenological immediacy. It is especially through the registers of infra- and ultrasound that the more-than-human targets of sonic governance are implicated.

Governing the imperceptible

The relationship of sound to the imperceptible is aporetic in that it hides and reveals phenomena. The often acousmatic character of sound – where the source is heard but unseen – is endemic to this relationship. So is the frequency range of sound. Beyond registers perceptible to human hearing are those that transmit on ultrasonic (above 20 kHz) and infrasonic (lower than 20 Hz) frequencies often used in medical and military contexts. These frequencies, while inaudible or barely audible to humans can cross sensory thresholds through their vibratory qualities – evidenced in discussions about low frequency noise (LFN) such as what has been called 'the hum'.[4] They are also discernable to animals, plants, viruses, and bacteria (see Yusoff, 2013).

Ultrasound is most commonly associated with deterrence, commercial and military detection and sensing (such as motion sensing), and biomedical procedures such as sonography. Through sonic pressure, ultrasound registers beyond the upper limits of human hearing, sending and receiving millions of pulses each second. The frequencies at which ultrasound transmits are audible to animals and insects – a vulnerability capitalised on in the development of domestic and wild animal repellents, which can generate a sonic barricade up to 600 metres around the site of broadcast. Some ultrasonic frequencies can also breach the threshold of human hearing; devices perceptible only to those below their mid-20s, such as the 'mosquito', have been used to disperse gatherings of teenagers and young people in public places (Townsend, 2010). In the biomedical industries, the use of ultrasound has been primarily directed toward cell disruption, reducing particle sizes, and vaporisation; however, it has also been shown to render inactive bacterial spores, microbes, and parasites. In part, this has been of particular interest to researchers in the USA, hoping to arrest biological warfare such as anthrax – an experiment conducted by scientists placing spores in an envelope showed a 99.9% elimination rate through the use of high-intensity focused ultrasound at a frequency of 70–200 kHz for 30 seconds. The acoustic energy generates heat, along with the mechanical effects associated with acoustic pulses, to cause cell death (Zhou, 2011); this advancement has also been deployed as a key non-invasive and extracorporeal treatment of cancer tumours.

The deterrent and lethal effects of ultrasound for detection, monitoring, and nontoxic control have been explored by artist David Dunn through his work in bioacoustic ecology. In collaboration with physicist Jim Crutchfield, Dunn has proposed the use of ultrasound to intervene in a feedback loop of beetle infestation and deforestation. According to Dunn and Crutchfield (2009) correlations can be drawn between expanding beetle habitation and the mass decimation of trees through a fungus, which the beetles carry; this decimation leads to the release of thousands of tonnes of carbon into the atmosphere, significantly altering climate. In their work, Dunn and Crutchfield studied the ways bioacoustic communication affects infestation dynamics, namely, how tree-eating beetles use ultrasound to identify trees already vulnerable due to drought, and to further communicate this within the insect colony. While entomologists have argued that pheromones are more commonly critical to insect communication and identification of infestation sites, Dunn and Crutchfield found, through initial field research using ultrasonic signals in close range to disturb the beetles' sense, that acoustic signals were successful in breaking up or slowing down infestation. This, they argued, could help to lure insects away from damaged forests and, importantly, mitigate the further spread of insects into new territory.

From the ultrasonic to the infrasonic, acoustic technologies are used to explore, demarcate, and discipline biospheres (Hedlin et al., 2012). Increasingly, underwater drones are being used to for geo-exploration, using traditional sonar but with the added benefit of deep underwater mobility. Because of the mid to low frequency range of these pulses, and the vast underwater distances travelled by sound, a range of detrimental effects to sea life have been isolated. Physical changes in whale and dolphin populations such as tissue and organ damage and the growth of microbubbles, and behavioural changes including avoidance of sound emitting boats, cessation of singing, altered migration routes, and high levels of agitation have been noted, and scientists have directly linked mass beach strandings of sea mammals to sonar testing.

Further, offshore oil and gas exploration by geophysics companies use sonar devices, sending highly compressed air pulsing through the water and penetrating the sea floor, where sound levels, claimed to be 100,000 times that of a jet engine, are detonated every 10 seconds, 24 hours a day for weeks on end. Recent research has also shown that fish that communicate acoustically are having to produce louder sounds in order to be heard over manmade sounds – which threaten their existence (Holt and Johnston, 2014) and demand more robust regulation of anthropogenic ocean sound (Merchant et al., 2012).

Voice as biometric signature: warehouses, prisons, and borders

From ultrasound to sonic booms, buried microphones to acoustic screams, sound is used to control and monitor human movement. Surveillance and discipline through vocal technologies is a ubiquitous iteration of this. The use of voice-based biometrics to track and determine the movements of individuals has been thoroughly incorporated into varying levels of nation-state and corporate governance. This has been loosely consolidated around voice and speech capture and recognition software and hardware, which identify an individual biometric signature from the combination of vocal tract physiology and behavioural speaking aspects. From drones armed with super-sensitive microphones (under consideration by the NYPD for use in 'crime hotspots'), to 'smart' CCTV cameras able to eavesdrop on conversations (rumoured to have been used during the London 2012 Olympics), advances in voice biometrics are mapping new fields in surveillance capabilities. Such biometric technology is being deployed both in public and private spaces, perhaps most aggressively at border zones, and on the warehouse floor.

Voice biometrics have been used on asylum seekers and refugees for over a decade to impede movement outside of detention centres. Between the 2004

Asylum and Immigration (Treatment of Claimants, etc) Act and the 2006 *Immigration, Asylum and Nationality Act*, compliance with electronic tagging and monitoring moved from consensual to a conditional requirement in the UK.[5] Although electronic tracking has not been made compulsory, as this would constitute criminal abuse (though cases can be made to implement compulsory consent), compliance has been encouraged by authorities as beneficial to asylum seeker claims. This means that electronic monitoring has been assimilated into the suite of bureaucratic and technological apparatuses determining, and often severely and unjustifiably constraining the everyday mobilities of asylum seekers. Corporate service providers operating across the prison-detention spectrum, such as G4S and Serco, although are developing more accurate voice verification systems capable of accommodating multiple dialects, accent variations, and voice and condition changes.

Working in conjunction with the UK Border Agency (UKBA), such systems sit alongside security protocols designed to confirm, through accent and dialect, asylum seekers places of origin. Amidst controversy around the nature of language and accent acquisition, in 2003 the UK ratified the use of forensic linguistics in language analysis, to examine dialectological features relevant to the individual's geographical and social origin.[6] According to a 2011 report prepared for the UK Border Agency: New Asylum Model team (NAM +) by Home Office Science (2011: 5), language analysis is deployed

> to assist in establishing whether an asylum applicant is from their claimed country of nationality in cases of doubt; and to deter individuals from making fraudulent claims purely because particular countries have a perceived advantage – such as a high grant rate for asylum or humanitarian protection.

During the pilot of the programme the UKBA focused more closely on applicants from Afghanistan, Eritrea, Kuwait, Palestine, and Somalia, for whom Removal and Return Agreements were available. In 2013 claimants from Syria were added (Harper, 2013). While not compulsory, as for electronic monitoring, the refusal by an asylum seeker to participate in testing may have a detrimental effect on their asylum case. The profiling inherent to these systems of linguistic analysis confirm and intensify racial and ethnic discriminations, further illuminating the fundamentally racist and violent nature of the asylum processes.

The discrimination enacted along nation-state border zones also takes place within warehouses, already sites of precarious working conditions, labour exploitation, and biopolitical control. Over the last decade voice directed efficiency mechanisms have become ubiquitous along the logistics supply chain (Kanngieser, 2013). In warehouses and distribution centres, voice picking is used to manage the passage and pace of workers through the workplace. Voice picking is a system for supervising workers via headsets and microphones, consisting of a series of automated verbal directives issued from a company's warehouse management system, which recognises the response from the worker through speech recognition and converts it into productivity data. In a workforce that is significantly migrant and on flexible and temporary contracts, the ability for software to accommodate diversities of speech and language is imperative. Unfortunately, as commercial voice recognition softwares have shown, this accommodation is limited at best[7]; in the logistics industries this has severe consequences for worker's pay and roster, which are contingent on workers meeting targets conditional on accurate communication with speech recognition platforms. Two kinds of recognition systems are used in warehouse operations: speaker-dependent, which require speakers to 'train' the application to identify their voice; and speaker-independent, which rely on a pre-existing archive of voice patterns from which statistical models are derived. Both are based on assumptions that may conflict with the realities of the distribution centre labour force. Speech-independent systems, while marketed as being adaptable to any voice within minutes of activation, are fundamentally restricted in their capacity to accommodate vocal or sonic 'anomalies' (including accents, dialects, speech impediments, external noise) that fall outside of software parameters. Speaker-dependent systems, while being more exact in their ability to assimilate pathologies, accents, dialects, and even multiple languages, require time for their programming and are thus incompatible with high and fast turnover rates (Klie, 2009).

The use of voice to determine geographical mobility shows how deeply embedded racism is within biometric profiling. More than fingerprinting or iris scanning, voice announces the ethnic, social, geographical and cultural characteristics of the speaker. Arguments for the impartiality of technological 'security and monitoring' apparatuses obfuscate the parameters upon which algorithms are built, leading to the applications of norms and standards that are always deficient. When these seemingly 'objective' technologies are then deployed to enact decision-making protocols, it is clear how violent these inbuilt deficiencies can be. The invisibilisation of prejudice that technologies afford are in part connected to a lack of engagement with what lies behind the interface. This disconnection breeds an ignorance of the substantial role of "computerized code in shaping the social and geographical politics of inequality" (Graham, 2005: 562)

Conclusion

The myriad techniques of sonic governance over bodies and forms of life outlined here remain predominantly disconnected from one another, under-examined, and outlier to 'visible' iterations of control – police and military formations and their armouries of weapons, drones, ships, tanks, horses, gases, dogs, robots, barricades, and bullets. Sound is largely insensible when one is not within the requisite spatial and temporal proximity. The activity of sound can thus only be relayed through traces and effects, and these bodily responses, the narratives told through words, articulations and flesh, behaviour and memory, can only ever marginally convey the trauma of sonic warfare. Perhaps it is for this reason that concerns about sound weapons often end up relegated to the speculative, conspiracy-theory, edges of public knowledge and debate.

Beyond what has been introduced within this chapter, it is critical to note that sound as a threshold-crossing war dispositif is not limited to the biotic. Sound is becoming even more prolific within digital systems. A 2014 study on acoustic cryptanalysis demonstrated the viability of sound to break highly secure encryption algorithms. Deploying a side channel attack (a non-direct and unconventional attack line), a computer was able to listen via a microphone to the 10–150 KHz sounds the target computer made when decrypting encrypted data. Using both a high-quality parabolic microphone (at a distance of 4 metres), and a smartphone (at a distance of 30 centimetres), researchers showed how such information could be gleaned from different kinds of computing hardware at varying distances with equipment of even negligible sophistication (Genkin et al., 2014). The implications of this are enormous, especially given the ease at which such infiltrations can be undertaken in any setting to both retrieve information from, and instal information on to, personal computers. Similar in consequence has been the claim made by security researcher Dragos Ruiu to the infection of his lab computers by 'badBIOS' system-to-system malware carried by sound waves from computer microphones to speakers (Marks, 2013). What sound deployed in the sphere of the digital shows is just how ubiquitous and far-reaching its development as a means for contagion, constraint, and control may actually be, re-inscribing the perimeters of governance across spatial scales from the minute to the expansive.

This movement across different spatial and temporal scales necessarily requires a conceptual framework like Guattari's concept of transversality for its articulation, one that not only speaks to human relations and pathologies, but also the assemblages that concatenate human, animal, energetic, material, and technological phenomena. As Guattari sought to make clear, transversality is an inherently political concept, attendant to flows of power and multiple vectors of communication. It is also a concept that refuses to adhere to fixed categorical delimitations. By bringing together specific geopolitical concerns, this chapter has aimed to show how sound de- and reterritorialises geographies by looking at the bodies, objects, and infrastructures brought into new formations through it. It has traced out historical and contemporary practices of colonisation and militarisation of not only humans but of biospheres more broadly.

These sonic technologies have helped to comprise apparatuses of everyday discipline; however, they also hold the potential to, reshape and escape such administrations. In his final work *Chaosmosis* (1995), Guattari elaborated the movement of transversality as processes of subjectivation. For Guattari, ideal activities were made up of transversal lines that affectively engender "unprecedented, unforeseen and unthinkable qualities of being" (1995: 106). Sound cannot help but be transversal, and thereby be creative of new modalities and relations. These are not always connective, or substantive of cultural and social norms (as is often focused on in research on sound) – they can also be disruptive, alienating, and profoundly damaging (see Gallagher, 2013). Whatever the repercussion, sound is a commoning force, pushing proximate bodies into shared and environments to engender different ways of perceiving and apprehending the world. It is this geopolitical element of sound that makes it critical for further exploration, in terms of its development by state, military, and corporate organisations for territorial securitisation, and for the potential it holds for building lines of evasion and solidarity.

Notes

1 The use of sound and music for psychological warfare and torture in conflict environments has already been extensively documented and theorised, most significantly by Suzanne G. Cusick (2013). See also Hill (2012) and Pieslak (2009).
2 The SHOPAR is named after the traditional ram's horn instrument blown during Jewish religious ceremonies (Rawnsley, 2011).
3 To date, the LRAD is used extensively in maritime, law enforcement, military, and commercial activity. As for military and policing, its application in commercial contexts is primarily in deflection/dispersal – notably in the use of crop protection from avian predators and in logistics, ocean travel, and piracy.
4 Low frequency noise, and its effects, has had a contentious history, with individual experiences conflicting with results from scientific and audiological testing. "While only a relatively small number of people are affected, those who are tend to suffer severe distress . . . and they may suffer various symptoms such as depression or even feel suicidal. In some cases a source of LFN is found and can be dealt with. However, in many cases . . . no environmental sound that could account for the sufferer's reaction can be found, and the cause of the disturbance remains a mystery" (Moorhouse et al., 2011: 2). This conflict emphasises the importance of acknowledging not only the limitations of current testing parameters for low frequency sound, but also the strong affective and emotional responses sound as physical and psychological entity generates.
5 The Thirteenth Report (2004) of the Joint Committee of Human Rights states: "1.134 We accept that clause 22 does not strictly speaking authorise compulsory electronic monitoring against an individual's will, and that to this extent there is no risk of interferences with Article 8 rights arising from compulsory monitoring as such. However, an individual's agreement to co-operate with electronic monitoring can under this clause be made a condition of their admission to the country or release from detention. Consent to monitoring in those circumstances cannot be said to be freely given, and the imposition of such a condition therefore amounts to an interference with Article 8 rights which requires justification in each case".
6 From its inception, the use of language testing by border agencies has been duly criticised by human rights lawyers and linguists, who argue that "the assumption that language can be equated to nationality is problematic and assumes an essentialized model of nationality. Languages and dialects have permeable borders, they change over generations and people who grow up over several areas often have mixed accents or lose their 'mother tongue'. Language tests also depend heavily upon the expertise of the translator" (Griffiths, 2013: 290). See also Diana Eades' (2005) exemplary criticisms of language analysis and asylum claims.
7 This was seen in the case of the Apple iPhone automated voice recognition software (Siri) not recognising particular accents and dialects, such as the Scottish accent (Wade, 2011).

References

Altmann, J. (2001) Acoustic weapons: A prospective assessment. *Science and Global Security*, 9: 165–234.

Altmann, J. (2008) *Millimetre Waves, Lasers, Acoustics for Non-Lethal Weapons? Physics Analyses and Inferences*. Osnabrück: Deutsche Stiftung Friedensforschung.

Arkin, W. (1997) Acoustic anti-personnel weapons: An inhumane future? *Medicine Conflict and Survival*, 14: 314–326.

Bloomberg Businessweek. (2014) *Company Overview of Electro-Optics Research & Development Ltd*. Available online: http://investing.businessweek.com/research/stocks/private/snapshot.asp?privcapId=46428024. Accessed 4 May 2014.

Bogue, P., Sullivan, R., Anonymous and Grandinetti, G.C. (2014) Settlements and separation in the West Bank: Future implications for health. *Medicine, Conflict and Survival*, 30(1): 4–10.

Brunner, C. and Rhoades, T. (2010) Transversal fields of experience. *INFLeXions*, 4. Available online: www.inflexions.org/n4_introhtml.html. Accessed 1 May 2014.

Cox, C. (2011) Beyond representation and signification: Toward a sonic materialism. *Journal of Visual Culture*, 10(2): 145–161.

Cusick, S.G. (2013) Acoustemology of detention in the 'global war on terror'. In: Born, G. (ed.) *Music, Sound and Space: Transformations of Public and Private Experience*. Cambridge: Cambridge University Press. pp. 275–291.

Deleuze, G. and Guattari, F. (2004) *A Thousand Plateaus*. London, New York, NY: Continuum.

Department of the Army. (2009) *Grenades and Pyrotechnic Signals Field Manual*. Washington, DC: Department of the Army.

Dunn, D. and Crutchfield, J.P. (2009) Entomogenic climate change: Insect bioacoustics and future forest ecology. *Leonardo*, 42(3): 239–244.

Eades, D. (2005) Applied linguistics and language analysis in asylum seeker cases. *Applied Linguistics*, 26(4): 503–526.

Elpam Electronics. (2014) Available online: www.elpam.com/products/tunnels-disclosure/. Accessed 1 October 2014.

Feigenbaum, A. and Kanngieser, A. (2015) Response to 'Air Affinities'. *Dialogues in Human Geography*, 5(1): 80–84.

Gallagher, M. (2013) Listening, meaning and power. In Carlyle, A. and Lane, C. (eds.) *On Listening*. Devon: Uniformbooks. pp. 41–44.

Genkin, D., Shamir, A. and Tromer, E. (2014) RSA key extraction via low-bandwidth acoustic cryptanalysis. *Crypto*, 1: 444–461.

Goodman, S. (2009) *Sonic Warfare: Sound, Affect and the Ecology of Fear*. Cambridge, MA: The MIT Press.

Graham, S. (2005) Software-sorted geographies. *Progress in Human Geography*, 29(5): 562–580.

Griffiths, M. (2013) 'Establishing your true identity': Immigration detention and contemporary identification debates. In About, I., Brown, J. and Lonergan, G. (eds.) *Identification and Registration Practices in Transnational Perspective: People, Papers and Practices*. London: Palgrave Macmillan. pp. 281–302.

Guattari, F. (1989) The three ecologies trans Turner, C. *New Formations*, 8: 131–147.

Guattari, F. (1995) *Chaosmosis: An Ethico-Aesthetic Paradigm* (trans. P. Bains and J. Pefanis). Sydney: Power Publications.

Gurr, N.J. (1997) Non-lethal weapons: Lifting the lid on the issue of human rights. *The International Journal of Human Rights*, 1(4): 1–17.

Harper, M. (2013) *Equality (Language Analysis – Palestinian, Syrian & Kuwaiti Testing) Authorisation 2013 – WMS*. Available online: www.publications.parliament.uk/pa/cm201213/cmhansrd/cm130214/wmstext/130214m0001.htm. Accessed 1 May 2014.

Hedlin, M.A.H., Walker, K., Drob, D.P. and de Groot-Hedlin, C.D. (2012) Infrasound: Connecting the solid earth, oceans, and atmosphere. *Annual Review of Earth and Planetary Sciences*, 40: 327–354.

Hill, I. (2012) Not quite bleeding from the ears: Amplifying sonic torture. *Western Journal of Communication*, 76(3): 217–235.

Holt, D.E. and Johnston, C.E. (2014) Evidence of the Lombard effect in fishes. *Behavioral Ecology*, 25(4): 819–826.

Home Office Science: Migration and Border Analysis. (2011) *Language Analysis Testing of Asylum Applicants: Impacts and Economic Costs and Benefits September 2011. A Report Prepared for the UK Border Agency: New Asylum Model Team (NAM +)*. Available online: www.gov.uk/government/uploads/system/uploads/attachment_data/file/257177/language-analysis.pdf. Accessed 1 May 2014.

Ingold, T. (2011) *Being Alive: Essays on Movement, Knowledge, and Description*. London: Routledge.

International Committee of the Red Cross. (2014) Available online: www.icrc.org/customary-ihl/eng/docs/v1_rul_rule103. Accessed 1 October 2014.

International Solidarity Movement. (2014) Video: 15 tear gas grenades and 5 stun grenades fired at school children. *International Solidarity Movement*. Available online: http://palsolidarity.org/2014/08/video-15-tear-gas-grenades-and-5-stun-grenades-fired-at-schoolchildren/. Accessed 28 December 2014.

Joint Committee on Human Rights. (2004) *Thirteenth Report*. Available online: www.publications.parliament.uk/pa/jt200304/jtselect/jtrights/102/10204.htm.

Kanngieser, A. (2013) Tracking and tracing: Geographies of logistical governance and labouring bodies. *Environment and Planning D: Society and Space*, 31(4): 594–610.

Kanngieser, A. (forthcoming) Listening to the anthropocene: Sounding the 'natural' world. In Maitland, P. (ed.) *Object Emotions*. Cambridge: Harvard University Press.

Klie, L. (2009) Speech declares its independence. *Speech Technology*, 14(8): 28–33.

LaBelle, B. (2010) *Acoustic Territories: Sound Culture and Everyday Life*. London: Continuum.

Lewis, O. (2014) Top Palestinian cleric says Israel threw stun grenades into al Aqsa; Israel denies. *Reuters*. Available online: http://uk.reuters.com/article/2014/04/20/uk-palestinian-israel-violence-idUKBREA3J0MQ20140420. Accessed 5 May 2014.

LRAD. (nd) Available online: www.lradx.com/site/content/view/15/110/. Accessed 4 May 2014.

Ma'an News Agency. (2014) Israeli forces fire stun grenades at Palestinian wedding. *Albawaba News*. Available online: www.albawaba.com/news/palestinian-wedding-attacked-581836. Accessed 28 December 2014.

Marks, P. (2013) Can a computer virus communicate via your speakers? *New Scientist*. Available online: www.newscientist.com/article/dn24582-can-a-computer-virus-communicate-via-your-speakers.html#.VOSDIbCsXpA. Accessed 1 May 2014.

McGreal, C. (2005) Palestinians hit by sonic boom air raids. *The Guardian*. Available online: www.theguardian.com/world/2005/nov/03/israel. Accessed 4 May 2014.

Merchant, N.D., Blondel, P., Dakin, D.T. and Dorocicz, J. (2012) Averaging underwater noise levels for environmental assessment of shipping. *Journal of the Acoustical Society of America*, 132(4), EL343–EL349.

Moorhouse, A., Waddington, D. and Adams, M. (2011) *Procedure for the Assessment of Low Frequency Noise Complaints*. Manchester: University of Salford.

Nancy, J.L. (2007) *Listening* (trans. C. Mandell). New York: Fordham University Press.

Pieslak, J. (2009) *Sound Targets: American Soldiers and Music in the Iraq War*. Bloomington, IN: Indiana University Press.

Rawnsley, A. (2011) 'The Scream': Israel blasts protesters with sonic gun. *Wired*. Available online: http://goo.gl/Efl6De. Accessed 4 May 2014.

Revill, G. (2014) El tren fantasma: Arcs of sound and the acoustic spaces of landscape. *Transactions of the Institute of British Geographers*, 39(3): 333–344.

Roads, C. (2001) *Microsound*. Cambridge, MA: The MIT Press.

Russollo, L. (1986) *The Art of Noises*. New York, NY: Pendragon Press.

Scrimshaw, W. (2013) Non-cochlear sound: On affect and exteriority. In Thompson, M. and Biddle, I. (eds.) *Sound, Music, Affect: Theorizing Sonic Experience*. New York, NY: Bloomsbury, Academic. pp. 27–43.

Stengers, I. (2005) The cosmopolitical proposal. In Latour, B. and Weibel, P. (eds.) *Making Things Public: Atmospheres of Democracy*. Cambridge, MA: The MIT Press. pp. 994–1003.

Thibaud, J.P. (2011) The sensory fabric of urban ambiances. *Senses and Society*, 6(2): 203–215.

Townsend, M. (2010) Teenager-repellent 'mosquito' must be banned, says Europe. *The Guardian*. Available online: www.theguardian.com/society/2010/jun/20/teenager-repellent-mosquito-banned-europe. Accessed 1 May 2014.

Volcler, J. (2013) *Extremely Loud: Sound as a Weapon* (trans. C. Volk). New York, NY: New Press.

Wade, M. (2011) Scottish accent leaves iPhone lost for words. *The Times London*, 26 October.

Who Profits Research Centre. (2014) *Proven effective: Crowd control weapons in the occupied Palestinian territories*. Available online: http://whoprofits.org/content/proven-effective-crowd-control-weapons-occupied-palestinian-territories. Accessed 28 December 2014.

Wright, S. (2001) The role of sub-lethal weapons in human rights abuse. *Medicine, Conflict and Survival*, 17(3): 221–233.

Yusoff, K. (2013) Insensible worlds: Postrelational ethics, indeterminacy and the (k)nots of relating. *Environment and Planning D: Society and Space*, 31(2): 208–226.

Zhou, Y.F. (2011) High intensity focused ultrasound in clinical tumor ablation. *World Journal of Clinical Oncology*, 2(1): 8–27.

Index

FSC
www.fsc.org
MIX
Papier | Fördert gute Waldnutzung
FSC® C083411

Kolibri 360 GmbH
Ferdinand-Jühlke-Straße 7
99095 Erfurt, Deutschland
produktsicherheit@kolibri360.de